DEVELOPMENTAL AND CELL BIOLOGY SERIES

EDITORS

P. W. BARLOW   P. B. GREEN   C. C. WYLIE

Genome multiplication in growth and development
Biology of polyploid and polytene cells

GENOME MULTIPLICATION IN GROWTH AND DEVELOPMENT

Biology of polyploid and polytene cells

V. Ya. BRODSKY & I. V. URYVAEVA

Department of Cytology, Institute of Developmental Biology

USSR Academy of Sciences, Moscow

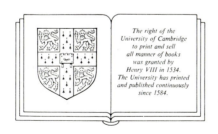

*The right of the
University of Cambridge
to print and sell
all manner of books
was granted by
Henry VIII in 1534.
The University has printed
and published continuously
since 1584.*

CAMBRIDGE UNIVERSITY PRESS

Cambridge

London New York New Rochelle

Melbourne Sydney

Published by the Press Syndicate of the University of Cambridge
The Pitt Building, Trumpington Street, Cambridge CB2 1RP
32 East 57th Street, New York, NY 10022, USA
10 Stamford Road, Oakleigh, Melbourne 3166, Australia

First published 1985

Printed in Great Britain by the University Press, Cambridge

Library of Congress catalogue card number: 84–12685

British Library cataloguing in publication data

Brodsky, V. Ya.
Genome multiplication in growth and
development – (Development and cell
biology series; no. 15)

1. Chromosomes
I. Title   II. Uryvaeva, I. V.   III. Series
574.87'322   QH600

ISBN 0 521 25323 3

Contents

Preface

The diploid chromosome set is the usual carrier of the genome in eucaryotes. At the same time, polyploid and polytene cells are widespread in somatic tissues of animals and plants and protozoans.

Genome multiplication, the reproduction of the genetic material in the cell during its polyploidization or polytenization, is only just beginning to be evaluated in developmental biology, although the phenomenon itself has long been known. For the geneticist, polyploidy is usually associated with plants, many of whose cultivated forms are polyploid. In animals, by contrast, this zygotic (generative) polyploidy is exceptionally rare. This book deals with the genome multiplication in somatic tissues of diploid organisms.

For a long time the polytene cells of salivary glands of dipteran larvae were considered the outstanding example of genome multiplication in somatic cells and numerous classical genetic studies were made using the giant polytene chromosomes of Diptera. Even now, they remain one of the main subjects of studies in molecular genetics. Many other differentiated cells also have an increased nuclear content. For example, the highly specialized cells of trophoblast of some mammals, the cells of the silk gland of *Bombyx mori* (the silkworm), the neurons of molluscs, and the antipodal cells in plants all have gigantic sizes and thousands, or even millions of copies of the gametic DNA. The phenomenon of genome multiplication is not restricted to a few exotic examples. Suffice it to say that almost all the mature functioning cells in the human myocardium are polyploid, likewise in the liver of rats and mice, and in the megakaryocyte series of the bone marrow in all mammals.

Among the aspects of polyploidy and polyteny upon which we will focus attention are those that relate to developmental and cell biology rather than those traditionally associated with genetics and cytology.

Studies of cell polyploidy began in our laboratory in the late 1950s. After a few years, the data obtained on the occurrence of polyploid cells were summarized (Brodsky, 1966). Our joint reviews beginning in 1970 and especially our book *Cell Polyploidy*, published in Russian in 1981, form the basis of the present book. The present book, however, is not a word-for-word translation of the earlier Russian book. It is, in fact, a totally new book. New chapters have been written, some revised and some rewritten. Changes have been introduced into all the remaining sections in the light of the most recent findings that have appeared since that time, and a number of new ideas have been put forward. We wrote this book after Nagl's book *Endopolyploidy and Polyteny in Differentiation and Evolution* (1978) had been published. Both books deal with the same problem but the material and discussion contained in them complement rather than overlap one another.

A considerable part of the material in the book consists of the results obtained by researchers in the authors' laboratory. We would like to thank Drs V. M. Faktor, A. M. Arefyeva, T. L. Marshak, T. B. Aizenshtadt, G. V. Delone, N. N. Tsirekidze, as we are well aware that, without the constant flow of new data and the productive discussion thereof, there would be no general ideas on this subject. We would like to thank Drs V. B. Ivanov, N. G. Khrushchov, I. I. Kiknadze, B. N. Kudryavtsev, I. B. Raikov, O. G. Stroeva and E. V. Zybina for their valuable remarks on the text of the book. We are also grateful to Drs A. P. Anisimov, G. E. Onishchenko and V. A. Vorobyev who supplied their original microphotographs. Our thanks also go to Mrs Glenys Ann Kozlov for her careful translation of the manuscript from Russian into English.

We cordially appreciate the initiative of Dr Peter W. Barlow in publishing our book in English; we also thank him and Dr J. Valerie Neal for their valuable comments on its structure, language and content.

I

Polyploidy and polyteny as phenomena of normal development

1

Introductory notes

A A historical survey and terminology

The first observations of a double diploid, i.e. a polyploid, chromosome set in somatic cells were made shortly after the discovery of chromosomes themselves. In his well-known book *The Cell in Development and Heredity*, E. B. Wilson (1925) mentions work dating from the beginning of the twentieth century in which examples of polyploidy are given. Thus, in epithelial cells of the pupal intestine of the *Culex* mosquito, metaphases with 6 (diploid set), 12, 24 or 48 chromosomes were found. Noting the rarity of these finds, Wilson suggested that they applied to old or degenerating tissues. Information on somatic polyploidy, the doubling of the number of chromosomes, was continually being accumulated during the study of mitosis (Mazia, 1961). Now, however, we can see that neither could the occurrence of polyploid cells be evaluated by this method of research, nor could the reasons for their emergence be determined. Mitoses in many differentiating cells are rare, and even more so in differentiated ones. Moreover, polyploid cells may enter the cell cycle less frequently than diploid ones. A study of only mitotic cells excludes from analysis polyteny (the multiplication of the number of chromonemata of the diploid, or conjugated, pairs of interphase chromosomes) which is a phenomenon no less common than polyploidy.

The first method especially aimed at studying somatic polyploidy was the measurement of the size of the nuclei. W. Jacobj (1925) was the first to note that, in mammalian hepatocytes, nuclear volumes fell into categories that were proportional to the doubling of the number of chromosomes. This volumetric method made it possible to reveal polyploidy by studying interphase nuclei. Many nuclei of rat and mouse liver cells were established as being polyploid. The liver is, however, almost the only example of a vertebrate tissue where an adequate evaluation of polyploidy can be made according to nuclear size. Attempts to apply volumetric methods to other tissues have not been successful. Even in the case of the liver, the size of the nucleus characterizes ploidy only in mammals. For example, rat hepatocytes with a diploid number of chromosomes have a nucleus with a volume of 10–30 units, while the tetraploid ones have a volume of 30–60 units (Bachmann & Cowden, 1965*a*). But the sizes of

3

diploid nuclei of hepatocytes in the frog, *Rana pipiens*, vary from 20 to 100 units. If the criterion of size were used to evaluate the ploidy of frog hepatocytes, not only would diploid nuclei be found, but also tetra- and even octaploid nuclei. In starved frogs, the size of the hepatocyte nuclei decreases, and when the temperature is lowered it increases. In this case, the extreme values of the volumes differ twofold (Alvarez & Cowden, 1966). If it were not known that the quantity of DNA in the nuclei actually remains constant (diploid), one might come to the erroneous conclusion that polyploidization occurs.

The general principles of tissue growth have been determined by the research on nuclear and cell sizes, but a reliable method of studying cell polyploidy has not thereby been discovered. It has repeatedly been found that the volume of the nucleus may double or more, while the chromosome set remains constant (as during growth of the diploid cell, at different phases of secretion, etc.). The criterion of size can be successfully used at present to study polyploidy in the liver and other tissues only where there is an additional cytophotometric control.

The concept of endomitosis was first put forward by L. Geitler (1938, 1953) and later developed by the cytologists of the Viennese school (Tschermak-Woess, 1956, 1963, 1971; Nagl, 1962, 1972a). Endomitosis (the doubling and separation of the chromosomes inside the nuclear envelope without a spindle being formed or the nucleolus being destroyed) would appear to explain the mechanism of formation of polyploid cells. When the mitotic cycle was not yet known, the description of endomitosis helped to advance the understanding of mitosis. It appeared logical to separate mitosis, the mode of cell fission, from endomitosis, the mode of polyploidization. The idea of endomitosis was accepted immediately and remains unchallenged to this day. The widespread occurrence of endomitosis is today thought doubtful but, in its time, research on this topic was important since it revealed the extent of polyploidy in animal and plant tissues (Geitler, 1953, Tschermak-Woess, 1956).

The main contemporary method used in studying cell polyploidy was established in 1948 as a result of the discovery of the rule of relative constancy of DNA. This important principle (due to Boivin, Vendrely & Vendrely, 1948) stipulated that the quantity of DNA in the nucleus is proportional to the number of chromosome sets. The rule of the relative constancy of DNA is also historically significant because it was an early indication of the genetic function of DNA. What is important for us is that a really reliable and precise method of determining the number and composition of the polyploid cells in any tissue had been discovered; and polyploidy can now be studied in interphase cells. Different tissues in the organism and in different species can be compared, from the onset of development to old age, as can various physiological and pathological states. DNA cytophotometry makes it possible to study polyteny as well;

in the polytene nucleus (just as in the polyploid one) the DNA content changes in proportion to the number of chromonemata.

The discovery of the rule of the relative constancy of DNA was made at the same time as the bases of cytophotometry were established. Cytophotometric methods developed by T. Caspersson (1951) were furthered considerably by many researchers. The main task was that of improving the accuracy of quantitative determinations of DNA in a single nucleus. The cytophotometry of DNA in cells stained by Feulgen's reaction revealed a huge number of polyploid cells. These cells were also discovered in tissues where endomitosis had not formerly been recorded (for instance, in those of mammals).

In time, certain deviations in the constancy of DNA were exposed by cytophotometry and later by the methods of molecular biology. These deviations, however, do not apply to most cases of polyploidy and polyteny (when the DNA of the diploid genome doubles completely and sometimes doubles 15–20 times). Non-multiple changes in the quantity of DNA do not always signify a loss of some fraction of DNA: in most cases when the entire diploid genome is preserved (or even after it has doubled several times) such changes indicate that a small part has not been re-produced. Losses of DNA have also been described, however (Beerman, 1977; see also Chapters 3A and 4).

The discovery of the mitotic cycle has expanded our present knowledge of cell polyploidy. It is now understood that the main event in proliferation is not so much mitosis itself, but the preparation for division. Autoradiography of DNA has become a valuable method of studying the mitotic cycle, as well as the cycle leading to polyploidy.

Over the last 20 years, polyploidization of cells has been shown to be a regular event in the development of many animal and plant tissues. Frequently, cell polyploidization accompanies the postnatal growth of tissues. Polyploidy or polyteny have been established as being needed for certain tissues to perform their specific functions to the full. Well-founded ideas have also been put forward on the significance of cell polyploidy in the evolution of organisms (Nagl, 1978).

The long and complicated history of the study of cell polyploidy has given rise to certain difficulties in ascertaining the terminology to be used, which it is not worthwhile going into here. The terms we use in this volume are the same as those used in our previous reviews (Brodsky, 1966; Brodsky & Uryvaeva, 1970, 1974, 1977).

It is obvious that the starting point in any polyploid or polytene series is the diploid cell – 2n, 2C, where n is the haploid chromosome set, and C is the quantity of DNA in the unreplicated haploid set. The following transformations of the diploid genome (Fig. 1.1) are described below.

Polyploidy (in botanical literature, especially in old works, this is termed 'polysomaty') is the doubling of the number of chromosomes in the

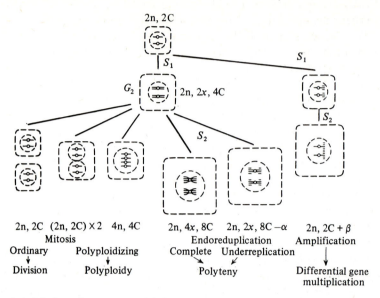

2n, 2C

G_2 2n, 2x, 4C S_1 S_1

S_1

S_2

S_2

2n, 2C	(2n, 2C) × 2	4n, 4C	2n, 4x, 8C	2n, 2x, 8C −α	2n, 2C + β
	Mitosis			Endoreduplication	Amplification
Ordinary		Polyploidizing	Complete	Underreplication	
↓		↓	↘	↙	↓
Division		Polyploidy		Polyteny	Differential gene multiplication

Fig. 1.1. Modes of genome multiplication. *n*, Haploid chromosome set; *x*, the number of chromatids in one chromosome; C, haploid DNA amount.

somatic cell. We emphasize that it is the cell which is considered and not the individual nucleus. It has been established that, in terms of the properties of its genome, the binucleate cell is the genetic and functional analog of the mononucleate cell with the same number of chromosomes (see Chapter 7B). These types of polyploid cells are formed in both animals and plants as a result of *polyploidizing mitosis*, a variant of ordinary mitosis, having a block at one stage or another.

Polyteny (in the botanical literature the term 'cryptopolyploidy' is sometimes used) is the repeated doubling of the number of chromonemata (chromatids) in the diploid, or conjugated diploid, set of the chromosomes without subsequent mitosis. This shortened version of the mitotic cycle is frequently designated endocycle, endoreproduction, endoreplication or endoreduplication.

If the interphase diploid cell is characterized by the number of chromosomes (2n) and the quantity of DNA (2C) proportional to it, and the polyploid cell by the doubling of these parameters (for example, 4n, 4C or 32n, 32C), then the number of chromatids per chromosome must also be defined. This number, which will also be the number of daughter chromosomes in the nucleus, can be designated by '*x*'. Thus, the G_2 or premitotic diploid cell is described by 2x and also 2n and 4C. If mitosis does not occur and the cell again replicates its DNA before the next mitosis, the chromatids double once more (4*x*), but the chromosome number remains diploid (2n), and the nucleus contains 8C of DNA.

The essence of both polyteny and polyploidy is the reproduction of the chromatids. In this respect, polyteny and polyploidy do not differ from one another. However, the mode of genome multiplication, the morphology of the nucleus and the physiology of the cell differ during polytenization and polyploidization. Therefore we distinguish between these phenomena although, in general terms, they can be considered as *genome multiplication*.

A rare variant of the polytene cycle is the *underreplication* of DNA. This phenomenon usually begins after several complete endocycles.

Amplification is the selective replication of a particular gene, when the number of the other genes does not change. Amplification has been clearly demonstrated in the ribosomal genes of certain oocytes. Some data on the amplification of genes in some somatic cells have also been obtained (see Chapter 6F).

Also noteworthy is yet another form of quantitative change in DNA (the only example where the rule of the relative constancy of DNA is really broken): the *diminution* of chromatin, or the elimination of chromosomes, in somatic cells (see Chapters 3G and 6F).

B The occurrence of polyploid and polytene cells and its relation to genome mass and multiplication

Most of the data on the occurrence of polyploid and polytene cells come from the tissues of mammals, insects and plants, and also Protozoa. In addition to these systems, we shall mention examples of the extensive polyploidization in cells of certain worms, molluscs, and echinoderms. In addition there is evidence for the doubling of DNA content in cells of Coelenterata (Campbell, 1973; Aizenshtadt & Marshak, 1974), Ctenophora (Pylilo, 1975), Oligochaeta (Christensen, 1966), Crustacea (Jost & Mameli, 1970; Bocquet-Védrine, 1970), and Tunicata (Fenaux, 1971). Polyploid cells are found in higher plants (see Chapter 4), and in many lower ones, too (for review see Nagl, 1978).

Cell polyploidy is a general biological phenomenon. It should, however, be noted that in some groups of organisms, polyploid or polytene cells have not yet been found although it would be more correct to say that no really serious attempts have been made to find such cells in these organisms. For example, insufficient research has been done on reptiles and birds, and fish have been studied even less. In the only species studied of these classes of vertebrates, mainly diploid nuclei were discovered in the liver, whereas in the livers of mice and rats the majority of cells are polyploid. In some mammals, however, diploid cells predominate in the liver (Shima, 1980).

It is not known whether birds or mammals with livers composed of diploid cells have polyploid cells in other organs. Nor is it known whether the livers of all fish, reptiles, and birds have diploid cells. Recently, however, 4C nuclei have been revealed in the liver of some fish, and 4C and 8C nuclei in the turtle's liver (Brasch, 1980). The livers of many amphibian species have been assumed also to consist of diploid cells, and this has been borne out by cytophotometric studies on nine species of toads by Bachmann (1970). A 2C quantity of DNA was also determined in isolated nuclei of the hepatocytes of two species of frogs, *Rana pipiens* and *Rana catesbiana* (Bachmann & Cowden, 1965*b*), but in the liver of a third species (the little chorus frog *Pseudacris*), 2C, 4C and 8C nuclei were found (Bachmann & Cowden, 1967). These are the same classes of nuclei as are found in the livers of mice and rats. Moreover, it is not known whether some (and even perhaps many?) of the diploid nuclei belong to binucleate cells 2C × 2, i.e. to cells that have a tetraploid genome. Fish tissues are known to contain small diploid cells. But there are giant polyploid nuclei in the yolk-sac syncytium of some teleosts (Bachop & Schwartz, 1974).

A comparison of eight organs in two species of plants, in the pea (*Pisum sativum*) and the sunflower (*Helianthus annuus*), revealed polyploid cells in all organs of the former whereas, in the same organs of the second species, there were only diploid cells (Evans & Van't Hof, 1975).

It would be interesting to find a purely diploid organism. It would likewise be interesting to compare in related species similar tissues with a considerably different expression of polyploidy. The very fact that there is a great species-specificity of polyploidy in analogous tissues points to the conclusion that polyploidy has a functional significance in certain species.

A link between polyploidization of cells and the mass of the organism's diploid genome, the 2C quantity of DNA, should be considered. Before examining this possible relationship, we shall discuss data concerning genome mass in different organisms.

Genome mass should theoretically correspond to the number of genes, which may intercorrelate with the number of proteins synthesized by the cell and hence (in the non-secreting cells) to cell mass. Such proportionality is not usually observed. The terminal differentiation of metazoan cells leads to the intensification of synthesis of a single or a few special proteins. Not only the complement of proteins but also their production and secretion (turnover) varies greatly in cells of different types; consequently, the cells of the same organism (having one and the same genome) show a manifold difference in size. The discrepancy between genome mass and the mass of cellular proteins is also affected by the fact that in the nuclei of eucaryotes, but not those of ciliates, two functions coexist: the metabolic and the genetic. Protein production depends on the number of active genes, i.e. the metabolic function of the nucleus, while the remainder of the DNA reflects the preservation of genetic information. In ciliates, the metabolic and genetic functions are divided between the macronucleus and the micro-nucleus, and here the quantity of protein correlates with the content of macronuclear DNA, rather than with micronuclear DNA (Ammermann & Muenz, 1982).

The larger size of the genome of eucaryotes as compared with that of procaryotes has often been noted (Sparrow, Price & Underbrink, 1972). Among multicellular organisms, the largest quantity of DNA is found in vertebrates, especially in some fish and amphibians, and also in some higher plants (Mirsky & Ris, 1951; Britten & Davidson, 1969; Nagl, 1978). It might be thought that the level of sophistication of an organism's functions is a consequence of the informational content of the genome. The complexity of organisms – the number of cell types, the optimization of cell and tissue interaction – definitely increases during the course of evolution. The number of genes should correspond to the degree of perfection of form and function. However, this dependence can only be discussed in very general terms. A considerable fraction of the DNA does not carry direct genetic information. However evolution does not always seem to be the result of morphophysiological complexity or of genome complexity. In many cases, speciation has resulted from a decrease rather than an increase in total DNA quantity in the genome.

Comparison of genome mass between species is complicated by the fact that repeated sequences may form a considerable part of the total DNA. The repeated sequences, located in the heterochromatic areas of the chromosomes, do not increase the amount of information possessed by the genome and many of them do not code for proteins at all. The functions of this type of DNA are not clear. Repeatedly multiplied genes are known to code the structure of ribosomal, transfer and histone messenger RNA.

The percentage of repetitive DNA may correlate with the overall DNA content. Thus, in fungi (where the genome mass is extremely low) repeated sequences make up a negligible part of the DNA; in insects where there is a greater DNA content, but whose genome mass is still low, the repeated sequences account for not more than 20%; in mammals, repeated sequences represent as much as 30–40% of total DNA content, while in amphibians and some higher plants (organisms with some of the highest genome masses) repeated DNA sequences form up to 80% of this mass (Flavell *et al.*, 1974).

Examples are known where the genome grows almost exclusively due to increases in repetitive DNA. Thus, in nine species of *Lathyrus*, the absolute 2C quantities of DNA form a series from 12.5–20.3 pg. Repeated sequences formed approximately half the DNA in the species with the smallest genome, while 70% of the repeated sequences were in the largest genome. The percentage of repeated sequences in the remaining species varied between these two values, but the amount of unique (non-repeated)

DNA in the nine species was practically the same (Narayan & Rees, 1976). In six species of the black larder beetle *Dermestes* (where the overall quantity of DNA differed twofold, from 1.7 to 3.8 pg), repeated sequences accounted for almost the whole of the difference (Rees, Fox & Maher, 1976).

Many examples are known, however, where the number of repeated sequences does not correspond to the weight of the genome. Thus, for example, the same quantity of repetitive DNA – some 30% – occurs in the genomes of *Drosophila melanogaster* (the total quantity of DNA in the haploid set (1 C) is 0.12 pg), *Xenopus laevis* (1 C = 3.1 pg), and humans (1 C = 2.9 pg). In two plants with an approximately equal proportion of repeated sequences (*c.* 55%), the content of DNA in the haploid set of one species was 1.6 pg (*Nicotiana tabacum*) and in the other 16.0 pg (*Anemone blanda*). In the onion, *Allium cepa*, which has approximately the same haploid quantity of DNA as the anemone (16.7 pg), 95% of the genome consists of repeated sequences.

The functions of most of the repeated sequences are not known; the significance of their variation between related species and even between higher taxonomic groupings is, therefore, not clear. In the case of copies of known function, they cannot be proportional to the genome size. So, the number of rDNA copies varies fourfold in *Vicia* species while the genome size (total DNA content) varies sevenfold (Lamppa, Honda & Bendich, 1984). No trend was revealed in the repeat length, precursor size or intron presence as the total DNA content increased. The major difference among ribosomal repeats occurs in the non-transcribed spacer regions.

When commenting on variation in the overall mass of the genome, it should be noted that many of the unique sequences do not carry information for protein structure. It is thought that they have a regulatory function (Davidson & Britten, 1979; Schmidtke & Epplen, 1980).

An interesting series of 2C quantities of DNA is obtained when comparing the major taxonomic groups. If the minimal genome for each group is selected from each of them (i.e. the genome that is supposedly least burdened with non-informative sequences), a curious, but as yet unexplained, regularity comes to light: the quantity of DNA in the minimal genome of organisms, beginning with viruses (RNA) and ending with vertebrates and higher plants, comprises a power series of numbers, 2^n, where n is an integer (Fig. 1.2). The procaryotes correspond to the series 2^0–2^{15}, the eucaryotes produce the series 2^{18}–2^{26}. We reproduce here the diagram in its original form (Sparrow & Naumann, 1976). Its biological purport is apparently increased if the organisms are brought together according to type, for example, where the minimum genome of arthropods is typical of insects, and the genome of nematodes is typical of worms. Then, besides the double quantity of DNA being preserved in

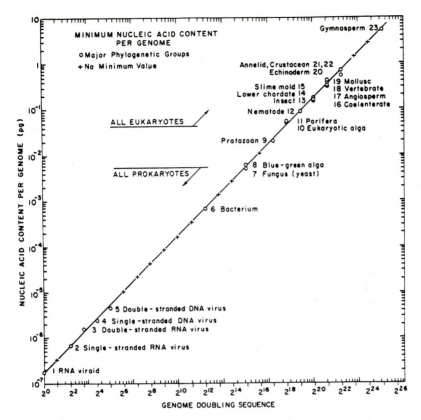

Fig. 1.2. Minimal genome mass (pg of DNA or RNA) in viruses and other major phylogenetic groups of organisms. The points indicated by a cross are doubling points not represented by a known minimum value for a major group (after Sparrow & Naumann, 1976).

a structure of this type, the correspondence of the mass of the genome to the complexity of the organisms can be determined: viruses → bacteria → blue-green algae → protozoans → seaweeds and sponges → arthropods → lower chordates → molluscs, vertebrates, higher plants. Naturally, contradictions arise in this series, too, which evidently demonstrate that, even when grouping organisms into large taxons, it is not always the minimum genome that has been selected. Making this reservation, the division into types reveals a general (and expected) regularity, while the division into classes (Fig. 1.2) puts slime molds in the same group as insects, and annelid worms in the same group as higher vertebrates.

The study of the genomes of related species of plants, insects and echinoderms reveals a correspondence between the quantity of DNA and the size of the chromosomes (see Rees & Jones, 1972; Sparrow &

Table 1.1. *Variation in the quantity of DNA (2C) in the diploid genome (chromosome number = 2n). Each species of each genus has the same number of chromosomes*

Genus	Number of species studied	2n	2C[a]	References
Drosophila	3	8	1.0–1.4	Laird, 1973
Dermestes	6	18	1.0–2.7	Fox, 1969
Chorthippus	5	17	1.0–1.4	Kiknadze & Vysotskaya, 1970
Bufo	3	22	1.0–1.5	Ullerich, 1966
Bufo	5	22	1.0–1.5	Bachmann *et al.*, 1978

[a] In arbitrary units, comparable only within each investigation.

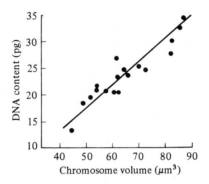

Fig. 1.3. Characteristics of the chromosome complement in 18 different *Lathyrus* species represented as a plot of 2C DNA content of each species against the respective total metaphase chromosome volume (after Rees, 1972).

Naumann, 1976; Nagl, 1977). In these cases, it may be suggested that endoreproduction or even amplification of the genes occurs in the gametes. A correlation between the sizes of the chromosomes and the weight of the genome has been observed, for example, in species of *Lathyrus* (Fig. 1.3), and also in pulses, onions, *Lolium* (Rees, 1972), Triticinae (Pegington & Rees, 1970), *Luzula* (Barlow & Nevin, 1976), and *Crepis* (Jones & Brown, 1976). The differences in quantities of DNA in similar species of animals (Table 1.1) can evidently be treated in the same way. Recently, a close relationship has been revealed between individual metaphase chromosome volumes and their DNA content in barley, *Hordeum vulgare* (Bennett *et al.*, 1982). Barlow & Nevin (1976) examined nucleolar volume and the number of chromocenters in the different species of *Luzula*; these show a relationship to the DNA content. These authors suggested that chromosome fragmentation could have generated the range of variation in chromosome number in the genus. Variation also occurs in the total DNA content.

Variation in the volume of the chromosomes need not be linked with variation in the quantity of DNA. Thus, when the karyotypes of cells from different tissues (and even from within one tissue at different stages of development) are compared, the sizes of the chromosomes may change, even though the DNA content is quite stable (Bennett, 1970, 1974). For example, in meristems of *Vicia faba* chromosomal size and protein content decrease approximately twofold in the course of development, the absolute values differing for the root and for the shoot. The size of the interphase nuclei also changes, but their DNA content remains the same.

It frequently happens, especially in plants, that chromosome volume remains constant within a genus or even within a group of related genera. The genome mass (the quantity of DNA) then depends on the number of chromosomes (see, for example, Bennett & Smith, 1976).

Changes in the absolute mass of the genome may be accompanied by changes in the size of the cells. The 'small cell' and 'large cell' phenomenon in certain species of animals and plants has long been known (Szarski, 1976). For example, the size of lymphocytes in bats (six species studied) are on average half the size of those in primates (18 species); the 2C quantity of DNA differs by approximately the same amount (Manfredi Romanini, 1974). Changes in cell volume, also proportional to genome mass, occur in 19 species of grasses (Sparrow & Naumann, 1974). Both parameters vary tenfold. The same correlation has been recently revealed for other animal species (Brasch, 1980).

Genome mass and, accordingly, the size of the cells, may have some ecological significance. It is known that the rate of genome reproduction depends on its mass (Grosset & Odartchenko, 1975 a, b; Ivanov, 1978). A study of many British plants revealed a certain dependence between genome mass and seasonal variation in division and growth of cells (Grime & Mowforth, 1982; see also Fig. 1.4).

Polyploidy, or even endoreproduction of DNA, as a factor in speciation has repeatedly been discussed (Ohno, 1970; Bachmann, 1972; Comings, 1972; Nagl, 1978). In some groups of angiosperms up to 70–80% of the species are polyploid (unlike gymnosperms, where there are few polyploid species). The regular doubling of DNA was discovered when different species of insects were compared. Polyploidy, and its converse, the reduction in the number of chromosomes, is believed to have occurred during the evolution of fish (Kirpichnikov, 1982) and birds (Bachmann, Harrington & Graig, 1972). Polyploid forms of amphibians have also been found (Bachmann & Cowden, 1967; Wasserman, 1970; Beçak & Goissis, 1971); one species is apparently allotetraploid (Beçak & Beçak, 1974).

Changes in part of the genome may occur in evolution, whereby slight changes take place in the quantity of DNA. One way of checking this

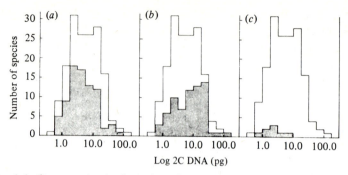

Fig. 1.4. Genome size in flowering plants occurring in the British Isles. Open histograms represent the DNA content distribution of all the 162 species whose DNA content has been estimated; the shaded histograms represent (*a*) the DNA contents of species with a widespread distribution (80 species), (*b*) those with a southern distribution (69 species) and (*c*) those with a northern distribution (8 species (after Grime & Mowforth, 1982).

possibility is to evaluate the DNA content in genomes of related species or, better still, in isolated populations of the same species. One example of such an investigation is the comparison of *Chironomus thummi thummi* and *Chironomus thummi piger*; this revealed a difference in the quantities of DNA in the spermatocytes (or in the cells of the salivary gland) of about 27–8% (Keyl, 1965). Changes in the DNA may, however, be smaller and so cannot be reliably determined by cytophotometry. Thus, research on genome masses in some species of green toad from disjunct geographical regions revealed average quantities of DNA from 98 to 120 units (Bachmann *et al.*, 1978). The extreme forms do perhaps vary, but, to be sure of that, the degree of error in the cytophotometric method used must be known.

Let us note the alternative (or supplement) to quantitative DNA changes in evolution. Evolution may occur due to the decreased variability of the genome masses. If it does, separation of the light and heavy genomes may result. It is obvious that this principle cannot be a general one for it would signify the finiteness of evolution. It may, however, be discussed with regard to a limited number of groups. A reduction in the variation in mass of the genome may possibly occur in vertebrates (Figs. 1.5 and 1.6).

In mammals and birds variation in genome mass is small (Figs. 1.5 and 1.7): approximately twofold in the species studied. By comparison, the 2C DNA contents in flowering plants vary 500-fold. So far, there are few data available for higher vertebrates (data for only several dozen species in each class). There are, however, data on species of lower orders.

Some direct data exist showing insignificant changes in DNA and the number of chromosomes in mammal speciation. Thus, when a comparison was made of the chromosome sets of the goat and the sheep, the quantity of DNA was found to be the same. There was also visual identity of karyotype, as judged by Giemsa (G-) band distribution (Sumner &

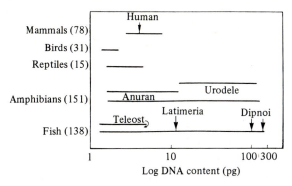

Fig. 1.5. Genome mass variations in vertebrates. The number of species studied is recorded in brackets (summarizing data of Bachmann, 1972; Rees, 1964; Thomson, Gall & Coggins, 1973; Manfredi Romanini, 1974; Morescalchi, 1974).

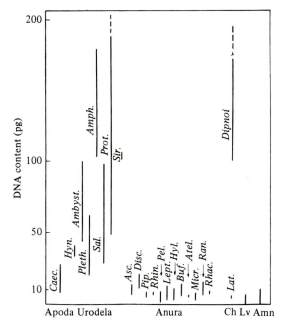

Fig. 1.6. Maximum and minimum amounts of nuclear DNA in Amphibia (orders Apoda, Anura, and Urodela) and other vertebrates. Ch, choanichthyes; Lv, other lower vertebrates; Amn, amniotes; Lat, Latimeria. Other abbreviations refer to amphibian families (after Morescalchi, 1974).

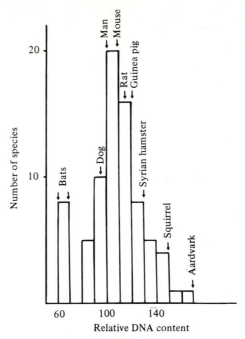

Fig. 1.7. Relative genome mass (with respect to the diploid human genome which is set at 100%) in mammals. Arrows indicate the DNA content of some particular species studied (summarizing data of Vendrely & Vendrely, 1956; Capanna & Manfredi Romanini, 1971; Bachmann, 1972; Manfredi Romanini, 1974; Formenti, 1975; Hatch *et al.*, 1976).

Buckland, 1976). Even in those rare cases where the diploid chromosome sets differ considerably within a genus, such variation seems to be unrelated to polyploidy. Thus, in three species of *Ellobius* with a diploid chromosome number of $2n = 17$, 36 and 54 (i.e. almost multiple changes in the number of chromosomes), the quantity of DNA in each was the same (Lyapunova *et al.*, 1980). In two species of mice (*Mus musculus* and *Mus poschiavinus*), where there was a considerable difference in the chromosome number ($2n = 40$ and 26), the diploid DNA content hardly differed at all – 5.9 and 6.3 pg (Manfredi Romanini, Minazza & Capanna, 1971). In two species of the salmon genus – the salmon and the trout – with $2n = 80$ and 60, the quantity of DNA did not differ (Rees, 1964). In all of these cases there are clearly changes within the initial karyotype (fragmentation or fusion of the chromosomes), which do not involve polyploidy. Some of these changes may involve heterochromatin and the number of rapidly reassociating sequences since these differ markedly in four species of gopher; in this example, however, the number of chromosomes ($2n = 36$) and G-bands in them remain constant (Lyapunova *et al.*, 1980).

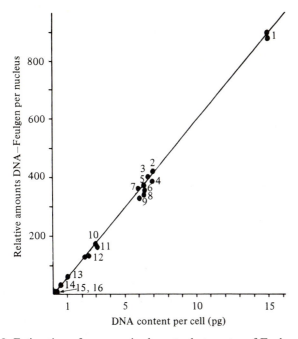

Fig. 1.8. Estimation of genome size by cytophotometry of Feulgen-stained DNA (arbitrary relative units) and by biochemical analysis of DNA (pg) in the same cell types. 57.20 ± 1.25 Feulgen arbitrary units correspond to 1 pg of DNA. The following cells were used: 1, erythrocyte nuclei of *Rana pipiens* (15 pg); 2, buccal epithelium from human female and, 4, from human male (7.0–7.3 pg); 3, granulocyte nuclei from human female (7.0 pg); 5, kidney nuclei from CF1 male mouse (6.0 pg); 6–9, erythrocyte nuclei from *Xenopus laevis* (6.3 pg); 10, spermatids from CF1 male mouse (3.0 pg); 11, sperm nuclei from *Xenopus laevis* (3.2 pg); 12, erythrocyte nuclei from white rock chicken (2.5 pg); 13, diploid hemocytes from *Bombyx mori* (1.0 pg); 14, maturing spermatids from *B. mori* (0.5 pg); 15, maturing sperm from *Drosophila melanogaster* (0.2 pg); 16, yeast cells *Saccharomyces cerevisiae* (0.05 pg). (After Rasch, 1974, who summarized data from many authors.)

Is somatic polyploidy connected with absolute genome mass? Let us examine the data in Fig. 1.8 (Rasch, 1974). The value of this material lies in the fact that it testifies to the complete comparability of cytochemical and biochemical studies of DNA. From the quantities of DNA measured by the authors, a series is obtained (Fig. 1.8): cells of *Drosophila* and silkworm (2C approximately 1 pg), chickens (3 pg), mice, humans, and clawed frog (6–7 pg), and tiger frog (15 pg). If one selects from these species the silkworm, the mouse, and the tiger frog, one discovers a 15-fold increase in the diploid genome mass and a decrease in the extent of polyploidy (polyteny) from 10^6 in the silk gland of the silkworm, down to 10^3 in the trophoblast (and 10^2 in the liver) of mouse; polyploid cells have not yet been found in the frog. However, when all the seven species cited in Fig. 1.8 are considered, no such correlation is revealed. There is less DNA in the nuclei of the salivary gland of *Drosophila* than in the cells of the trophoblast in mouse; the clawed frog has exactly the same total genomic DNA content as the mouse, but no polyploid cells have yet been found in it (whereas they are abundant in the mouse).

The diploid quantity of DNA in the mollusc *Aplysia californica* is 2 pg and DNA contents of up to 300000C have been found in its nerve cells. Certain plants have similar 2C DNA values, e.g. *Phaseolus coccineus* (runner bean, 2.7 pg), *Tropaeolum majus* (nasturtium, 2.7 pg) and *Capsella bursa-pastoris* (shepherd's purse, 1.7 pg). Highly polytene cells are found in their suspensor cells, which although not as large as the mollusc nerve cells, contain considerable quantities of DNA all the same: 8192C, 2048C, and 1024C, respectively, for the three plant species (see Nagl, 1978). In birds, however, with the same genome mass as *Aplysia* or *Tropaeolum* not even 8C cells have been found. It would be interesting to study the ploidy of somatic cells in fish. Some species of fish have genomes equal in mass to those of *Aplysia*, while others have some of the heaviest genomes known. Among mammals, variation in the 2C quantity of DNA is small, as noted above. However, the number of polyploid cells differs greatly even in the species of a single order (for example, in mice and guinea pigs). Thus, the dependence between genome mass and somatic polyploidy is manifest in only a very general form: cells of high ploidy are found more frequently in species with small genomes than in species with large 2C DNA contents.

The significance of somatic polyploidy should evidently be seen not so much as a compensation for a low absolute mass of the genome (a considerable part of which does not carry direct genetic information) but rather as a way of multiplying the number of structural genes in the tissues where the number of cells cannot increase for some reason or other. The particular advantages of polyploid cell lines as compared with diploid ones have already been discovered (see Chapter 7E); this may facilitate the selection of polyploid cells in organisms with any genomic mass.

Conclusion

At the moment, some 300 cases of polyploidization or polytenization of cells of different types have been discovered in animals, plants, and protozoans. The number of examples will undoubtedly grow considerably as research continues. Thus, we do not consider the numerous descriptions of polyploidy based on the single criterion of the nucleus and cell size valid. Re-examination of all the morphological observations by new methods represents an interesting task. For example, dozens of examples of polyploidy may be found in old manuals on cytology, beginning with the classic book by Wilson (1925), and these could be re-assessed.

Although there is considerable information available on quantitative changes in DNA and the sizes and number of chromosomes in different species of organisms, it is not yet sufficient to formulate broad generalizations. In different groups of animals and plants, speciation may occur in different ways. One possibility is that of quantitative changes in the genome and the subsequent acquisition or elimination of certain proteins. Another way is by means of changes in the regulation of gene activity by rearrangement of the available genes. The reasons for the tremendous differences in the number of chromosomes within certain genera (e.g. *Crocus*) are not clear; conversely, it is perhaps less surprising that some species of a given genus (e.g. bats) have practically identical diploid chromosome sets. Nor has it been explained why generative polyploidy is such a common feature in angiosperms, and yet polyploidy hardly occurs in gymnosperms. The absolute mass of the genome in mammals, and especially in some plants and amphibians is so great that it would be sufficient for a million genes (Neyfakh & Timofeeva, 1984). Genetic analysis has revealed no more than 5×10^4 active genes in mammals, and 5×10^3 in *Drosophila*. The functions of most of the DNA sequences therefore are not known, at present; hence, the significance of the transformations which occur in the mass of the genome during evolution remains unclear.

2

The cells of vertebrate animals

Polyploidy is common in the tissues of mammals (Table 2.1), but this may in part be a consequence of them having been more thoroughly studied than any other vertebrates. When evaluating the distribution and number of polyploid cells, it should be remembered that binucleate and multinucleate cells are also polyploid with respect to their total genome. Binucleate cells occur in many mammalian tissues and sometimes they are the predominant component in a tissue. Binucleation is especially characteristic of glandular tissues, for example, the pancreas, salivary glands and also in gastric glands. Binucleate cells are common in the cervical ganglion and also in some other sympathetic ganglia. When evaluating the contribution of binucleate cells to a tissue, one has to reckon with the fact that only one half to two thirds of any binucleate cell is present in a histological section (and morphological studies are usually carried out on sectioned material).

The highest degree of polyploidization of a mature mammalian cell type is the bone marrow megakaryocytes; the level of ploidy in mature megakaryocytes ranges from 8n to 64n (Table 2.1). Many polyploid cells are found in the liver, the exorbital gland, the transitional epithelium of the urinary bladder, and in decidual cells of the uterus (Table 2.1). The trophoblast of the placenta of some mammals is unique both in the amount of DNA accumulated – up to 4906C – and in the mechanism of its reduplication (the endocycle).

Cell polyploidy is more widely distributed in the adult tissues of certain mammals (mice and rats, for example) than in any other organism. However, the amount of DNA accumulated and the degree of multiplication of chromosome number in mammals is not large compared with some other animals and plants. The highest ploidy classes – 32n and 64n – are characteristic of only a few cells of certain mature mammalian tissues and, even in these tissues, 4n cells usually predominate. In other tissues, the 4n class (or its binucleate analogue, 2n × 2) is practically the only class of polyploid cells. In mammals, polyploidization is exact: each doubling of the DNA corresponds to a doubling of the number of chromosomes.

Many mammalian cells which are diploid in normal development *in vivo*,

Table 2.1. *Occurrence of polyploid cells in mammalian tissues*

Species	Type of cell	Level of polyploidy (DNA content)			References
		%	Usual	Maximum	
Mouse	Hepatocytes	99	4C × 2	32C	Epstein, 1967
Rat	Hepatocytes	90	4C	16C	Nadal & Zajdela, 1966
Cat	Hepatocytes	60	4C	8C	Grygoryeva et al., 1970
Syrian hamster	Hepatocytes	40	2C × 2	4C	Wheatley, 1972
Guinea pig	Hepatocytes	15	2C × 2	8C	Dubovaya, Kushch & Brodsky, 1977
House shrew	Hepatocytes	4–9[a]	4C	4C	Shima, 1980
Human	Hepatocytes	30[a]	4C	4C	Ranek, 1976 a, b
Human	Hepatocytes	12	2C × 2	4C	Kudryavtsev et al., 1982
Mouse	Cardiomyocytes	90	2C × 2	16C × 2	Brodsky, Tsirekidze & Arefyeva, 1984
Rat	Cardiomyocytes	95	2C × 2	4C	Kuhn, Pfitzer & Stoepel, 1974; Katzberg, Farmer & Harris, 1977
Pig	Cardiomyocytes	80	2C × 2	2C × 32	Pfitzer, 1971b; Gräbner & Pfitzer, 1974
Rhesus monkey	Cardiomyocytes	50[a]	4C	8C	Pfitzer, Knieriem & Schulte, 1977
Human	Cardiomyocytes	80	4C	8C × 2	Moubayed & Pfitzer, 1975
Mouse	Megakaryocytes	100	16C	32C	Odell et al., 1969
Rat	Megakaryocytes	100	16C	32C	Odell, Jackson & Friday, 1970
Guinea pig	Megakaryocytes	100	16C	32C	De Leval, 1964
Rabbit	Megakaryocytes	100	32C	64C	Garcia, 1964
Human	Megakaryocytes	100	16C	64C	Weste & Penington, 1972
Rat	Cells of the exorbital gland	80[a]	4C	32C	Desaive, 1967
Rat	Retinal melanocytes	80	2C × 2	4C × 2	Marshak & Stroeva, 1974
Mouse	Bladder epithelium	90	4C	16C	Levi, Cowen & Cooper, 1969
Human	β-cells of islets of Langerhans	–	4C[a]	8C	Ehria & Swartz, 1974
Rat	Endothelium of the aorta	50	2C × 2	4C × 2	Maljuk, 1970
Mouse	Mesothelium	–	2C × 2	2C × 4	Zaborskaya, 1965
Mouse	Fibroblasts	–	2C × 2	4C × 2	Brodsky & Krushchov, 1962
Rat	Smooth myocytes	20	4C	4C	Owens, Rabinovitch & Schwartz, 1981
Rabbit	Neurons of cervical ganglion	–	4C	2C × 2	Yarygin & Felichkina, 1977
Mouse	Decidual cells	60[a]	4C	32C	Ansell, Barlow, McLaren, 1974
Rat	Decidual cells	100	16C	32C × 2	Zybina & Grishchenko, 1972
Human	Amnion cells	50[a]	4C	8C	Klinger & Schwarzacher, 1960
Mouse	Giant cells of the trophoblast[b]	100	256C	1024C	Barlow & Sherman, 1972
Rat	Giant cells of the trophoblast	100	512C	4096C	Zybina, 1970; Nagl, 1972b
Rabbit	Giant cells of the trophoblast	100	256C	1024C	Zybina, Kudryavtseva & Kudryavtsev, 1973

[a] In the total nuclear population, binucleate cells are not taken into account. [b] Polyteny.

are polyploidized in culture. Unlike tissues *in vivo*, where cells arrested in G_2 are extremely rare (if they are there at all), some cultured cells have doubled or even quadrupled chromosome sets. Polyploid cells are usually found in malignant tumors of humans and animals. Giant, evidently polyploid, cells are, however, observed in normal diploid tissues after treatment with ionizing irradiation.

We shall examine examples of polyploid cell populations in mammals and some other vertebrates, paying special attention to the origin of these cells.

A The liver: the mitotic origin of polyploid cells

Polyploidy is a characteristic feature of the mammalian liver, although this phenomenon is expressed to various extents in different species (Table 2.2). Polyploidization of hepatocytes has been studied in greatest detail in mice and rats where almost the entire parenchyma of the adult liver is formed of polyploid cells. Hepatocytes constitute about 60% of all cells in the liver, and contribute up to 90% of its weight (Bucher, 1963; Jype, Bhargava & Tasker, 1965; West, 1976). The liver of only a few other species of mammals have been studied. Thus, it is only recently that the ploidy distribution of parenchymal cells in human liver has been determined. On average, 88% of human hepatocytes are diploid, and 9% are binucleate cells with diploid nuclei ($2C \times 2$). When hepatitis occurs, the number of $2C \times 2$ cells increases, sometimes up to 40% (Kudryavtsev *et al.*, 1982). Lack of certainty regarding the ploidy of hepatocytes in other vertebrates was noted in Chapter 1. We recall that in the frog, one species with polyploid nuclei has already been found (Bachmann & Cowden, 1967).

The ploidy of murine hepatocytes is usually no more than 16n; very occasionally 32n cells are found in old animals, or after regeneration. One mouse strain (NMRI) is known, however, to possess a considerable number (*c.* 13%) of 16n hepatocytes with occasional 64n and even 128n cells (Gerhard, Schultze & Maurer, 1971). The doubling of the DNA in hepatocytes is exact and corresponds to the doubling of chromosome number (Fig. 2.1), i.e. these are genuinely polyploid cells (Sun & Chu, 1971; Uryvaeva & Faktor, 1971*b*).

The liver parenchyma in new-born mice or rats contains diploid cells; these divide intensively, increasing their numbers. Binucleate $2n \times 2$ hepatocytes appear in the first few days of postnatal life, and they have increased markedly in number by 2 weeks of age. Then mononucleate tetraploid cells emerge, followed by binucleate $4n \times 2$ cells and finally by mononucleate 8n cells (Fig. 2.2). This succession of hepatocyte cell classes has been established by many cytophotometric studies (Alfert & Geschwind, 1958; Inamdar, 1958; Epstein, 1967; Evans, 1976). The main features of postnatal polyploidization have been confirmed by recent research carried

Table 2.2. DNA ploidy classes (%) of hepatocytes in adult mammals

Order	Species	2C	2C×2	4C	4C×2	8C	8C×2	References
Monotremata	Tachyglossus aculeatus	89.2	10.8	–	–	–	–	
Marsupialia	Macropus agilis	3.3	1.1	88.2	4.7	2.7	–	Gahan & Middleton, 1982[a]
	Macropus canguru giganteus	45.3	13.4	29.8	6.2	5.2	0.1	
Insectivora	Erinaceus europeus	86.5	12.2	2.3	–	–	–	
Rodentia	Microtus fortis	26-40	18-19	29-40	8-9	2-3	1-2	Kudryavtsev et al., 1984[a]
	Microtus sahalinensis	40-60	47-58	2-5	0-1	–	–	
	Microtus subarvalis	31-55	40-56	2-9	0-3	0-1	–	
	Cavia porcellus	85-95	5-15	0-1	–	–	–	James, Schopman & Delfgaauw, 1966
	Rattus rattus	6-17	13-27	53-75	1-2[c]	–	–	
	Mus musculus	0.1-4	10-30	25-40	30-45	3-9	2-5[a]	Our laboratory, summary data
Carnivora	Nycteretus procyonoides	91-96	3-8	0-2	–	–	–	Kudryavtsev et al., 1984[b]
	Alopex lagopus	82-88	11-18	0-0.5	0-0.6	–	–	
	Vulpes vulpes	68-84	15-30	2-3	–	–	–	
	Mustela vison	78-96	4-21	0-1	–	–	–	
	Acinomyx jubatus	0.7	0.9	80.9	17.1	0.4	–	Gahan & Middleton, 1982[a]
Perissodactyla	Equus zebra	91.7	5.4	3.5	–	–	–	Kudryavtsev et al., 1984[b]
	Equus caballus	94-98	2-5	1-2	–	–	–	
Artiodactyla	Sus scrofa	77-82	16-21	1-3	–	–	–	
	Bos taurus	93-95	3-5	1-3	–	–	–	
	Cervus albirostris	79.1	7.7	12.6	0.6	–	–	
	Oryx dammah	83.2	7.8	8.9	0.1	–	–	Gahan & Middleton, 1982[a]
	Giraffa camelopardalis	91.1	5.3	3.6	–	–	–	
	Tayassu tajacu	87.4	10.7	1.9	–	–	–	
Primates	Pan troglodytes	37.5	50.3	8.8	2.5	0.7	0.2	
	Gorilla gorilla	81.3	12.2	5.6	0.6	0.3	–	
	Pongo pygmaeus	91.2	5.3	2.9	0.3	0.3	–	
	Macaca mulatta	63.5	32.1	3.4	0.9	0.1	–	
	Homo sapiens	83-91	6-13	1-7	0-1	0-0.2	–	Kudryavtsev et al., 1982[b]
	Homo sapiens	58	12	24	2	4[e]	–	Koike et al., 1982

[a] Only one animal was sampled.

[c] Range of values computed from the mean ± three standard errors

[e] These data are from the diagram given by these authors.

[b] Variability between animals also stated.

[a] 0.5% and 0.2% of the cells were 16C and 16C×2, respectively.

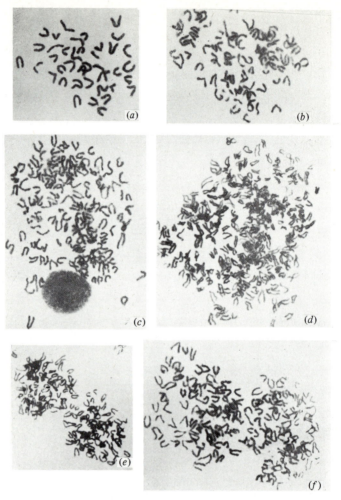

Fig. 2.1. Diploid and polyploid chromosome sets of hepatocytes from regenerating mouse liver. (*a*) 2n, (*b*) 4n, (*c*) 8n, (*d*) 16n, (*e–f*) binucleate 2n × 2 and 4n × 2 (after Uryvaeva & Faktor, 1971*b*).

out using DNA cytophotometry, flow cytometry, and also by volume estimation with the help of a Coulter Counter (Shima, 1980; Böhm & Noltemeyer, 1981*a*; Deschênes, Valet & Marceau, 1981; Steele *et al.*, 1981).

A more in-depth analysis has led to the building of a model of hepatocyte transformation during the growth of rat liver (Fig. 2.3) and, more recently, of mouse liver. The former is due to the work of Nadal and coworkers (Nadal & Heyman-Blanchet, 1965; Nadal & Zajdela, 1966; Nadal, 1970)

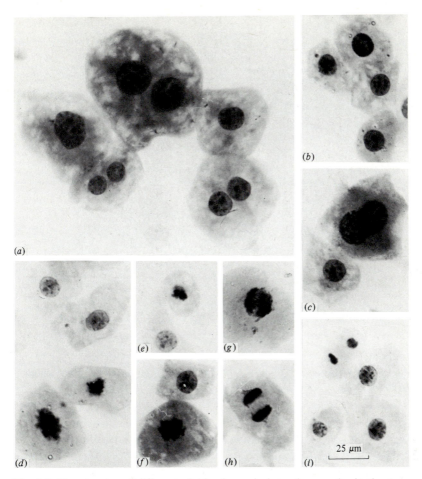

Fig. 2.2. Hepatocytes of different ploidy classes in interphase and mitotic stages isolated from mouse liver. (*a*) Cell types from an intact liver of adult mouse, (*b–i*) mitoses in hepatocytes of regenerating liver. (*b,c*) Prophases, (*d–f*) metaphases, (*g,h*) anaphases, (*i*) telophase.

and the latter, in the main to work in our laboratory. In the diagram of Nadal & Zajdela (see Fig. 2.3) attention is concentrated on the formation of binucleate cells which serve as precursors of mononucleate ones. In this system, the main feature of the polyploidization process is also the alternation of mononucleate and binucleate generations. In the mouse, a typical transformation of murine hepatocyte classes is: $2n \rightarrow 2n \times 2 \rightarrow 4n \rightarrow 4n \times 2 \rightarrow 8n \rightarrow 8n \times 2$ (Brodsky & Uryvaeva, 1977). According to both models, the polyploid series begins with binucleate $2n \times 2$ cells. Therefore, the mode of their formation is of principal importance.

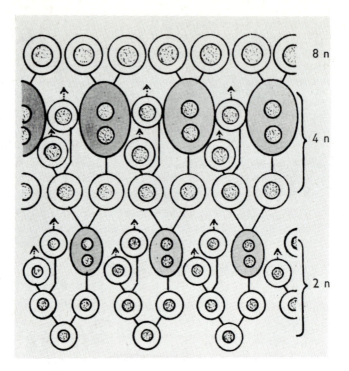

Fig. 2.3. Transformation of cell ploidy classes in the rat liver (after Nadal & Zajdela, 1966).

It has long been suggested that binucleate cells of the liver are formed as a result of acytokinetic mitosis (Beams & King, 1942; Wilson & Leduc, 1948). However, at that time, other possible modes of binucleate-cell formation, such as amitosis of the nucleus or cell fusion were discussed. Recently, evidence for the mitotic origin of binucleate hepatocytes has been obtained. [³H]thymidine was injected into 7–13-day-old CBA/C57BL mice and was incorporated into the diploid nuclei of hepatocytes at the moment of DNA synthesis. Later, the label was found in the G_2-nuclei and then in diploid nuclei again (Fig. 2.4). Thus, in mice of this age, all nuclei which have entered the mitotic cycle divide. Tetraploid nuclei do not accumulate. Part of the diploid nuclear population was found, after mitosis, to consist of binucleate $2C \times 2$ cells. The accumulation of binucleate cells was connected with the passage through mitosis, since the first labelled binucleate cells were found after the completion of the first labelled mitoses. With time the number of labelled mitoses diminishes and the number of labelled binucleate cells no longer increases (Uryvaeva & Lange, 1971). The mitotic origin of binucleate rat hepatocytes (Fig. 2.5), of

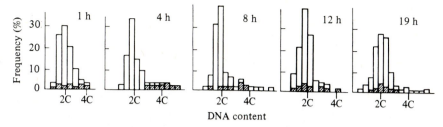

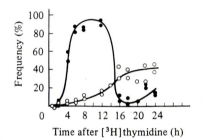

Fig. 2.4. DNA content (C) of labelled (hatched columns) and of unlabelled hepatocyte nuclei at different times after a single [³H]thymidine injection into 7–13 day-old CBA/C57BL mice (after Uryvaeva & Lange, 1971).

Fig. 2.5. Relationship between appearance of labelled mitoses (filled circles) and accumulation of labelled binucleate cells (open circles) in the liver of 3-week-old rats after a single injection of [³H]thymidine (after Marshak, 1974).

cardiomyocytes and of retinal pigment cells was subsequently demonstrated by the same method (see chapter 2B and 2C).

A numerical method, coupled with autoradiography can also demonstrate the mitotic origin of binucleate cells in the rat liver. 3–16 h after an injection of [³H]thymidine, the proportion of newly formed, labelled, binucleate cells gradually increased up to 30% (Fig. 2.5). Therefore, the rate of binucleate-cell formation in the labelled population was 2.3% h⁻¹. The frequency of the labelled interphase cells was *c.* 3.3%. This means that the rate of binucleate-cell formation in the entire hepatocyte population was $(2.3 \times 0.033) = 0.076\%$ h⁻¹. According to morphological observations, the frequency of acytokinesis (telophase unaccompanied by cell cleavage) was 70% in the 3-week-old rats studied. On average, the mitotic index was 0.12%, i.e. acytokinetic mitosis occurred at an approximate rate of 0.084% h⁻¹. This figure (0.084) is close to the rate of binucleate-cell formation calculated from autoradiographic data (0.076).

The appearance of mononucleate polyploid hepatocytes and the further transformation of their classes have been studied at later ontogenetic stages in mice of the above-mentioned strain. All determinations were carried out

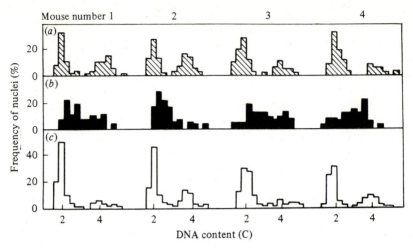

Fig. 2.6. DNA content (C) of (*a*) post-mitotic hepatocyte nuclei; (*b*) nuclei at the time of synthesis and (*c*) unlabelled nuclei in 20–30-day-old CBA/C57BL mice weighing 10–12 g. Data are presented for each of four mice separately. Nuclei shown in (*a*) were labelled by a [^{14}C]thymidine injection 3–6 days before fixation; in (*b*), nuclei were labelled by a [^{3}H]-thymidine injection 1 h before fixation. Each histogram represents 100–200 nuclei (after Brodsky, Faktor & Uryvaeva, 1973).

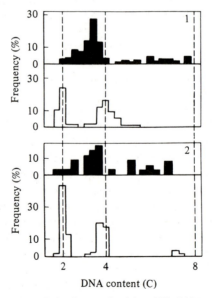

Fig. 2.7. DNA content (C) in the synthesizing (filled histograms) and unlabelled (open histograms) hepatocyte nuclei of 1.5-month-old CBA/C57BL mice weighing 15–17 g. Each paired set of histograms represents cells from the same mouse. Mice 1 and 2 were injected with [^{3}H]thymidine, 1 h before killing (after Brodsky *et al.*, 1973).

on cells of the same mouse. The main method used consisted of measuring the quantity of DNA in the hepatocytes with a double-thymidine label (^{3}H and ^{14}C). Cells in the act of synthesizing DNA (S phase) were labeled with one of the isotopes. The second isotope, given earlier, marks cells which have had time to complete mitosis. Combined autoradiography and cytophotometry on the same nuclei proves that mononucleate polyploid cells arise from binucleate cells of lower ploidy. It is known that the proliferation of hepatocytes is of a rhythmic nature; there may be a pause, with almost zero mitotic activity in the growing liver, after a wave of divisions (Muljutina, Kogan & Nekhljudova, 1978).

In mice 3–4 weeks old, just as in younger ones, mainly diploid nuclei enter into mitosis. Many mitoses are found in binucleate cells. After mitosis, not only 2n nuclei but also 4n nuclei are labelled (Fig. 2.6). In 5-week-old mice, the entire population of nuclei and the mononucleate and binucleate classes of hepatocytes were studied (Fig. 2.7 and Table 2.3). In mice of this age, 2n × 2 cells predominated; there were also quite a lot of mononucleate polyploid cells. After mitosis, the level of ploidy rose and the number of 2n and 2n × 2 hepatocytes fell.

In adult mice, the diploid hepatocytes practically cease to maintain themselves. Only 0.1–4% of parenchymal cells were diploid. Among the newly formed cells, there were hardly any 2n × 2 hepatocytes, and octaploid (4n × 2 and 8n) cells predominated. Hepatocyte ploidy increases still more in old mice, aged 1.2–2 years (Table 2.4), the main classes of cells being 4C × 2 and 8C. The majority of hepatocytes are thus octaploid, though cells with 32 chromosome sets have been seen. But some small groups of diploid hepatocytes are conserved, even in old mice.

Polyploidization of hepatocytes is the result of the alternation of acytokinetic and complete mitoses. Neither amitosis nor cell fusion plays any substantial part in polyploidization (compare Le Bouton, 1976). Direct observations of cell transformation resulting from mitosis demonstrate the mitotic origin of binucleate hepatocytes and, likewise, of the rest of the polyploid series. Furthermore, in a system such as liver parenchyma which is heterogeneous in the ploidy of its cells, fusion, if it occurred, would lead to the appearance of cells with nuclei of different ploidy. This would also be the outcome of accidental nuclear amitosis. According to modern cytophotometric data, practically all binucleate hepatocytes have nuclei of equal ploidy. After a prolonged period following the injection of radioactive thymidine, both nuclei of a binucleate cell are labelled; this need not be the case if accidental fusion occurred.

The role of mitosis in the polyploidization of hepatocytes is particularly well illustrated when the number of binucleate cells is altered, for instance, following damage to the liver, or when there is an excess (of lack) of hormones, or by the nutritional regime. Thus, an arrest to growth at a time

Table 2.3. *Polyploidization of hepatocytes in young 15–17 g CBA/C57BL mice[a]*

Mouse number	No. of cells examined	Non-labelled (non-cycling) cells (%)						No. of cells examined	Labelled (newly formed) cells (%)					
		2C	2C×2	4C	4C×2	8C	8C×2		2C	2C×2	4C	4C×2	8C	8C×2
1	104	12.5	51.9	22.1	13.5	–	–	60	–	11.7	21.7	56.7	10.0	–
2	124	12.1	46.0	19.4	20.2	1.6	0.8	81	–	4.9	40.8	44.4	7.4	2.5
3	98	14.3	45.9	28.6	9.2	1.0	1.0	87	2.3	18.4	10.3	59.8	6.9	2.3
4	84	21.4	34.5	32.1	9.5	2.4	–	88	8.0	18.2	51.1	20.5	1.1	1.1
5	110	10.0	34.5	39.1	14.5	1.9	–	77	3.9	2.6	42.8	46.8	3.9	–
6	114	11.8	41.6	25.7	18.8	1.4	0.7	35	–	25.7	37.2	37.1	–	–

After Brodsky, Faktor & Uryvaeva, 1973.
[a] Cells were fixed 40 h after six injections of [³H]thymidine given at intervals of 8–10 h. Results were obtained by a combination of autoradiography and Feulgen–DNA cytophotometry of the same nucleus.

Table 2.4. *Classes of hepatocytes in the liver of old mice of the hybrid strain CBA/C57BL*

Mouse number	Age (years)	Weight (g)	Ploidy class (%)									Average ploidy (n)	
			2C	2C×2	4C	4C×2	8C	8C×2	16C	16C×2	32C	Nucleus	Cell
1	1.2	33	1.15	7.70	11.15	50.33	16.07	11.48	2.13	–	–	4.9	8.3
2	1.2	42	0.76	10.13	15.87	43.98	17.78	9.56	1.72	0.19	–	4.8	7.9
3	1.2	34	0.34	5.28	10.39	50.09	15.33	16.35	2.21	–	–	5.1	8.8
4	1.5	55	–	10.76	10.13	38.61	12.03	20.89	5.70	0.63	–	5.6	9.8
5	2.0	56	0.85	14.90	9.50	47.80	14.18	9.65	2.55	0.14	0.43	4.7	8.1
6	2.0	55	–	2.39	11.23	32.40	26.15	20.99	5.52	0.55	0.74	6.3	9.9

After Faktor & Uryvaeva, 1975.

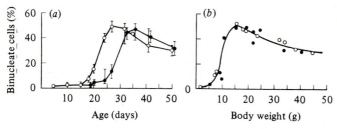

Fig. 2.8. Binucleation in the liver of adolescent rats weaned at two different ages. (*a*) Rats weaned early at 15 days after birth (open circles); rats weaned at 25 days of age (filled circles). At least six rats are represented at each time point. (*b*) Relation between the percentage of binucleate cells in early- and late-weaned rats and body weight (after Wheatley, 1972).

when acytokinetic mitosis is normally occurring reduces the proportion of binucleate cells. If the ensuing mitosis (which transforms $2n \times 2$ cells into 4n cells) is inhibited, the relative number of binucleate cells is then higher than normal. One such example from the rat is cited in Fig. 2.8 (from Wheatley, 1972). In this figure, the accumulation of binucleate hepatocytes correlates with the increase in body weight. In the group of animals where growth is arrested, the number of binucleate cells increases later than in the control group. In other experiments, the growth of young rats and their liver was slowed by thyroidectomy; here the polyploid hepatocytes did not accumulate. Injection of thyroxine stimulated mitoses and several days later the number of tetraploid cells had risen to almost the normal level (Mendecki *et al.*, 1978). The old observations on the arrest of polyploidization of the liver in growth-hormone-deficient animals may be interpreted in the same way (see Leuchtenberger, Helweg-Larsen & Murmanis, 1954; Geschwind, Alfert & Schooley, 1958; Swartz, 1967; Carrière, 1969). Here adult dwarf mice had a liver composed of diploid cells; the injection of a growth hormone (especially combined with thyroxine) normalized growth. At the same time, polyploid cells appeared in the liver parenchyma. Mitosis is also inhibited in the liver by the impact of various stresses such as hypokinesia and dietary restrictions. The arrest of growth of the organ halts polyploidization of hepatocytes (Faktor *et al.*, 1979; Enesko & Samborsky, 1983).

In some papers, circadian variation in the relative number of various ploidy classes in rat liver has been described (Philippens, Rover & Abbecht, 1981; Bhattacharya, van Noorden & James, 1983). Apparently, in growing liver, a time-dependent transition from mononuclearity to binuclearity, and vice versa, may be the result of circadian synchronization of mitosis.

In adult mice and rats, hepatocytes have a low mitotic index. However, massive damage to the liver stimulates almost all the surviving hepatocytes to enter the division cycle (Stöcker *et al.*, 1972; Uryvaeva & Faktor, 1975).

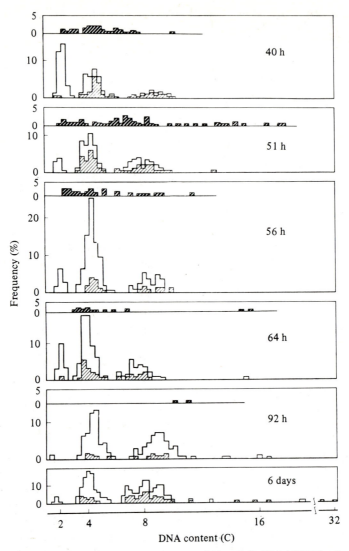

Fig. 2.9. Polyploidization of hepatocyte nuclei in adult CBA/C57BL mice (weight 26–8 g) during liver regeneration. DNA–Feulgen cytophotometry and autoradiography with double [³H] and [¹⁴C]thymidine-pulse label. Each graph represents the hepatocytes of one animal at a particular time (from 40 h to 6 days) after partial (two-thirds) hepatectomy. Lightly hatched histograms, DNA in nuclei labelled with [¹⁴C]thymidine injected 28 h after the operation; heavily hatched histograms, DNA in nuclei labelled with [³H]thymidine injected into the same mouse 1 h before killing; open histograms, unlabelled nuclei of hepatocytes of the same mouse. 200–300 nuclei were scored for each mouse. The [¹⁴C]label is confined mainly to postmitotic and G_2 cells. The [³H]label marks cells in S phase (after Uryvaeva & Brodsky, 1972).

Table 2.5. *The average DNA content* ($\bar{C}$) *in cycling* (*labelled*) *and non-cycling* (*unlabelled*) *hepatocytes.*

Mouse number	Time after partial hepatectomy (h)[a]	Average DNA content per cell ($\bar{C}$)		
		Labelled (l)	Unlabelled (u)	l/u
1	36	8.00	4.59	1.74
2	42	6.88	5.12	1.34
3	42	7.50	4.90	1.53
4	50	6.33	4.66	1.36
5	50	5.45	4.08	1.34
6	58	6.81	4.60	1.48
7	66	8.02	5.23	1.53
8	66	6.66	4.27	1.56
9	72	6.75	4.34	1.55
10	81	6.28	4.73	1.33

After Uryvaeva & Brodsky, 1972.
[a] There was a continuous infusion of [^{3}H]thymidine in CBA/C57BL mice at different periods after partial hepatectomy.

The formation of cells of high ploidy during regeneration of the liver has long been noted in karyometric and cytophotometric works (Bucher, 1963; James, Schopman & Delfgaauw, 1966). In Fig. 2.9, the results of DNA cytophotometry and autoradiography of double-thymidine labelled hepatocyte nuclei are shown at different times of regeneration after partial hepatectomy. Among the DNA-synthesizing nuclei (^{3}H-labelled), 2C and 4C nuclei predominate. After mitosis (^{14}C-label) 4C and 8C nuclei were found; 16C and 32C nuclei appeared later.

According to Gerhard, Schultze & Maurer (1973), the number of nuclei after regeneration of mouse liver increases by only 1.1-fold, while the average ploidy of the nuclei increases by 1.66-fold. According to our data on liver regeneration (Uryvaeva & Brodsky, 1972), the average ploidy of the post-mitotic cells rises by *c.* 1.5-fold (Table 2.5). If all the cells were to divide, their ploidy would not increase. If all the cells were to be polyploidized as a result of mitosis, the average ploidy of the post-mitotic cells would grow twofold. Consequently, during regeneration, part of the cell population undergoes complete mitoses and the other part undergoes polyploidizing mitoses.

A characteristic result of regeneration is a decrease in the relative number of binucleate cells while the number of mononucleate polyploid cells increases considerably (Fig. 2.9; Table 2.6). This is particularly noticeable in the change in the number of 8n and 16n hepatocytes, of which there are very few before hepatectomy. Some hepatocytes undergo two mitotic cycles during regeneration, but most of them undergo just one cycle

Table 2.6. *The decrease in the number of binucleate hepatocytes during regeneration of liver in CBA/C57BL mice*

	Ploidy class (%)					
	2C	2C × 2	4C	4C × 2	8C	8C × 2
Control[a]						
Mean	8.9	48.4	19.6	21.6	1.0	0.5
Range	4.2–12.4	37.2–55.6	12.1–39.2	15.8–26.4	0–1.7	0–1.5
40 h after partial hepatectomy[b]						
Mean	16.4	2.1	78.9	1.1	1.5	–
Range	9.3–22.0	0.4–6.5	68.0–85.3	0–1.8	0.4–1.8	–

After Brodsky & Uryvaeva, 1977.
[a] Eight young mice, each weighing 14–17 g.
[b] Four mice each weighing 14–16 g were operated. The percentage of the various nuclear classes (DNA C-values) are a selected population of labelled post-mitotic cells. Cells were fixed 15 h after the injection of [³H]thymidine (following the wave of labelled mitoses). The methods are described in the notes to Table 2.3.

(Fabrikant, 1967; Liozner & Markelova, 1971; Maurer, Gerhard & Schultze, 1973). A sharp drop in the number of binucleate cells and a simultaneous rise in average ploidy are indicators of differences in the mechanisms of polyploidization during regeneration (compared with undisturbed ontogenetic growth) in the liver. During regeneration, direct transformation of the mononucleate classes – $4n \to 8n$, $8n \to 16n$ – probably occurs, by-passing the binucleate stage. In regenerating liver, arrested metaphases were found, in addition to acytokinetic and other polyploidizing mitoses (Klinge, 1968). In accordance with the ideas of Altmann (1966), incomplete mitoses, which lead to polyploidy, may be the result of a defect in the achromatic division apparatus. Endomitosis, of the form described by Geitler (1953), was not observed, either in normal liver or during regeneration (Klinge, 1968).

Aside from completely ordinary mitosis (and the deviations from it, which are quite normal in liver development), pathological mitoses are also observed, which lead to chromosomal aberrations; some of them may be caused by chromosome aberration that occurred during the previous interphase. The number of aberrant mitoses (with anaphase and telophase bridges, chromosome fragments, or multipolar mitoses) is especially large in old mice (Curtis, 1963).

Polyploidization of hepatocytes is an irreversible process. A decrease in the level of ploidy, 'depolyploidization', has not been observed. Stimulation of hepatocyte proliferation leads to their further polyploidization (Fig.

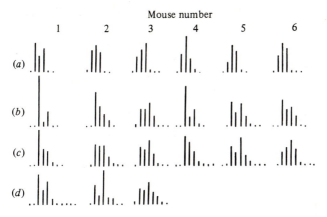

Fig. 2.10. The progression of polyploidy in the mouse liver after repeated partial hepatectomy. Each vertical column represents samples from the liver of the same animal (1–6), removed by sequential operations. (*a*) Before partial hepatectomy; (*b*), 1 month after first partial hepatectomy; (*c*) 1 month after second partial hepatectomy; and (*d*) 1 month after third partial hepatectomy. The height of the vertical bars in each graph shows the relative frequency of a hepatocyte class. Each group of vertical bars, read from left to right, represents 2n, 2n × 2, 4n, 4n × 2, 8n, 8n × 2, 16n, 16n × 2, 32n, 32n × 2, 64n (after Faktor & Uryvaeva, 1975).

2.9). After three partial hepatectomies, the average ploidy of the hepatocytes increased from 5.3–6.2n per cell (at the moment of the first operation) to 7.5–10.4n 1 month after the third operation in the same mice (Faktor & Uryvaeva, 1975). The age of the mice, after three operations, was 5–6 months. The distribution of the hepatocyte ploidy classes in normal animals of this age differs little from that of 2-month-old animals (the control at the first operation). However, after three operations, the DNA histograms (Fig. 2.10) were similar to those for liver parenchyma of normal 2–3-year-old mice (see also Table 2.4). A substantial decrease in the number of binucleate cells was observed: from 60–73% before the operations and, in more adult normal mice, to 24–43% after three operations.

Stimulation of proliferation, caused by inducers of adaptive liver growth, for example, gestagens and phenobarbital, also leads to further hepatocyte polyploidization (Schulte-Hermann *et al.*, 1980; Böhm & Noltemeyer, 1981*b*).

It was formerly thought that among hepatocytes with a multiple quantity of DNA there were cells which were arrested for a long period before mitosis and then capable of beginning division without new DNA synthesis – the so-called G_2-population (Gelfant, 1963*a*,*b*; Perry & Schwartz, 1967). To reveal the cells of the G_2-population, resting cells are usually stimulated to proliferate. An analysis of cell kinetics, DNA synthesis and mitosis during regeneration of the liver did not reveal the existence of a G_2-population of any substantial size (Grisham, 1969). The proliferative pool in the liver parenchyma of young rats and mice reaches almost 100% during regeneration. In experiments with labelled thymidine, three different results may be obtained: (i) G_2-cells begin mitosis without synthesizing DNA and, accordingly, without incorporating thymidine; (ii) G_2-cells do not begin mitosis but, after proliferative stimulus, they undergo subsequent endoreproduction, followed by mitosis; (iii) as with second possibility, but with no mitosis even after a second endoreproduction. The result of the first type of behaviour of the G_2-population is unlabelled mitoses, that of the second is labelled mitoses with diplochromosomes, and that of the third is an excess of labelled nuclei relative to mitoses. None of these manifestations of the G_2-population were found in the liver of mice or rats. All the thousands of hepatocyte mitoses that have been observed in regenerating liver were labelled by continuous infusion of labelled thymidine (Stöcker, 1966; Brodsky, *et al.*, 1969). Diplochromosomes were never found. The number of S-phase cells corresponded well to the number of mitoses (Polishchuk, 1967; Schultze *et al.*, 1973).

Another method of revealing cells blocked G_2 is to study mitoses in the tissue when it is cultured *in vitro*. Some authors have managed to induce a high level of proliferation in cultured adult rat hepatocytes (Koch & Leffert, 1979). About 37% of cells were thereby stimulated to DNA synthesis and all the observed mitoses were labelled. This means that all the mitotic cells also undertook an earlier synthesis.

The number of hepatocytes capable of entering the mitotic cycle decreases as the animal ages. Continuous [³H]thymidine labelling of young rats or mice after a partial hepatectomy (or following the toxic effect of carbon tetrachloride) revealed up to 99.8% labelled hepatocytes (Stöcker *et al.*, 1972; Uryvaeva & Faktor, 1975). In old animals, no more than 74–75% of the hepatocytes were labelled after similar treatment. Cells of high ploidy predominate among the unlabelled cells. The question arises whether cell properties change with age or whether the cells do not enter the cycle because of a changed internal environment in the old organism.

Hepatocytes synthesize serum proteins, glycogen and bile, deaminate amino acids, metabolize xenobiotic substances, and perform other special functions; in the course of these processes a large number of substances is synthesized (see Tsanev, 1975; Salganik, 1979). Indeed, many of these functions are carried out within one and the same hepatocyte. During stimulation of hepatocyte proliferation after a partial hepatectomy, some of the specialized functions become less active (for instance, the drug-metabolizing function), but others are maintained or even amplified.

The reasons for polyploidization of hepatocytes in many ways lie in the question of why the diploid hepatocyte does not complete mitosis. One of the reasons for acytokinetic mitosis is the incomplete preparation of the cell for division owing to the combination in it of premitotic and tissue-specific processes. It is hardly likely, however, that this is the only or even the main reason for polyploidization. The size and the composition of the polyploid hepatocyte population varies considerably between mammals (Table 2.1). The establishment of polyploidy in some species may be caused by the beneficial properties conferred by multiple genome copies (see Chapters 7D and 7E).

When summarizing current knowledge of polyploidy in the liver, the following main points may be noted (see also Chapters 7B, 7C and 8B). The first polyploid hepatocytes – binucleate cells with diploid nuclei – arise through the non-completion of ordinary mitoses. The mechanism of polyploidization is the alternation of acytokinetic and complete mitoses. During regeneration, other types of polyploidizing mitoses frequently occur. The emergence of polyploid cells is probably connected with the special relationship between proliferation and differentiation in developing cells. Their fixation during evolution is apparently caused by the beneficial properties of polyploidy compared to diploidy. This will be discussed in detail further on.

B The myocardium: polyploidy and hypertrophy

Many polyploid cells have been discovered in the myocardium of mammals (Table 2.7). In humans there is a particularly large number of mononucleate polyploid cardiomyocytes; some of these cells reach 16C and 32C ploidy level (Sandritter & Scomazzoni, 1964; Moubayed & Pfitzer, 1975). The frequency of binucleate cardiomyocytes in humans amounts to about 13–14%, but in adult rats and mice they comprise up to 90% of cells of the myocardium (Kogan, Belov & Leontjeva, 1976; Katzberg, Farmer & Harris, 1977). Almost all murine binucleate myocytes are tetraploid $2C \times 2$ cells, although some other ploidy classes have been shown (Fig. 2.11). An interesting form of cell polyploidy develops in the pig's myocardium. In addition to large numbers of mononucleate and binucleate polyploid cardiomyocytes, there are multinucleate cells that have from 4 to 32 nuclei (Gräbner & Pfitzer, 1974, see also Fig. 2.12).

Polyploidization of the myocardium was not properly appraised until very recently. Almost all the early studies were carried out on sectioned material. The observed number of binucleate cells is known to be greatly underestimated in sections and the ploidy of mononucleate cells cannot therefore be evaluated. Conclusions drawn from studies of isolated cardiomyocyte nuclei are also restricted. Although very clear histograms of the DNA content in the nuclei have been produced by these studies, the conclusion that cardiomyocytes are diploid in most of the animals studied is false. The number of polyploid *nuclei* in the myocardium of mice and rats does not, in fact, exceed 5–10% (Kuhn, Pfitzer & Stoepel, 1974; Brodsky, Arefyeva & Uryvaeva, 1980*a*), though 80–90% of the *cells*

Table 2.7. *Polyploid cardiomyocytes in the normal myocardium of adult mammals*

| Species | Polyploid cells (% of total population) | | References |
	All	Binucleate only	
Human	Not less than 80	—[a]	Sandritter & Adler, 1978
Human	Not less than 80	13–14	Schneider & Pfitzer, 1973
Human	c. 70	13–14	Moubayed & Pfitzer, 1975
Human	Not less than 50	—[a]	Eisenstein & Wied, 1970
Rhesus monkey	60–80	—[a]	Pfitzer, Knieriem & Schulte, 1977
Pig	—[a]	> 10 (35% tetranucleates, 13–20% multinucleates)	Pfitzer, 1971a; Gräbner & Pfitzer, 1974
Guinea pig	—[a]	c. 90	Katzberg, Farmer & Harris, 1977
Guinea pig	—[a]	81	Korecky, Sweet & Rakusan, 1979
Rat	—[a]	95	Kogan, Belov & Leontyeva, 1976
Mouse	—[a]	c. 80	
Mouse	> 90	≥ 80	Brodsky, et al., 1980a
Dog	—[a]	85	Fischman, 1979
Rabbit	—[a]	78	Korecky & Rakusan, 1978
Cat	—[a]	76	Korecky & Rakusan, 1978
Cow	—[a]	54	Korecky & Rakusan, 1978

[a]No information is available.

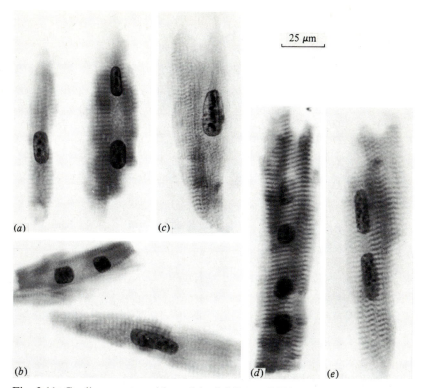

Fig. 2.11. Cardiomyocytes with nuclei of different DNA contents isolated from 3–4-week-old mice. (*a*) 2C and 2C × 2, (*b*) 4C and 2C × 2, (*c*) 8C, (*d*) 2C × 4, and (*e*) 4C × 2.

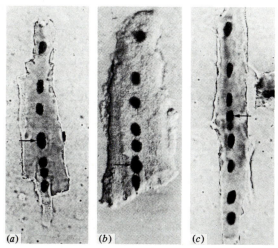

Fig. 2.12. Multinucleate myocytes in pig heart. (*a,b*), A tetraploid nucleus (arrows) between diploid nuclei; (*c*), a nine-nucleated cell with seven diploid nuclei, one hypodiploid nucleus (arrow) and a micronucleus (after Pfitzer. 1980).

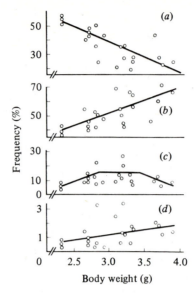

Fig. 2.13. Changes in the cell population of heart myocytes in CBA/C57BL mice growing from 2.5 to 4 g. (*a*) 2C, (*b*) 2C × 2, (*c*) 4C, and (*d*) 4C × 2 and 8C. Each point represents one animal (after Brodsky *et al.*, 1980*a*).

(cardiomyocytes) are binucleate and therefore polyploid. The functional similarity of the binucleate cells with mononucleate polyploid ones will later be demonstrated for various cell types, including cardiomyocytes (see Chapter 7B).

Polyploidization of the cardiomyocytes is a regular event in early postnatal development. We discuss here our data on the ventricular cardiomyocytes of mice. In 1-day-old mice, approximately 5% of cells are binucleate (2C × 2) ones, the remainder are diploid. Substantial changes occur between the third and sixth day after birth (Fig. 2.13). During this short period, the proportion of diploid cells falls to 20–30%, and the binucleate (2C × 2) cell population increases to 60–70%. Later, the rate of polyploidization of the myocardium slows down (Table 2.8). By 3 weeks of age, polyploid cells comprise more than 90% of the population.

It is noteworthy that the number of the cardiomyocytes increases by approximately 20% in mice between the first and third days after birth. Then, during the main polyploidization period (3–14 days), cell numbers remain constant. A slow rate of polyploidization continues, until 21 days after birth; the population size increases by another 5–7% and then does not change (Brodsky *et al.*, 1983*b*).

In rat myocardium, the first polyploid cells are also binucleate. Their numbers grow and (just as in mice) binucleate cardiomyocytes eventually make up almost the entire cardiac muscle (Fig. 2.14). Recently, Kasten,

Table 2.8. *Changes in the ploidy classes of the ventricular cardiomyocytes during post-natal life of CBA/C57BL mice*

Age (days)	No. of cells examined	Ploidy class (%)								
		2C	2C×2	4C	4C×2	8C	2C×4	2C+2C+4C	8C×2	16C
1	1592	90.1	7.4	2.4	–	0.1	–	–	–	–
3	2079	92.7	4.8	2.3	0.1	0.1	–	–	–	–
7	1488	31.0	67.0	1.2	0.8	–	–	–	–	–
21–30	3203	1.7	82.9	2.8	9.4	0.8	1.3	0.7	0.2	0.2
90	926	4.2	81.4	4.8	7.2	0.9	1.0	0.3	0.1	0.1
360	892	4.1	83.0	4.5	6.5	0.1	0.9	0.9	–	–

After Brodsky, Tsirekidze & Arefyeva, 1984.

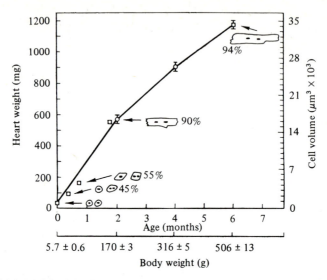

Fig. 2.14. Weight of heart (circles), mean volume of a myocyte (squares) and frequency of binucleate myocytes (expressed in %) in rats from birth to 6 months of age (after Katzberg *et al.*, 1977).

Kudryavtsev & Rumyantsev (1982) revealed by cytofluorometric estimation that 75–80% of the binucleate (mainly 2C × 2) cardiomyocytes appear between birth and 7–10 days, while DNA synthesis ceases by day 14. Binucleate cells also appear early in human myocardium (Schneider & Pfitzer, 1973). In the left ventricle of new-born babies, binucleate cardiomyocytes comprise approximately 14% of the total. The number of 4C mononucleate cells rises from 5% at 4.5 months after birth to 50–60% at 14 years of age. In adult humans, 8C cells are common, and quite a few cells are of higher ploidy.

In those species studied, binucleate cardiomyocytes appear earlier than do the binucleate cells of the liver and are formed at a considerably higher rate. In CBA/C57BL mice, by the sixth day after birth, approximately 70% of cardiomyocytes are binucleate (2C × 2) (Table 2.8) while in the liver only 2% of hepatocytes are binucleate (2C × 2), and there are not yet any other polyploid cells. According to the estimates of Uryvaeva & Lange (1971), the rate of formation of the binucleate hepatocytes in 1–2-week-old mice is 0.025% h^{-1}. The rate of binucleate-cardiomyocyte formation is higher by approximately one order of magnitude: 0.23% h^{-1} (Brodsky *et al.*, 1980*a*).

Polyploid cardiomyocytes, like the polyploid cells of the liver, are formed as a result of incomplete mitosis. In the heart of 3–4-day-old mice, DNA is synthesized almost exclusively by diploid cells (Fig. 2.15). After

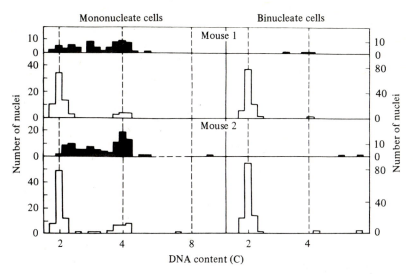

Fig. 2.15. DNA content (C) in DNA-synthesizing (filled histograms) and unlabelled nuclei (open histograms) of the mononucleate and binucleate heart myocytes in 3–4-day-old mice after a single [¹⁴C]thymidine injection (see also Fig. 2.16).

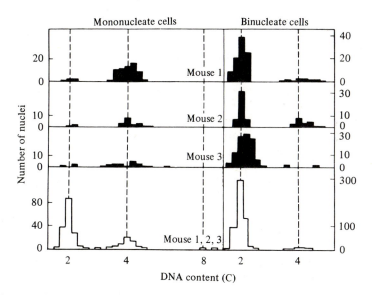

Fig. 2.16. DNA content (C) in postmitotic newly formed (filled histograms) and unlabelled nuclei (open histograms) of mononucleate and binucleate heart myocytes 32–36 h after [¹⁴C]thymidine injection into 3–4-day-old mice (after Brodsky *et al.*, 1980*a*).

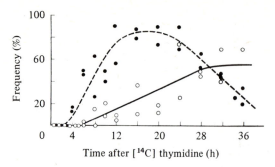

Fig. 2.17. Relationship between the appearance of labelled mitoses (filled circles) and the accumulation of labelled binucleate cells (open circles) in murine heart ventricles after a single injection of [^{14}C]thymidine into 3–4-day-old mice (after Uryvaeva, Arefyeva & Brodsky, 1980).

mitosis, the label of radioactive DNA precursor is found primarily in the 2C×2 and 4C cardiomyocytes (Fig. 2.16). The labelled binucleate cardiomyocytes accumulate as labelled mitoses proceed (Fig. 2.17). Simultaneous with the accumulation of binucleate cells, the number of diploid cells decreases (Fig. 2.13).

Acytokinetic mitoses in the myocardium have frequently been observed, both in fixed preparations (Klinge, 1970, 1971; Rumyantsev, 1972; Przybalski & Chlembowski, 1972) during studies of the heart *in vivo*, and also in living cardiomyocytes *in vitro* (Chacko, 1973; Goode, 1975).

Mononucleate 4C cardiomyocytes in mice, just like binucleate ones, are formed as a result of incomplete mitosis. But the mode of their formation differs from that in the liver. As noted earlier, in the liver acytokinetic and complete mitosis alternate and binucleate hepatocytes are the ancestors of mononucleate polyploid ones. The situation in the myocardium is quite different. Binucleate cardiomyocytes hardly ever enter the mitotic cycle in the period studied (Fig. 2.15). 4C cells are formed directly from diploid ones, as can be seen from the post-mitotic labelling pattern (Fig. 2.16).

The main cell transformations in the ventricles of mouse myocardium are the following: 2C → 2C×2 and 2C → 4C. In rare cases, other changes are also possible.

In the myocardium of 5–6-day-old mice, a few post-mitotic diploid cardiomyocytes were found, i.e. cell division has continued at a low rate. In rare cases, binucleate cells synthesizing DNA were also seen (Fig. 2.15). Hence an increase in the ploidy of the nuclei of binucleate cells (2C×2 → 4C×2) was occurring; or the number of the nuclei in the cardiomyocyte was increasing, i.e. 2C×2 → 2C×4 (in mice and rats approximately 0.5% tetranucleate cardiomyocytes were observed); or fusion of metaphase plates is possible, together with the formation of mononucleate tetraploid cells (2C×2 → 4C+4C). All these variants of cardiomyocyte transforma-

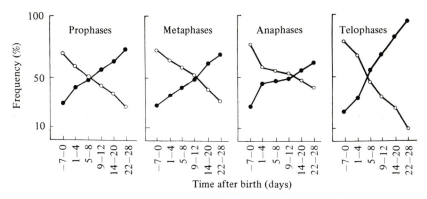

Fig. 2.18. Changes of the proportion of normal (open circles) and abnormal mitoses (filled circles) in rat cardiomyocytes up to 28 days after birth (after Klinge, 1971).

tions have been substantiated not only by cytochemical but also by morphological data.

O. Klinge (1970) observed modified ('kollabierte') prophases and sometimes atypical metaphases in rat cardiomyocytes. We have also seen unusual prophases. From time to time, pictures of paired prophases and metaphases are encountered. The number of normal mitotic figures decreases almost linearly from 70–80% in new-born rats to 20–30% after 2–3 weeks (Fig. 2.18). Klinge noted spindle defects in dividing cardiomyocytes. Should this be confirmed, this important observation may allow us to understand the specific mechanism of polyploidizing mitosis in the heart muscle cells, just as in the liver (Altmann, 1966). Among the pictures of mitoses (Fig. 2.19), cell cleavage has very seldom been observed (only twice among thousands of cells examined).

In evaluating cell transitions, it should be taken into account that by 2–3 weeks after birth in mice, and possibly also in rats, the number of myocytes in the heart ventricles has stabilized. DNA synthesis in cardiomyocytes may be observed at 10–15 days, but is sharply diminished, although it is not completely blocked (see Rumyantsev, 1981). It is interesting that during the first months of life differentiation of the heart myocytes continues. The organization of the contractile apparatus and the intermedial-filament system changes in both the atria and the ventricles, (Carlsson *et al.*, 1982).

Thus, the main precursor for all types of polyploid mouse cardiomyocytes is a diploid cell. The same is probably true of the rat, though, in this species, there are even fewer 4C cells than in mice (and most of the mature cardiomyocytes are binucleate $2C \times 2$ cells). In humans, mononucleate polyploid cells predominate and the number of binucleate cells decreases with age. During hypertrophy of the human myocardium (e.g. after an

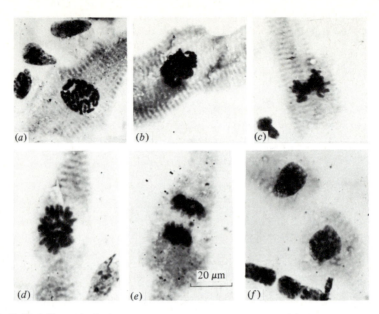

Fig. 2.19. Mitoses in isolated cardiomyocytes in 3–6-day-old mice. (*a*) Prophase, (*b*) 'collapsed' pro-metaphase, (*c,d*) metaphases, (*e*) anaphase, and (*f*) a rare telophase with cell cleavage.

infarction) the number of binucleate cardiomyocytes again increases considerably (Schneider & Pfitzer, 1973). As already mentioned (p. 37), a characteristic of pig myocardium is the high percentage of multinucleate cells. There may be as many as 32 nuclei in a single cardiomyocyte, although cells with 4–8 nuclei predominate among the multinucleate ones (Gräbner & Pfitzer, 1974). The nuclear DNA content of a single cell may vary from 2C to 16C, but the main class consists of 2C nuclei (80%). Sometimes, non-doubling variants (such as 6C or 12C nuclei) are encountered in a chain of nuclei. In these rare cases, multipolar mitoses may contribute to the formation of the multinucleate cell. These mitoses are observed in the myocardium of pigs, rats and mice.

Rumyantsev (1963, 1966) was the first to note that mitoses can occur even in embryonic cardiomyocytes which have accumulated many myofibrils. Therefore, tissue-specific protein synthesis in the myocardium – i.e. the synthesis of contractile proteins and the formation of myofibrils – occurs concurrently with preparation for cell division (Weinstein & Hay, 1970; Chacko, 1973). In this respect, cardiac muscle differs from skeletal muscle. In the latter, mitosis ceases before the myoblast fusion and myofibril formation. On the other hand, in smooth muscle cells, mitosis occurs concomitantly with differentiation and also in differentiated cells (Cobb & Bennett, 1970; Chamley & Campbell, 1974). Smooth muscle cell contraction ceases during mitosis; likewise, during metaphase and anaphase, cardiomyocytes do not contract (Mark & Strasser, 1966; Kasten, 1972; Pollinger, 1973; Goode, 1975). Differentiation of the cardiomyocytes causes the number of mitoses to drop sharply, but it does not inhibit them completely, even in old animals. Thus, according to the data of Rumyantsev (1963, 1966, 1981) and Klinge (1970) for rats, the mitotic index falls by *c.* 50% during the first few days after birth; during the next 10 days it falls another ninefold. Even in 1–2-year-old rats, a mitotic index of 0.002–0.005% is still observed in cardiomyocytes of the ventricles. If the labelling index is 0.02% and S-phase lasts approximately 13 h (Rumyantsev & Kassem, 1976), then barely 10% of cardiomyocytes enter the cell cycle during the period from the age of 6 months to the end of the rat's life span.

Where there is hypertrophy of the myocardium, as after infarction or in conditions of hyperfunction, the incorporation of [^{3}H]thymidine into the myocytes of certain regions of the myocardium is observed and there is also an increased quantity of DNA in some post-mitotic cells (Rumyantsev, 1966, 1979; Klinge, 1970; Steinert *et al.*, 1974; Pfitzer, Knieriem & Schulte, 1977; Oberpriller, Ferrans & Carroll, 1983). An interesting and as yet unexplained consequence of stimulation of proliferation in differentiated cardiomyocytes is the much more intense reaction of the atrial compared to the ventricular cardiomyocytes. Rumyantsev (1979) suggests that atrial myocardium cells, just like the cells of the conductive system of the heart, are less specialized than the ventricular myocardium cells. The latter are much bigger, contain many more myofibrils and have thickened intercalated discs. However, even the ventricular cells are capable of entering the mitotic cycle. After repeated injections of tritiated thymidine, labelled cells appear in the perinecrotic zone of the infarction. There are few (maximum 5%) labelled cells in the ventricles, but 60% of atrial myocytes are labelled. Even so, it seems that, in principle, the ventricular myocytes constitute a reversible post-mitotic population and not a static one.

A noteworthy characteristic of cardiomyocytes is their intensive growth after polyploidization. From birth to the age of 1 year in mice, the weight of the heart ventricles increases approximately 30-fold though the number of cells increases during this period by only 25%. The increase in mass due to polyploidization of the cells accounts for only one fifteenth of the growth. The main mechanism for the very considerable post-natal growth of the ventricles is the increase in mass (protein content per cell) of the cardiomyocytes, with the number of chromosomes remaining constant, i.e. the cells remain out of cycle (see also chapter 7C).

Polyploidization and subsequent cell growth is also considerable in the postnatal development of the human heart. The weight of the heart of an adult human is 16 times the weight of the heart of a new-born baby, and the average volume of the cardiomyocytes increases by 15-fold (Zak, 1974).

Thus, the cardiac muscle of mammals is a polyploid tissue. Polyploidization occurs in early post-natal development and is the result of incomplete mitosis. Myocyte growth continues after the cessation of polyploidization.

C The pigment epithelium of the retina: the growth of differentiating cells

The pigment epithelium of the retina in some mammals presents yet another example of large-scale polyploidization during ontogenesis. The attractiveness of the melanocytes as a subject of research lies firstly in the fact that it is easy to discern their differentiation (they contain melanin pigment granules) and secondly, unlike other polyploid cell lines, the pigment epithelium layer is a uniform cell population. The dynamics of polyploidization of the pigment epithelium have as yet been studied in the retina of rats only.

Compared with the liver and the myocardium, polyploidization of the pigment epithelium is less significant, both in its scale and in the extent of multiplication of the genome. Binucleate cells predominate, a fact first noted in retinal melanocytes of the rabbit (Vinikov, 1938) and later studied quantitatively in rats (T'so & Friedman, 1967; Stroeva & Brodsky, 1968; Marshak & Stroeva, 1973). The number of binucleate cells in the pigment epithelium of adult grey rats is 70–80%. Almost all binucleate melanocytes have diploid nuclei ($2C \times 2$) though some octaploid ($4C \times 2$) cells are also observed (Fig. 2.20). In rats, approximately 1% of cells are mononucleate polyploid or multinucleate ($2C \times 4$); in some animals this percentage is greater.

The pigment epithelium has been studied using tangential sections. Since the main part of the single-row layer of cells, and all their nuclei, can be accommodated in one section, DNA measurements and the evaluation of the number of binucleate cells can be made accurately.

Polyploidization of the pigment epithelium occurs in the first few days of post-natal life of the rat, being completed by day 11–15, when the animal's eyes have opened (Fig. 2.21). The number of polyploid – $2C \times 2$ – cells begins to increase 2 or 3 days after birth. Before that, the melanocytes are almost exclusively diploid.

Polyploidization of the melanocytes, as in other cell types studied, is an event dependent upon the mitotic cycle. The number of DNA synthesizing cells and mitoses coincides with the time of polyploidization of the melanocytes in the early post-natal life of rats. During normal development, from 2 to 9 days after birth, the proportion of binucleate cells increases from 10–20% to 80%. Experimental blocking of the cycle (e.g. by removal of the lens) halts polyploidization (Fig. 2.22). On the seventh and ninth days after lens removal there was no labelling with [³H]thymidine. It is not therefore surprising that the number of binucleate cells in the experimental

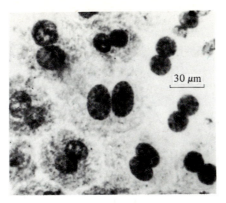

Fig. 2.20. Binucleate 2C × 2 and 4C × 2 cells in a tangential section of the pigment epithelium of the retina in an 18-day-old grey rat. Feulgen-stained, after partial depigmentation (after Marshak & Stroeva, 1973).

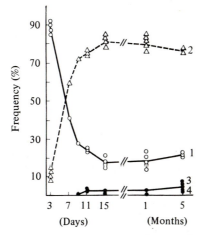

Fig. 2.21. Changes in mononucleate and binucleate melanocytes in the retina of grey rats from 3 days to 5 months after birth. Lines labelled 1, 2, 3 and 4 represent cells with that number of nuclei (after Stroeva & Panova, 1976).

microphthalmics remained at the same level as when proliferation ended (i.e. as at day 5).

Binucleate melanocytes form by acytokinetic mitosis. In tissue where 2C × 2 melanocytes are formed, there are hardly any mononucleate tetraploid cells. Labelled binucleate cells accumulate after the completion of labelled mitoses (Fig. 2.23). The same mechanism has been observed in hepatocyte and cardiomyocyte binucleation (Fig. 2.5 and Fig. 2.17).

Acytokinesis was proposed long ago to be the mechanism of formation of binucleate pigment cells of the retina. The mitotic origin of binucleation

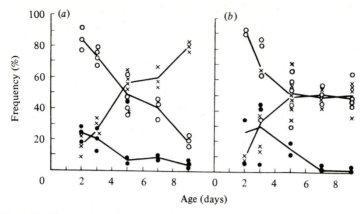

Fig. 2.22. The number of DNA synthesizing ([³H]thymidine-labelled) nuclei (filled circles), of mononucleate (open circles) and binucleate (crosses) melanocytes in (*a*) normal and (*b*) microphthalmic 2–8-day-old grey rats (after Panova & Stroeva, 1978).

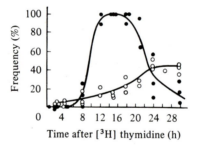

Fig. 2.23. Labelled mitoses (filled circles) and labelled binucleate melanocytes (open circles) after a single injection of [³H]-thymidine into grey rats (after Marshak, 1974).

is established by comparing quantitative cell kinetic data. In 4-day-old rats, binucleate melanocytes form at a rate of 0.11% h^{-1}, according to data on the accumulation of [³H]thymidine labelled binucleate cells and labelling indices. According to the frequency of acytokinetic telophase figures, the rate of formation of binucleate cells is 0.18% h^{-1}.

The relationship between proliferation and differentiation in melanocytes is probably similar to that in hepatocytes or cardiomyocytes with proliferation proceeding simultaneously with differentiation. Cells which have begun to accumulate pigment during embryonic development and after birth continue to take up [³H]thymidine (in rats) even when the intensity of melanin synthesis is still high. Electron microscopy reveals mitoses in cells with a considerable number of melanosomes at different stages of maturity (Marshak, Gorbunova & Stroeva, 1972). Differentiation during proliferation has been confirmed by the discovery of tyrosinase activity (an

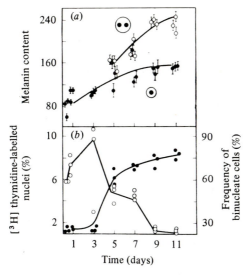

Fig. 2.24. Dynamics of melanocyte differentiation in the retina of grey rats from birth to 11 days. (*a*) Changes in melanin content (arbitrary units determined by cytophotometry of unstained sections, $\lambda = 590$ nm) in mononucleate and binucleate melanocytes. (*b*) The accumulation of binucleate melanocytes (filled circles) together with changes in the frequency of DNA-synthesizing melanocyte nuclei (open circles). Each point in (*a*) and (*b*) indicates one animal (after Marshak, Stroeva & Brodsky, 1976).

enzyme specific to melanin synthesis) in dividing pigment cells (Jimbow, *et al.*, 1975). A few mitoses were also found in melanocytes of old animals. Just as in the myocardium, differentiation and accumulation of tissue-specific proteins does not block the ability of the cells to enter the mitotic cycle.

The dynamics of melanin accumulation (Fig. 2.24) have shown that the period of polyploidization – days 3–11 – coincides with tissue-specific synthesis. It is noteworthy that blocking DNA synthesis in differentiating pigment cells may intensify melanin synthesis (Zimmermann, 1975), and that stimulation of proliferation slows down tissue-specific synthesis (Whittaker, 1970; for more detail see chapter 8B).

Polyploidy (binuclearity) is characteristic of the pigment layer only in some mammalian species; in others, polyploid melanocytes are not found. Species-specific differences in melanocyte polyploidization are not yet clear.

The number of polyploid melanocytes is known, in rats, to depend on total eye weight, but not on age or genotype. Experiments with a mutant strain of rats (MSU$_{BL}$, in which eyes of small weight and size frequently develop) are especially interesting. Just as in normal rats, in the micro-

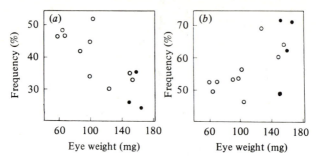

Fig. 2.25. Frequency of (*a*) mononucleate and (*b*) binucleate retinal melanocytes plotted against eye weight (mg) in normal (filled circles) and in mutant microphthalmic MSU$_{BL}$ (open circles) rats (after Marshak & Stroeva, 1974).

phthalmic ones all ploidy classes of melanocytes are found: there are mainly 2C and 2C × 2, with a few 4C and 2C × 4 cells. The less the weight of the eye, the fewer binucleate melanocytes there are (Fig. 2.25).

Melanocytes represent only a very small part of the cellular population and therefore, of the mass of the eye. Hence, direct calculations on the contribution of cell polyploidization to the growth of the tissue are impossible here (at any rate, of the type performed for the myocardium and liver). It has been shown, however, that changes in the pigment epithelium should correlate with those in the scleral sector of the eye. During polyploidization of melanocytes, the area of the scleral sector nearly doubles. The increase in the area of the melanocytes (during transformation of 80% of the diploid cells into binucleate ones) is of the same order (Stroeva & Panova, 1976, 1983). It has been suggested that melanocyte binucleation may be a mode of surface control in the epithelial layer as well as a mode of growth coordination between the epithelial and the scleral areas (Stroeva & Mitashov, 1983).

Thus, in rats, retinal melanocyte growth is achieved in the first 10 days by means of polyploidization. The size of the melanocytes at each ploidy level does not change in that time (Marshak, Stroeva & Brodsky, 1976). The melanocytes do not then enter into the mitotic cycle but rather they become distended under the influence of intraocular pressure (Coulombre, 1956). From the 15th day after birth for several months, the area of the scleral sector expands fourfold. Does the cytoplasm of the non-cycling melanocytes become hypertrophied? This is a feasible suggestion. It has already been noted that, during the preparation of tangential sections, the melanocytes of young rats do, in the main, occur only in one section 5 μm in thickness. If the cells were only to distend after that, without increasing their mass, their thickness in the layer would be approximately 1 μm.

Thus, 70–80% of the rat pigment epithelium layer consists of binucleate 2C × 2 cells. Polyploidization is completed during the first 2 weeks of

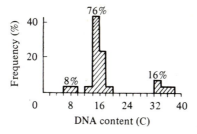

Fig. 2.26. DNA content (C) in the thrombopoietic megakaryocytes (after Paulus, 1968).

postnatal life. Polyploidy is a function of eye growth and not of the age of the animal. During polyploidization, growth of the pigment epithelium is probably caused solely by this process, and not by cell distension or cytoplasmic hypertrophy. Further growth then begins on the basis of a constant genome.

D Megakaryocytes: the differentiation of the polyploid cell

Megakaryocytes are the cells of the hemopoietic tissue that produce blood platelets. In the mammals studied – rats (Odell, Jackson & Gosslee, 1965), mice (Odell *et al.*, 1969), guinea pigs (De Leval, 1964), rabbits (Garcia, 1964; Dupont *et al.*, 1983), and humans (Weste & Penington, 1972) – they are always polyploid. The extent of genome duplication in megakaryocytes is greater than any other known to occur in mammalian tissues. The nuclei of some human megakaryocytes have 64 chromosome sets; in other mammals, 16n and 32n cells are usual (see, for example, Fig. 2.26).

As megakaryocytes mature, three types of cells can be discerned (Odell & Jackson, 1968; Odell *et al.*, 1969; Odell, Jackson & Friday, 1970; Odell, 1972). These are: type I, immature cells with basophilic cytoplasm; type II, maturing cells with acidophilic granular cytoplasm; and type III, mature cells which produce blood platelets. In mice, the ratio of the three types is 10:86:4 respectively, i.e. differentiating cells predominate. The maturing promegakaryocytes (type II) are also the most numerous in rats, comprising 67% of the entire population. In rats, as in mice, terminally differentiated megakaryocytes form the smallest population compartment; they perish following the fragmentation of a considerable part of the cytoplasm.

The immature megakaryoblasts of type I are capable of DNA replication. Type II megakaryocytes contain label 4 h after the introduction of [³H]thymidine; type III megakaryocytes are labelled after 72 h.

The megakaryocytes become cells of high ploidy even before the main marker of specific differentiation (the release of platelets) has become

Table 2.9. *Types and ploidy classes of megakaryocytes in rats*

Types	DNA classes (%)		
	8C	16C	32C
Immature proliferating (Type I) unlabelled[a]	23	66	11
Labelled[a]	54	42	4
Total	32	59	9
Differentiating (Type II)	15	68	17
Mature (Type III)	7	78	15
Total population[b]	19	66	15

After Odell *et al.*, 1970.
[a] 30 min after administration of [³H]thymidine.
[b] 1122 cells in total.

apparent. Even immature type I megakaryocytes can reach ploidy levels of 8, 16 and 32n (Table 2.9). The same ploidy classes are found in type II and III megakaryocytes. Thus, complete differentiation occurs in some of these cells as soon as they become octaploid. Other cells undergo one or two cycles of polyploidization before terminal differentiation. Consequently, it is hardly likely that multiplication of the genome in itself determines the course of differentiation; it is more probable that the two processes (polyploidization and differentiation) occur simultaneously.

   The description above, of the dynamics of maturation and polyploidization, refers to the compartment of morphologically recognizable megakaryocytes (i.e. cells from the 8n level of ploidy upwards; Fig. 2.27). The initial polyploidization (the transition from the diploid to the octaploid state) occurs in the precursor-cell compartment. These cells are not morphologically recognizable, and have therefore not been studied. The initial mechanisms can, to a certain extent, be interpreted by examining the behaviour of bone marrow cells *in vitro*, using microcinematography (Kinosita, Ohno & Nakasawa, 1959). Some diploid cells undergo ordinary mitosis and increase the number of diploid precursor cells. Other cells undergo mitosis up to cytokinesis. Then, rather than going through cytokinesis, the daughter nuclei fuse, forming a mononucleate tetraploid cell. Some cells undergo this type of mitosis four times, ultimately forming a 32-ploid immature megakaryocyte.

   There is another model for the initial events in megakaryocytopoiesis. Autoradiographic studies of rat spleen have demonstrated fusion of the lymphocyte-like precursor cells to form a multinucleate cell (Sklarew, *et al.*, 1971). A high-polyploid mononucleate cell is formed from the multinucleate cell as a result of a multipolar mitosis which is blocked during metaphase. This model requires the support of additional data. At

PRESENT MODEL OF MEGAKARYOCYTOPOIESIS

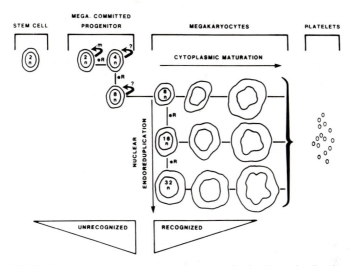

Fig. 2.27. Model of megakaryocytopoiesis; m, mitosis; R, reduplication (after Nakeff, 1980).

the moment, among the early precursor cells that can be discerned using acetylcholinesterase stain (Long & Williams, 1981), no multinucleate cells have been found.

Ideas on the ways of raising the level of polyploidy in a morphologically recognizable compartment are controversial, although there are sufficient data which should permit some conclusion to be reached. Hematologists are well aware that megakaryocytes are truly polyploid cells. Metaphase plates in megakaryocytes contain multiple copies of separated chromosomes. Diplochromosomes are not characteristic of these cells. Although different authors give different interpretations of morphological descriptions of mitosis, it is clear that the level of polyploidy ranges from 8- to 16-ploid and from 16- to 32-ploid, and rises by means of incomplete, or polyploidizing, mitoses. In more than 90% of cases, polyploid multipolar mitoses without cytokinesis are observed. Arrested mitoses (where there is dissolution of the nuclear membrane, but without separation of anaphase chromosomes) occur less frequently (Goyanes-Villaescusa, 1969). Depending on the distribution of centrioles, and also on spindle anomalies, mono-, bi- or multinucleate cells are formed with fragmented or segmented nuclei. An interesting circumstance noted during the analysis of more than 1000 mitotic megakaryocytes in rat bone marrow was the absence of cell cleavage (Odell, Jackson & Reiter, 1968). This means that recognizable megakaryocytes are not self-maintaining and do not increase their number.

The growth of individual megakaryocytes occurs as a result of poly-ploidization, on the one hand, and the accumulation of cytoplasmic mass during differentiation, on the other. The sizes of both the mature and the immature megakaryocytes are proportional to nuclear DNA content (Odell *et al.*, 1970; Dupont *et al.*, 1983), i.e. the cytoplasmic mass and the mass of platelets produced are strictly proportional to the ploidy level and the gene dosage. This is understandable because polyploidy develops as a result of only slight deviations in the mitotic cycle confined to the concluding phase of mitosis.

Unlike hepatocytes, cardiomyocytes and melanocytes, the degree of polyploidization of megakaryocytes is not age-dependent. Whereas the polyploidization of the former cell types is coupled to the growth of the corresponding organs, megakaryocytes are polyploid in both embryonic and postnatal life, and from the moment the platelet-producing system comes into being. The average ploidy level does not change significantly during the animal's lifetime either, at least so long as there are no strong influences on the specific function of the megakaryocytes. In experiments with animals treated with antiplatelet serum, the ploidy-class distribution of megakaryocytes tended towards a higher average ploidy (Penington & Olsen, 1970). The ability of immature megakaryocytes (type I) to undergo further replication cycles is a reserve mechanism for regulating the overall mass of platelets in the organism. The main source for the increase in the number of mature megakaryocytes is the premature differentiation of cells derived from the proliferating compartment of the haemopoietic cells.

Since terminal differentiation (platelet production) begins only after the megakaryocytes have become octaploid, the question arises whether this level of polyploidy is the minimum required for specific differentiation. As an alternative proposition, we may conjecture that polyploidy is not a decisive event in megakaryocyte differentiation, but only offers a number of advantages for the subsequent functioning of this cell lineage.

The polyploid megakaryocyte, which releases fragments of cytoplasm (blood platelets), has been acquired during the evolution of mammals. In other vertebrates, there are diploid nucleated cells called thrombocytes whose entire differentiation is accompanied by ordinary (not polyploidizing) mitoses just as in other types of blood cells. These diploid thrombocytes are both homologous and analogous to the highly polyploid megakaryocytes in mammals. The advantage of the latter is their greater functional economy: right up to its death one polyploid cell secretes numerous working elements, while the diploid thrombocyte in lower vertebrates is the functioning element and perishes immediately after completing its function. Intermediate systems are also known. Thus, in the salamander, *Batrachoceps*, the diploid thrombocytes release portions of cytoplasm which possess the properties of blood platelets. It is even more interesting that, during some diseases of the blood in humans (e.g. leukemia), diploid megakaryoblasts become megakaryocytes without polyploidizing. These diploid 'megakaryocytes' are capable of releasing blood platelets (Undritz & Nusselt-Bohaumilitzky, 1968).

Thus it seems that during the evolution of the megakaryocyte the cells have become progressively more polyploid but that this is not as a necessary condition for their differentiation. It is difficult to suggest which of the consequences of genome multiplication gives megakaryocyte polyploidization the advantage over the equivalent process of diploid-cell multiplication. It may be that the production of giant cells, with the consequent large cytoplasmic mass, favours the formation and release of platelets, or it may be that some other property of polyploid cells is the key to their advantage (see Chapter 7E).

E Giant specialized cells of the transitional epithelium of the bladder

The surface of the epithelium of the urinary bladder in mammals is covered by large polyploid cells whose luminal membrane has a complex structure. The cells increase in size and ploidy level from the basal to the superficial layer. The small basal cells are diploid, but the intermediate and upper ones contain 4-, 8- and 16-ploid nuclei. This was demonstrated in mice, using chromosome analysis (Walker, 1958) and by microspectrophotometry (Levi *et al.*, 1969*a*). Microflow fluorometry of a suspension of isolated mucosal nuclei has shown that the transitional epithelium in mice (of the hairless strain) is, on average, 39% diploid, 53% tetraploid and 4% octaploid, with some 3% of nuclei of an even higher ploidy (Farsund, 1975). During bladder tissue regeneration, the number of octaploid cells may increase three- or fourfold, but in time this is normalized (Farsund, 1976). Polyploidy in the transitional epithelium is established even in fetal life; in young mice it varies a little but is stable by the age of 2 months. Superficial cells rarely have a single polyploid nucleus; they are usually binucleate or multinucleate (Fig. 2.28). The nuclei of the multinucleate superficial cells in mice often seem, on the basis of a visual impression, to be 16n and 32n, the individual nuclei being usually octaploid. In a healthy human, mainly diploid nuclei (and a small quantity of tetraploid nuclei) have been observed in bladder mucosa, using flow cytometry (Klein *et al.*, 1982).

The processes of cell replacement and, consequently, the formation and maturation of new cells in a normal transitional epithelium, take place very slowly. The cell-turnover time is determined to be from several months to over a year (Stewart, Denekamp & Hirst, 1980). Migration of basal cells to the surface, and their differentiation, takes at least 12 weeks, and the lifetime of the differentiated superficial cells is in excess of 200 days (Martin, 1972). Because of this slow rate of development, it is understandable that knowledge of the mode of polyploid-cell formation has been

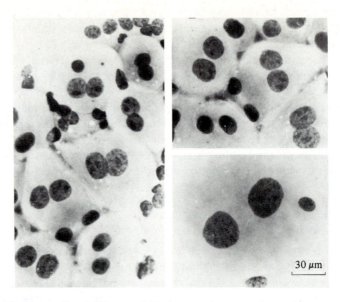

Fig. 2.28. Giant cells of different ploidy in squash preparations of the transitional epithelium of mouse bladder. All three photographs are presented at the same magnification.

obtained from tissues whose proliferation has been artificially accelerated. Enhanced rates of cell loss and replacement can be induced by means of mechanical damage, by the action of toxic substances, or by ionizing irradiation.

Martin (1972) made histological and autoradiographical studies of the kinetics of cell replacement in the urinary bladder of guinea pig after acute distension. He came to the conclusion that cell fusion is involved in the formation of polyploid cells. Soon after bladder distension, both the incorporation of [^{3}H]thymidine and mitoses were observed, but only in the basal layer. Label appeared in the intermediate cell layer after 10 days, and in the giant superficial cells only after 12 weeks. Thus, the labelled cells of the basal layer migrated to the surface very slowly and differentiated into the specialized superficial cells and these became multinucleate. Initially, binucleate cells were formed by the fusion of two adjacent cells; multinucleate cells were formed by repetition of this process. If the nuclei themselves fused, then tetraploid nuclei were formed. Martin has offered completely convincing autoradiographic and morphological evidence that fusion occurs not as an immediate sequel to mitosis, but much later, after the cells have already migrated from the basal layer. Labelled binucleate cells are produced after cell migration from the basal layer. They are not observed earlier than 11 weeks after the injection of [^{3}H]thymidine and, in most cases, only one nucleus was labelled. It is hard to imagine how this could happen in any way other than by the fusion of a labelled with a non-labelled cell. Moreover, during the migration of cells to the surface, links with the basement membrane are retained in the form of thin cytoplasmic processes. Even the giant superficial cells are, as it were, anchored to the basement membrane and, from the number of cytoplasmic processes, one can determine the total number of cells which took part in the fusion process.

The results of a study of the cell kinetics of the transitional epithelium in the mouse, following the effect of toxic substances (Levi, *et al.*, 1969*a*), do not agree with the results obtained using the guinea pig system. No newly formed binucleate cells with one labelled and one unlabelled nucleus were observed in autoradiographs. In experiments conducted by Farsund (1976), an injection of cyclophosphamide caused foci of cell loss and resulted in intensified DNA synthesis. DNA synthesis and mitosis were noted not only in the basal diploid cells, but also in those cells containing tetra- and octaploid nuclei. During the period of rapid formation of new polyploid cells, no evidence of cell fusion was found using serial microscopic sections (Farsund & Dahl, 1978). These authors believe that repeated cycles of DNA synthesis, which are not followed by mitosis, are responsible for the increased level of polyploidy.

However, descriptions of polyploid mitoses in the transitional epithelium are known. Walker (1958) found that 50% of metaphases displayed a tetra- or octaploid number of chromosomes in new-born mice. Levi *et al.* (1969*a*) described colchicine-arrested metaphases with 40, 80, and 160 chromosomes in the proliferating bladder of mice. This means that the polyploid cells in the upper layers of the epithelium are capable of mitosis and are replaced, not by the upward migration of the lower-lying diploid cells, but by self-maintenance. It is still not clear, however, whether the polyploid mitoses go to completion (and hence increase the number of polyploid cells) or are incomplete (and hence lead to a re-arrangement of the nuclear material and increase the level of cell ploidy). The anomalous forms of polyploid mitosis described in mice by Fleroff (1936) – multipolar, multifigure, acytokinetic – would lead to binucleate and multinucleate cells with different distributions of the chromosomal material among the nuclei (Fig. 2.29).

Cell fusion as the mechanism of formation of binucleate and multinucleate superficial cells may predominate over mitotic polyploidization in some species, e.g. the guinea pig (Martin, 1972). In any case, cytoplasmic connection between the superficial cells and the basement membrane and, what is more important, the existence of more than one of these cytoplasmic processes (the evidence of cell fusion), is a species-specific characteristic. They are found in the epithelium of the human bladder, but not in that of rats (Hicks, 1975).

The function of the superficial cells is that of ensuring the production,

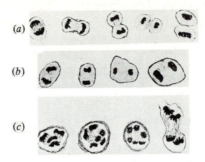

Fig. 2.29. Drawings of normal and abnormal mitoses, in cells of the bladder epithelium of fetal and neonatal mice. (*a*) Normal, (*b*) acytokinetic, and (*c*) multipolar mitoses (after Fleroff, 1936).

maintenance and renewal of the barrier-membrane elements. These cells are 20–100 μm in diameter and form a fairly regular polygonal plate on the luminal surface so producing a membrane with a unique structure. The morphology, chemical structure, and functioning of the barrier membrane has been well studied in a number of works (reviewed by Hicks, 1975). It is probable that small diploid cells are not able to produce the powerful Golgi apparatus where portions of the superficial membrane are synthesized. Moreover, a small volume of cytoplasm may be inadequate for the continuous translocation of membrane elements which accompanies each act of contraction and distension of the bladder wall. It is interesting that, after repeated doses of toxic substances, there is a shedding and complete disappearance of the superficial layer and all its polyploid cells. Initially, after this treatment, the remaining diploid cells proliferate without poly-ploidization; this leads to the formation of a multilayered epithelium consisting of diploid cells but lacking a membrane barrier. The superficial membrane appears only after a considerable time, once the normal polyploid cell types have been restored (Hicks, 1976; Farsund & Iversen, 1978).

The formation of polyploid cells appears to be a necessary part of the normal differentiation of the transitional epithelium. Depending on the mechanism involved, i.e. whether it is the fusion of the cells or polyploidizing mitosis, polyploidy here is evidently a means of attaining cells of the gigantic size necessary for the expression of tissue function.

F Salivary glands and the stimulation of their proliferation with isoproterenol

In mammals, there are known to be three pairs of salivary glands: the parotid, the submaxillary and the sublingual gland. The paired exorbital

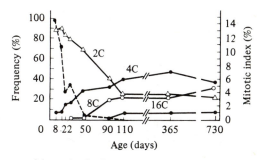

Fig. 2.30. Changes with age in the frequency of nuclei with different DNA content (C) and in the mitotic index (dotted line) in cells of the exorbital gland of rats (after Desaive, 1967).

glands (or Löwental gland) are morphologically (and apparently function-ally) similar to them. This gland is sometimes called the lacrymal gland, although there is evidence for its involvement in the digestive process in the buccal cavity (see Kühnel, 1972). The exorbital gland is especially well developed in rats.

The kinetics and mechanism of polyploidization of the cells of the salivary glands have not yet been studied. But many binucleate cells can be discerned in these glands as well as some cells with a large nucleus. Little is known about polyploidization in the exorbital gland. In cytophotometric works performed on thick sections (Desaive, 1965, 1967), changes in the ploidy classes of its cells were found during ontogenesis. It has become clear that, in mature rats, the acinar (i.e. the gland) cells of the exorbital gland form a polyploid line with a high percentage of cells with 8n and even 16n nuclei. In old rats there are many binucleate cells, some with a DNA content of 16C × 2. In the rat, the frequency of highly polyploid cells is apparently much greater in the exorbital gland than in the liver. Repetition of this research using a more precise cytophotometric technique would be of value. The cells of the exorbital gland, like those of the liver, are truly polyploid since a polyploid chromosome number is found in metaphase plates.

Polyploidization of the acini of the exorbital gland of rats occurs during the first 3 months after birth. Judging from the dynamics of mitoses in the diploid nuclei (Fig. 2.30) intensive multiplication of the diploid cells takes place during the first 3 or 4 weeks of life. It is not clear whether binucleate (2C × 2) cells accumulate at this time. Up to day 17, all mitoses in the exorbital gland occur in diploid cells but after 3 weeks, 2n as well as 4n and 8n mitoses are found. Mitoses still occur in the acinar cells of the exorbital gland of 2-year-old rats (although the mitotic index is less than 0.001%) and some of the metaphases and anaphases contain 32C DNA (Desaive, 1967).

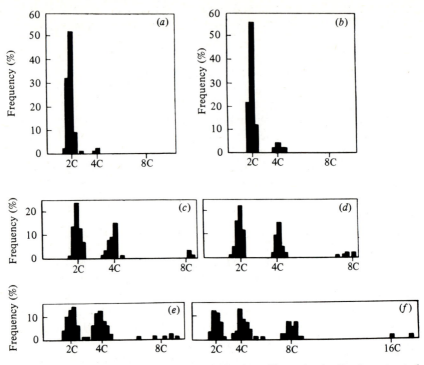

Fig. 2.31. DNA content (C) in the nuclei of submaxillary gland cells of rats treated with isoproterenol. (*a*) Control at 8 weeks old, (*b*) control at 7 months old, (*c–f*) after repeated injections of isoproterenol (starting at 8 weeks) for up to 20 days to demonstrate a rise in ploidy (after Radley, 1967).

The changes in the DNA content of the cells of the main salivary glands still need to be described. A cytophotometric determination of DNA content of the nuclei in the acinar cells of the rat submaxillary gland (Fig. 2.31) shows that almost all the nuclei are diploid. However, recalling the long-standing error with regard to the diploidy of rat and mouse cardiomyocytes, one should remember that these cells may have more than one nucleus.

More interesting is the study of ploidy of cells of the parotid salivary gland. Stimulation of mitosis in salivary gland cells by isoproterenol results in considerable polyploidization. Besides the salivary glands, the kidney, the liver, and the urinary bladder also respond to the injection of the chemical (Winter, 1974; Gerzelli & Barni, 1976). The increase in the number of mitoses in these organs cannot, however, be compared with that in the salivary glands. Among the salivary glands, the parotid reacts most strongly. 60–70% of the cells in the parotid gland of the adult rat synthesize DNA 28–38 h after a single injection of isoproterenol (in the control, the labelling index is only 0.3%; Baserga, Sasaki & Whitlock, 1969). Mitotic activity also increases dramatically in the submaxillary gland (Barka, 1967a,b, 1970). Acinar cells and the cells of the intercalary ducts are also stimulated.

Following the elevated level of mitotic activity induced by isoproterenol, the number of 4C and 8C nuclei increases considerably (Fig. 2.31) and approximately half of the metaphases show polyploid sets of chromosomes (Schneyer, Finley & Finley, 1967).

The characteristics of the growth induced by isoproterenol in the salivary glands have been generalized by R. Baserga (1970). The two most important conclusions are as follows:

1. Before the cells enter S phase, secretion of saliva is sharply increased (see also Simson, 1969). 2 h after the injection of isoproterenol, amylase activity in the salivary gland decreases tenfold. The cells are almost completely devoid of secretory granules. This low level of secretory proteins is maintained for almost 10 h. Then, resynthesis begins; 24 h after the injection, the control level of secretion is restored. Resynthesis of the secretory proteins takes place on stable, already present templates since inhibition of RNA synthesis by actinomycin D does not impede the restoration of the amylase levels (Byrt, 1966; Whitlock, Kaufman & Baserga, 1968). It may be important that the resynthesis of secretory proteins coincides with the completion of the gland cells' preparation for mitosis.

2. In cells preparing for mitoses, protein synthesis occurs in several intense 'bursts' over a short period of time. These 'bursts' are particularly marked in the 1–8 h period following the injection of isoproterenol. The previous blocking of RNA synthesis, or the slowing down of its translation, prevents further DNA synthesis and mitosis.

Thus, in the initial phase of the cells' preparation for mitosis, they cease to synthesize tissue-specific proteins. The mRNA for the secretory proteins is not destroyed, however, but nor is it translated until synthesis of those proteins necessary for the cycle occurs. Similarly, in epithelial pigment cells *in vitro*, after the induction of proliferation depigmentation occurs. This is because, during the preparation of the cell for mitosis, tyrosinase mRNA ceases to be translated (Whittaker, 1968a,b, 1970). The secretory granules themselves do not interfere with mitosis and secretion resumes before mitosis is completed.

The middle period of the cells' preparation for mitosis coincided in time with the resynthesis of the secretion. The replication enzymes – thymidine kinase, thymidylate kinase, thymidylate synthetase and DNA polymerase – accumulate in salivary gland cells during the peak period for amylase synthesis, 18–20 h after the injection of isoproterenol. It is from this time on that the proteins necessary for the G_2–mitosis transition, and for the realization of mitosis itself, are synthesized.

The manner in which isoproterenol works is not clear. Being a derivative of epinephrine, isoproterenol does not penetrate the cell. Consequently, its particularly powerful effect on the salivary gland is apparently caused by the fact that the acinar (and some duct cells) have the appropriate receptors; other cells lack these receptors, or possess only very few. The direct effect of isoproterenol is the result of its influence on the adenylate cyclase system. The effect of isoproterenol or, to be more correct, of cAMP, determines the readiness of the cell to intensify one process or another; the response is specific to the cell type itself and is not specifically determined by the stimulator. But, without stimulation, these processes are manifested poorly or not at all. Thus, in the differentiated gland (where few mitoses occur, but the cells are still capable of entering active cycle), isoproterenol stimulates mitoses. In the developing gland, isoproterenol accelerates differentiation, but mitoses are thereby slowed down (Schneyer & Shackleford, 1963; Schneyer, 1973).

The cells that have been stimulated to DNA synthesis by isoproterenol degenerate with time. New acini are formed from the remaining (unstimulated) cells, which become responsive to isoproterenol. The gland tissue may regress and develop several times (Domon *et al.*, 1978). Consequently, the induction of proliferation in the salivary gland by means of isoproterenol produces effects different from those described regarding the stimulation of mitosis in the liver or in the myocardium. The artificial nature of the situation in the salivary gland is obvious. Isoproterenol causes growth that is not needed by the organ. However, the mitotic cycle thus induced is sufficiently typical of polyploid cell populations, and further study of the effects of isoproterenol is highly desirable.

G The trophoblast: polytenization of a mammalian cell type

The trophoblast is a part of the mammalian placenta, and is formed from the blastocyst wall. The trophoblast participates in the attachment of the embryo to the uterus (implantation) and in sustaining the embryo. During placentation in some rodents, the trophoblast contains primary and secondary giant cells. Primary giant cells surround first the cavity of the blastocyst, and then the yolk sac; they remain viable up to the 12–13th day of development in the rat embryo. Secondary giant cells occur at the boundary between the maternal and embryonic parts of the placenta (Fig. 2.32). They are preserved in the placenta until the end of the pregnancy. Both the primary and secondary giant cells of mice and rats accumulate huge quantities of DNA – up to 4096C, although cells with 256–512C predominate (Zybina, 1963, 1970, 1983; Barlow & Sherman, 1972, 1974; Nagl, 1972*b*).

The progressive multiplication of DNA content in the nuclei of the secondary trophoblast cells is characteristic of the second half of rat embryonic development. On day 11, 16C and 32C cells already predom-

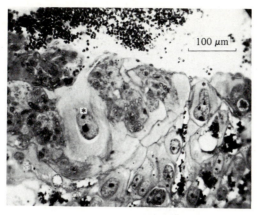

Fig. 2.32. Giant secondary trophoblast cells at the centre and below on the right (asterisks). Compare these with the low-ploid decidual cells above the central giant cell (photograph kindly provided by Dr E. V. Zybina).

inate, and very few octoploid cells are found. On day 13 mainly 64–256C cells are found. On day 15, 128–256C nuclei are dominant, but there are a few nuclei with 512C and 1024C. Thus, in approximately 1 week, 8–10 cycles of DNA replication occur, without division of the cell. The rate and scale of this cell growth has no parallel in any other mammalian tissue. There is a similar increase in DNA content in the trophoblast of mice (Barlow & Sherman, 1972), rabbits and field voles (Zybina, Kudryavtseva & Kudryavtsev, 1973, 1975).

A noteworthy feature of the colossal accumulation of DNA in the primary and secondary cells of the trophoblast is the complete absence of mitoses. Repeated endoreduplications occur which lead to polyteny. Study of the chromatin in giant trophoblast cells of four species of rodents revealed some similarities with the polytene chromosomes of Diptera. In trophoblast nuclei, bundles of chromonemata are visible, located under the nuclear envelope and around the nucleolus. Sometimes, a band-like pattern can be seen (Zybina, 1977). A characteristic of rodent trophoblast nuclei is the variable visual expression of the chromosome in the endo-reduplication cycle. Secondary giant cells of the rat trophoblast display two types of nuclei which seem to correspond to two stages of the endo-reduplication cycle. In nuclei of the first type, condensed chromonemata form bundles. In the nuclei of the second type, the chromonemata are decondensed over large portions of the chromosome, and separate from one another; union of chromonemata is preserved in the heterochromatic areas and in the nucleolus-organizer region. The nuclei with decondensed chromosomes incorporate [3H]thymidine, i.e. this morphological state corresponds to the stage of DNA synthesis (S); the nuclei with polytene

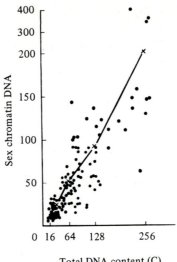

Fig. 2.33. DNA content (arbitrary units) of the sex chromatin body plotted against total DNA content (C) in the nuclei of giant trophoblast cells of rats (after Zybina & Mosyan, 1967).

chromosomes in the form of bundles are found during the non-DNA-synthesizing G-stage (Zybina, 1963).

In the nuclei of trophoblast cells, polytenization (multistrandedness), and not polyploidization, takes place. The chromosomes themselves are not spiralized, as in the case of mitosis, and there is no apparatus for chromosome separation. The nucleolus is preserved at both phases (S and G) of the nuclear cycle (Zybina, 1961, 1977). In the course of cell growth, although the chromonema bundles do not visibly increase in number, they do become thicker and there is a single sex chromatin body at all times (Zybina & Mosyan, 1967; Nagl, 1972b; Zybina, 1980). The DNA content in this body, which is a complex of sex chromosomes, increases proportionally with the total nuclear DNA content (at any rate, for six or seven endoreduplications; Fig. 2.33). The quantity of satellite DNA is proportional to the main band of DNA (Sherman, McLaren & Walker, 1972). The mean number of the constitutive centromeric heterochromatin bodies in murine giant trophoblast nuclei is 12, which is close to the value (9) found in the small diploid nuclei (Barlow & Sherman, 1974). As the nuclei grow to their giant size, the area occupied by these chromocenters remains a constant fraction of the total nuclear area in the giant cells (Fig. 2.34).

Actinomycin D was used to condense the chromatin in mouse trophoblast and chromosome-like structures were thereby caused to form. Their number was constant in nuclei with different DNA content (more than

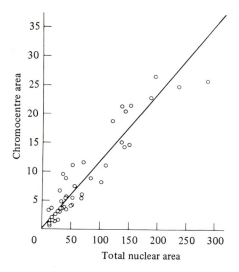

Fig. 2.34. Relationship between total nuclear area and chromocenter area (both arbitrary units) in the giant trophoblast nuclei of the mouse (after Barlow & Sherman, 1974).

512C) and corresponded to the diploid chromosome number (2n = 40) of the mouse (Snow & Ansell, 1974). The number of chromocenters in trophoblast cells also testifies to the diploid nature of the chromosome set contained in the giant nuclei with their huge quantity of DNA (Barlow & Sherman, 1974).

Thus, trophoblast cells which have accumulated enormous quantities of DNA – nuclei of more than 1000C – possess only two chromosome sets. These are obviously not polyploid nuclei. The trophoblast nuclei differ from the polytene nuclei of Diptera (of, for instance, *Drosophila*) in two ways: (i) somatic conjugation (the pairing of homologous chromosomes) does not occur in the trophoblast; and (ii) the polytene structure of the trophoblast nuclei is not stable: it does not occur in all cells, nor in all chromosomes, nor throughout the length of the chromosome.

Comparison of the mode of DNA reproduction in trophoblast cells with that in the best-known polytene cells (in Diptera) has given rise to doubt with regard to the polytene nature of trophoblast cells. However, chromosome conjugation and bands (the former absolutely, and the latter to a considerable extent) distinguish the larval cells of Diptera from all other cells studied in animals and plants. Moreover, a temporary polytene structure has been discovered, not only in the trophoblast, but also in many other cells which accumulate huge quantities of DNA without undergoing mitosis. Examples of such cells include nurse cells of the ovary in insects (Chapter 3B), the giant gland cells of the silkworm (Chapter 3C), and many plant cells (Chapter 4). In all these cells, a diploid number of chromosomes was found and the quantity of DNA was a multiple of that determined for the 2n set.

The giant trophoblast cells are formed as a result of endoreduplication cycles rapidly following one another. Two other possible mechanisms have been discussed with regard to the increase in mass of these cells: (i) the fusion of small diploid cells and (ii) the absorption of surrounding cells by the trophoblast cell, especially during the penetration of the trophoblast into the uterine endometrium. Both of these possibilities have been shown to be unlikely by results of experiments using mouse chimeras whose cells differed with respect to marker enzymes (Gearhardt & Mintz, 1972; Chapman, Ansell & McLaren, 1972; see also Chapter 6D).

The trophoblast cells develop to a giant size and accumulate huge quantities of DNA not only in their normal location *in vivo*, but also when transplanted to other regions of the body where they lack any kind of connection with their normal functions (Jollie, 1960; Hunt & Avery, 1971; Dorgan & Schultz, 1971; Avery & Hunt, 1972). The small diploid precursor cells of the trophoblast, when transplanted under the kidney capsule of syngeneic mice, transform into giant cells at the same time as those which remain *in situ*. The program of polytenization, and also the lifespan and the time of death of the cells, did not change in experiments with cultivated fragments of placenta *in vitro* (Jollie, 1960; Dorgan & Schultz, 1971). Just as *in situ*, the transplanted cells do not grow through the fusion of small cells. We note that growth to giant size is not, in itself, vital to the life of the cell. The blocking of nuclear DNA synthesis halts growth, and the resulting small cells survive in the explants. In the transplanted trophoblast, just as in the placental trophoblast, a specific enzyme complex is formed. This catalyzes the dehydrogenation of pregnenolone to progesterone (Salomon & Sherman, 1975). Recently, Ilgren (1981*a*) has established the influence of cell contact on the growth of trophectoderm-derived cells *in vitro*. Giant nuclei occur most frequently when intercellular contact is either minimal or absent.

The huge polytene cells of the trophoblast exist for just a few days in the placenta of mice, rats, and field voles. A week before the end of pregnancy, the nuclei of the giant cells (having accumulated huge quantities of DNA) cease endoreduplication. At the same time, the polytene chromosomes undergo gradual decondensation, after which the giant nucleus fragments into dozens of small nuclei (Zybina, 1963; Zybina, Kudryavtseva & Kudryavtsev, 1979; Zybina & Rumyantsev, 1980). As a result, multinucleate cells are formed; these are present in the placenta for a few more days and then degenerate. Ultrastructural and cytophotometric research has shown that, during fragmentation of the giant trophoblast nuclei, the chromosomal material tends to be segregated into its individual genomes. The chromosome markers of the interphase nucleus (nucleoli, hetero-

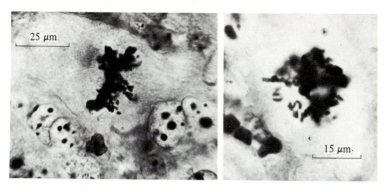

Fig. 2.35. Polyploid mitoses in the intermediate part of the rat placenta (photographs kindly provided by Dr E. V. Zybina).

chromatin blocks, including the sex chromatin body) are often distributed among the fragments almost in equal parts. DNA cytophotometry also indicates a tendency for genome segregation. It is, of course, difficult to imagine the mechanism whereby 512–1024C of DNA, the whole quantity of which is gathered into just two sets of condensed chromosomes, is segregated into single genomes. Zybina *et al.* (1979) have suggested that the chromosome material is transformed in the following manner. At the later stages of trophoblast differentiation, the chromonemata bundles loosen and (prior to nuclear fragmentation) complete disintegration of the bundles into individual chromonemata takes place. Thus, fragmentation is preceded by a complicated restructuring of the genetic material in the original, giant nucleus; as a result, the polytene material converts to polygenomes. The individual genomes then become isolated and are subject to segregation.

It should be stressed that this process has nothing to do with cell division and the replacement of the diploid cell population. The cell nuclei undergoing fragmentation have completely lost their ability to reproduce their chromosomes, nor can the fragments synthesize DNA.

Another mode of polyploidization is employed by cells of the trophoblast in the connective zone of the placenta, and by the tertiary giant cells of the trophoblast which migrate from the embryo to the maternal tissue (Zybina & Grishchenko, 1970). Polyploidizing mitoses occur which resemble those found in the liver. The first mitosis is acytokinetic, forming a binucleate cell ($2C \times 2$). In the next mitosis, the chromosome plates unite and complete division probably occurs, resulting in the formation of two mononucleate tetraploid cells (Fig. 2.35). The following sequence is then suggested: the third mitosis provides a binucleate cell ($4C \times 2$); the next mitosis produces an octaploid ($8C$) cell. By day 14 of pregnancy in rats, approximately half of the tertiary cells of the trophoblast are octaploid.

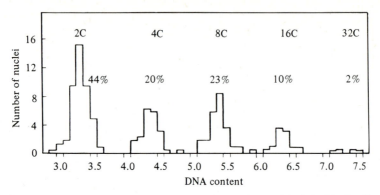

Fig. 2.36. DNA content (log₂ arbitrary units) in the nuclei of decidual cells of the mouse (after Ansell, Barlow & McLaren, 1974).

Further mitoses are not found, but the quantity of DNA in the cells doubles once or twice more, reaching 16C or 32C. This means that, after polyploidization, the cells undergo one or two endoreduplication cycles.

The decidual cells of the endometrium in the uterus become polyploid via a more usual mechanism (Zybina & Grishchenko, 1972). These cells are formed in rats on the fifth or sixth day of pregnancy at the site where the embryo will implant. The decidual cells form from mononucleate, diploid precursor cells of connective tissue. From day 8–10 of pregnancy they all become polyploid; half of these cells become binucleate. Among the mononucleate cells, 8C and 16C cells predominate. The binucleate cells are mainly 8C × 2, but huge 32C × 2 cells are also found. Polyploid cells are apparently formed by incomplete mitosis. Acytokinetic mitoses have been observed, and multipolar mitoses in large cells. In these cells, the number of nucleoli increases considerably (up to 10–12). The original connective tissue cells possess only one or two nucleoli. If the decidual reaction is evoked artificially, polyploid cells are also formed (Fig. 2.36). Results of research using the marker enzyme glucose phosphate isomerase in mouse chimeras, have completely ruled out the possibility of cell fusion during the formation of large polyploid decidual cells (Ansell, Barlow & McLaren, 1974).

Genome multiplication is perhaps a general feature of extraembryonic development, in mice at least. Ilgren (1980) has found nuclei containing more than a 4C amount of DNA, binucleate cells and metaphases with multiple sets of chromosomes in cells of the yolk-sac endoderm and of the amnion of mice.

Study of the trophoblast has thus produced new insights into cell growth in mammals (Fig. 2.37). The primary and secondary trophoblast cells are polytenized. Owing to polyteny, the cells achieve giant sizes in an extraordinarily short time. Apart from the increased cell growth rate, the

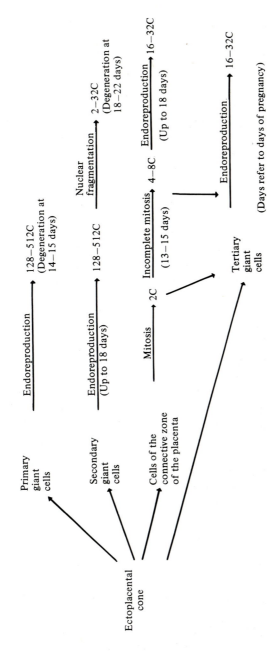

Fig. 2.37. Pathways of genome multiplication in different cell populations of the rat trophoblast (scheme kindly provided by Dr E. V. Zybina).

functional advantages of trophoblast-cell polyteny are not yet clear. But giant growth is evidently programed in the development of these cells. At the same time, the trophoblast illustrates a point which we shall note again many times: the total genome content of the tissue rather than the cell's individual genome is regulated during development. Giant mononucleate cells do not form in the trophoblast in all mammals. They are not found, for example, in the human trophoblast. However, one of the layers of the human trophoblast is, in itself, a syncytium with numerous nuclei. Hence, a tissue composed of a small number of highly polytene cells can be considered to be an analog of a tissue composed of multinucleate cells.

From the example of the trophoblast, it is understood that the concept of 'polyteny' is wider than would at first appear, on the basis of study of just the salivary glands of Diptera. The morphological criterion of having a polytene nucleus is secondary compared with the attendant biological significance of giant cell growth.

H The extraordinary increase in the DNA content of cerebellar Purkinje cells (hyperdiploidy)

According to early reports Purkinje cells of the cerebellum have a tetraploid DNA content (Brodsky, 1966; Sandritter *et al.*, 1967; Lapham, 1968). These results were later contradicted by results of observations employing the new cytophotometric and cytofluorometric methods (see Brodsky *et al.*, 1979, 1980*b*). Only 1–2% of rat Purkinje cells have a 4C amount of DNA, the majority of the cells being diploid. But, in Purkinje cells (and, for the moment, only in these nerve cells) unusual quantities of DNA have been discovered – amounts intermediate between the diploid and the tetraploid levels (Fig. 2.38).

First discovered in Purkinje cells of the rat cerebellum (Brodsky *et al.*, 1974; Bernocchi, 1975), these 'hyperdiploid' (H2C) amounts of DNA have since been found in Purkinje cells of chickens, cats and humans. In all species studied, H2C cells make up just a part (as a rule this is only a small part) of the Purkinje cell population. Most relevant data are available for rats where more than 100 animals, and several thousand cells have been examined. In some rats, H2C cells were not found; in others, the number of these cells ranged from 1 to 33% (Brodsky *et al.*, 1979).

The H2C cells were revealed by three cytochemical methods of quantitative DNA analysis: by cytophotometry using either the (i) Feulgen reaction or (ii) UV photometry and (iii) by DNA cytofluorometry.

Interesting differences have been found in the properties of the H2C cells from ordinary 2C Purkinje cells. H2C cells contain more condensed chromatin than do the 2C cells (Fig. 2.39). The surplus dense chromatin and the excess DNA is contained in the nucleolar zone, i.e. in the perinucleolar chromatin and in the nucleolar DNA.

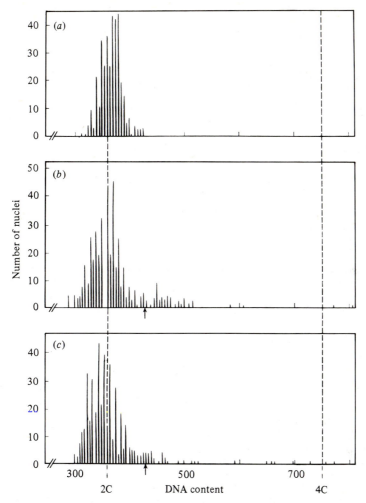

Fig. 2.38. Nuclear DNA content (arbitrary units and C) of (*a*) cerebellar granule cells, (*b*) Purkinje cells in rats infected with Kilham DNA virus, and (*c*) Purkinje cells in healthy rats. Dotted lines indicate the mean 2C and 4C level. Arrows indicate the upper limit of the 2C DNA distribution (after Brodsky *et al.*, 1979).

Of late, new data have been reported concerning DNA in Purkinje cells. Cameron, Pool & Hoage (1979) have described low levels of [³H]thymidine incorporation into mouse Purkinje cells. Thymidine labelling has been confirmed in other experiments (Marshak *et al.*, 1980; Brodsky *et al.*, 1983*a*). In our work, the number of labelled rat Purkinje neurons was very small and [³H]thymidine incorporation was usually restricted to the nucleolar zone. In rats, extra nucleic acid content was shown by UV

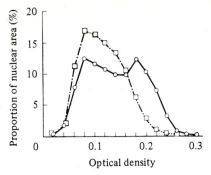

Fig. 2.39. The proportion of the nuclear area occupied by Feulgen-stained chromatin of different optical density in 2C (squares) and H2C (circles) rat Purkinje cells (after Brodsky *et al.*, 1979).

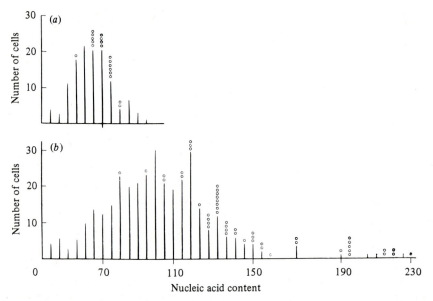

Fig. 2.40. The distribution of total nucleic acid content (arbitrary units from ultraviolet cytophotometry, $\lambda = 260$ nm) in nucleoli of (*a*) rat glial and (*b*) Purkinje cells. The lines relate to mononucleolate nuclei. Each circle indicates the total nucleic acid content in nucleoli of one binucleolate cell (after Mikeladze & Brodsky, 1980).

cytophotometry to occur in the nucleoli of a few Purkinje cells while most of the cells displayed a normal nucleic acid distribution (Fig. 2.40, compare with Fig. 2.38).

Three suggestions were at first made with regard to the causes of the H2C effect: (i) a viral infection of the nuclei; (ii) premature termination of the S period in the neuroblast mitotic cycle; (iii) gene amplification. The first possibility was ruled out by experiments using Kilham-virus infected rats (Fig. 2.38). Ribosomal gene amplification seems more likely in the light of studies of the Purkinje cell's nucleolus.

Naturally, H2C cells still need to be studied further. The rare occurrence and the significance of this phenomenon are both of interest. At the moment, gene amplification has only been conclusively demonstrated in oocytes and in some neoplastic and cultured cells (see chapter 6F).

I A G$_2$-block in erythropoietic cells during acute anemia

The development of erythroid cells is a classical example of the complete cessation of proliferation during the accumulation of a specific differentiation product. After several mitoses and definite changes in the cell properties, the main one of which is the accumulation of hemoglobin, proerythroblasts transform into polychromatophilic erythroblasts (Fig. 2.41). All the intermediate cell types of the erythroid series are diploid, as are the cells which form without divisions from the polychromatophilic erythroblast, i.e. reticulocytes and erythrocytes (if they preserve their nucleus).

It has been discovered (see Gazaryan, 1982) that, in pigeons, during phenylhydrazine-induced anemia, the rate of erythropoiesis increases and the differentiation of the erythroid cells is modified. Immature bone marrow cells enter the bloodstream and become enlarged. In addition, the basophilic erythroblasts transform into polychromatophilic reticulocytes without undergoing cellular division.

DNA cytophotometry of cells of the erythropoietic lineage, during phenylhydrazine-induced anemia, reveals an extraordinary phenomenon: erythroid cells develop with a doubled (4C) DNA content (Figs. 2.41 and 2.42). In the early stages of anemia, diploid basophilic erythroblasts enter the circulation. Meanwhile, 4C basophilic erythroblasts accumulate in the bone marrow. After 5 days, few diploid basophilic erythroblasts remain in the blood, almost all of them having been replaced by 4C cells. A histogram of DNA content for basophilic erythroblasts of bone marrow is similar to a corresponding histogram for erythroblasts in circulation. But diploid cells in the bone marrow signify another phenomenon. They mark the beginning of normalization. The 4C form of the basophilic erythroblast becomes a 4C reticulocyte. Then, the late reticulocytes undergo mitosis without additional DNA synthesis, and become diploid. The ultimate product of anemic erythropoiesis, like normal erythropoiesis, is the diploid erythrocyte.

The 4C erythroblast of anemic erythropoiesis is formed from the 2C erythroblast which can synthesize DNA but does not enter mitosis. The 4C form is larger than the diploid one. There can be no doubt that this is a G$_2$-cell which has been blocked before mitosis. The fact that they

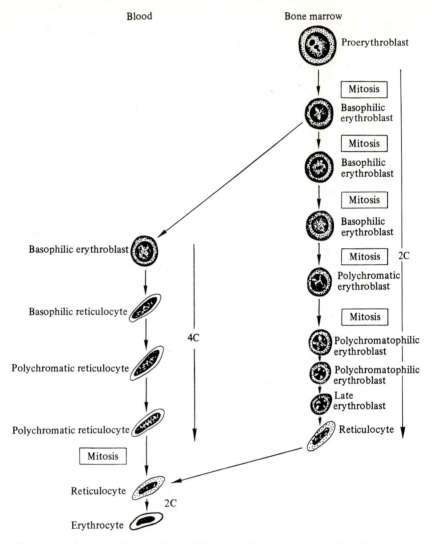

Fig. 2.41. Scheme of normal (right-hand side) and anemic (left-hand side) erythropoiesis in pigeons (after Kulminskaya, Gazaryan & Brodsky, 1978).

undergo differentiation in the G_2 state is something new and unusual for the erythroid cells.

The question that needs to be answered is the following: what is the mechanism of the G_2 block, and how is it maintained for several days? At the moment, we have no precise answer to this question. One possible avenue of investigation is an analysis of the normal events that prepare a cell for mitosis.

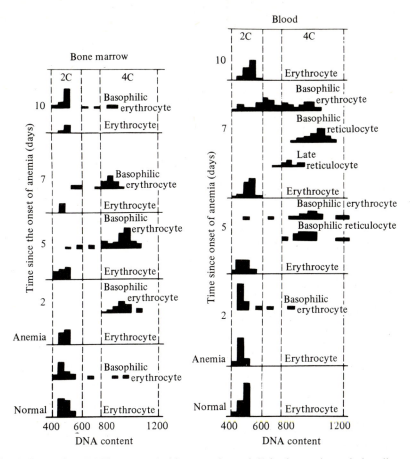

Fig. 2.42. Nuclear DNA content (arbitrary units and C) in the erythropoietic cells of anemic pigeons. Dotted lines represent the confidence limits ($\pm$two standard deviations) for 2C and 4C nuclei (after Kulminskaya *et al.*, 1978).

In mammals and amphibians, the earliest stage at which hemoglobin can be detected is the proerythroblast; there is already much hemoglobin in the basophilic erythroblast (Müller, 1972; Müller *et al.*, 1973; Mack, 1977). Amino acids are incorporated into these cells more intensely than into other erythroid cells – into globin and other proteins. The erythroid cells represent one of the many examples of simultaneous synthesis of DNA and of tissue-specific proteins (Jataganas, Gahrton & Thorell, 1970). Hemoglobin is formed throughout the preparation of the cells for mitosis (Fig. 2.43).

The hemoglobin content of the primary erythroblasts increases considerably in anemia. In particular, during phenylhydrazine-induced anemia, the synthesis of hemoglobin in basophilic erythroblasts occurs at double

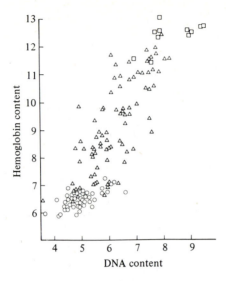

Fig. 2.43. Cytophotometric estimates of DNA (Feulgen staining, $\lambda = 566$ nm) and hemoglobin (unstained cells, $\lambda = 433$ nm) contents (both arbitrary units) in the same erythroblasts of chick embryo. Circles, G_1 cells; triangles, S cells; squares, G_2 cells (after Campbell *et al.*, 1971).

the rate observed in polychromatophilic erythroblasts (Borsook *et al.*, 1962). The hemoglobin may represent as much as one third of the protein synthesized in primary erythroblasts (Sussman, 1973). Primary erythroblasts of young chickens contain some 20% of their final hemoglobin content (Campbell *et al.*, 1971; Holtzer *et al.*, 1972).

There can be no doubt about the coexistence of premitotic and other (non-proliferative) protein synthesis in erythroid cells, including the primary erythroblast. Non-proliferative protein synthesis is expanded in anemia. It appears probable that incomplete preparation of the cell for mitosis blocks it in G_2; this prevents cell division from taking place even though there is active tissue-specific protein synthesis and cell growth.

Thus, the accumulation of the tissue-specific product in a diploid population in erythroid cells is combined with proliferation. In determining when cell proliferation will come to an end, the quantitative parameters of the cellular functions (primarily the premitotic and other, non-proliferative functions, including hemoglobin synthesis) are of considerable importance. The accumulation of the differentiation product in itself does not interfere with mitosis. In anemic erythropoieses, the reticulocyte undergoes mitosis with an almost complete complement of hemoglobin. Erythroid cells, blocked in G_2-phase, differentiate and then, after a further interval, undergo mitosis. The omission of several mitoses facilitates the accelerated differentiation of red blood cells and, therefore, the elimination of the anemia.

Conclusion

Cell polyploidy is a phenomenon common to normal development in mammals. Doubling of the DNA content as well as the chromosome set has been described in many tissues. Mammalian cells do not normally go beyond tetraploidy, and the binucleate variant of tetraploid cells ($2n \times 2$) is often the most common form. However, some highly polyploid cells (up to 64n) are observed. All mature megakaryocytes are highly polyploid. Many polyploid cells are observed in human myocardium.

Diploid cells are conserved in all polyploidized tissues. In the megakaryocytic and bladder lines diploid cells are the stem precursors of these differentiated types. In the liver, the myocardium, and the pigment epithelium, all the cells reach maturity and the number of diploid cells decreases sharply (but do not disappear) during tissue growth.

The expression of polyploidy differs between species, even when comparing the same tissue. The liver is known to be largely polyploid in mice and rats but largely diploid in humans and guinea pigs. Myocardiocyte-cell populations in all species studied consist of at least 90% polyploid cells. All mature bladder epithelial cells and megakaryocytes are highly polyploid.

The polyploid cells of mammals and other vertebrates studied are formed by a block in mitosis at one phase or another. Endomitosis, the division of the chromosomes within an intact nuclear envelope, has not been found in vertebrates. The usual forms of polyploidizing mitoses are acytokinetic mitosis (which results in binucleation), and metaphase arrest or anaphase fusion (which results in mononucleate polyploid cells). The cells of the trophoblast provide a unique example of genome reproduction in vertebrates; here polytenization of the nucleus, which is so typical of invertebrate and plant tissues, occurs. An unusual observation is that of the additional DNA accumulation in non-cycling Purkinje cells of the cerebellum. Another interesting discovery is the G_2-block in erythropoietic cells during anemia. All of these phenomena – partial genome multiplication, G_2-block and, especially, polytenization – are often noted in invertebrate and plant cells.

Genome multiplication is a characteristic phenomenon of ontogenesis, and it occurs at a precise time and is usually of short duration during tissue development. The rate of polyploidization evidently depends on the rate of cell differentiation and the growth of the tissue; this is related to the overall lifespan of the animal. After the cessation of cell division, many tissues in mammals continue to grow on account of cell polyploidization.

However, cytoplasmic growth has also been shown to occur in cases where the genome remains unchanged.

The widespread distribution of polyploid cells in mammalian tissues, and the regular time at which polyploidization occurs, indicate that cell polyploidization is not an accidental process but a programed part of development.

3

The cells of invertebrate animals

Polyploid and polytene cells have been found in all types of invertebrate animals. The extent to which DNA accumulates in single nuclei of some specialized cells in insects, molluscs and worms has no equal. The polytene cells of insects have been studied more than any other; literature reviews concerning polytene chromosomes of Diptera have been published by Beerman (1952), Geitler (1953), Bier (1959), and Kiknadze (1972). In this chapter, we concentrate on the developmental and physiological aspects of polyteny. The peculiarities of genome reduplication in some cells of molluscs, worms, and echinoderms will also be examined.

Table 3.1 cites examples of research on the cellular DNA content of insect cells, mainly in their larval tissues. Data on the maximal level of DNA in one cell type or another, however, are often limited. This is because although the doubling of DNA is, as a rule, synchronous in cell populations, and occurs at each larval instar, the final instars have not always been examined.

The polytene nucleus is very eye-catching and the number of instances in which polyteny occurs within the insects might be considerably increased simply by visual inspection of tissue preparations. Thus, although polytene chromosomes have been recorded in 20 species of dipterans from 10 families by inspection of salivary glands (Nagl, 1978), we have found polyteny in additional species through inspection of tissues such as trophocytes, cells of fat body and Malpighian tubules.

It is essential to note that in many cases where there are obviously huge increases in the number of chromonemata in the nucleus, polytene structures cannot be seen (or are only visible under special conditions or at certain stages of cell development). Sometimes, no polytene cells are ever seen in any tissue of some invertebrates even though the nuclei may have accumulated a huge amount of DNA. We shall present some of these examples.

In some early works, the extent of genome reduplication was often assessed on the basis of nuclear and cell size. Brilliant descriptions of nuclear and cell growth have been presented (see, for instance, Fig. 3.1). DNA cytophotometry of invertebrate cells may demonstrate many more

Table 3.1. *Examples of DNA multiplication in somatic cells of insects*

Species	Organ	Maximum DNA content (C)	References
Drosophila melanogaster	Salivary gland	512	Rudkin, 1972
Drosophila virilis	Salivary gland	2 048[a]	Rasch, 1970
Chironomus thummi	Salivary gland	8 192	Vlasova, Kiknadze & Sherudilo, 1972
Chironomus tentans	Salivary gland	16 384	Daneholt & Edström, 1967
Sciara coprophila	Salivary gland	4 096	Rasch, 1970
Drosophila melanogaster	Malpighian tubules	16 384	Rasch, 1970
Sciara coprophila	Malpighian tubules	1 024	Daneholt & Edström, 1967
Apis mellifera	Malpighian tubules	64	Merriam & Ris, 1954
Schistocerca gregaria	Malpighian tubules	8	Fontana, 1974
Mormoniella vitripensis	Malpighian tubules	32	Rasch, Cassidy & King, 1977
Triatoma infestans	Malpighian tubules	64	Mello, 1971
Dermestes sp.	Malpighian tubules	16[a]	Fox, 1970
Drosophila melanogaster	Ovary (trophocytes)	1 024	King, 1970
Chrysopa perla	Ovary (trophocytes)	512[a]	Zaichikova, 1976
Gerris najas	Ovary (trophocytes)	32	Won Chul-Choi & Nagl, 1977
Porphyrophora hameli	Ovary (trophocytes)	256	Magakyan, Karalova & Hachatryan, 1979
Schistocerca gregaria	Testicular wall	8	Kiknadze, Kolesnikov & Lopatin, 1975b
Dermestes sp.	Testicular wall	32	Fox, 1969
Bombyx mori	Silk gland	500 000	Rasch, 1974
Chironomus tentans	Hindgut	512	Daneholt, & Edström, 1967
Sarcophaga bullata	Foot pad	2 048	Roberts, Whitten & Gilbert, 1974
Bombyx mori	Muscles	256	Komarov, 1976
Apis mellifera	Pharyngeal gland	256	Merriam & Ris, 1954
Drosophila virilis	Fat body	128	Rasch, 1970
Gryllus maculatus	Fat body	128	Romer, 1972
Dermestes sp.	Fat body	16	Fox, 1969

[a] Not completely corresponding to the power series 2^n.

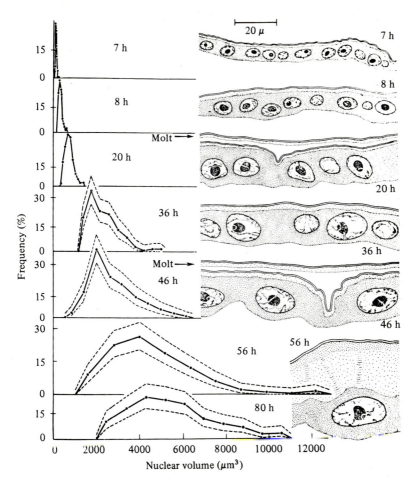

Fig. 3.1. Growth in volume of the epidermal nuclei in *Calliphora erythrocephala* during larval development. Dotted lines indicate 95.5% confidence limits about the mean. A sketch of the epidermis and its nuclei is shown on the right. h: Hours after hatching from the egg (after Wagner, 1951).

examples of genome multiplication, as occurred when this technique was applied to vertebrate material.

A Polyteny in the larval tissues of Diptera

In three families of the large order of Diptera – Drosophilidae, Chironomidae, Sciaridae – there are now classic examples of the development of polytene chromosomes and of huge accumulations of nuclear DNA. This feature is mainly characteristic of larval tissues. In adult insects, polyteny is well expressed only in the Malpighian tubules, and sometimes in the trophocytes of the ovary.

Nikolai K. Koltzoff was the first to propose that polytene chromosomes were multistranded structures (Koltzoff, 1934). This idea was developed in a number of subsequent publications (Bauer, 1936; Bridges, 1936; Frolova, 1936; Prokofyeva-Belgovskaya, 1937). Concrete evidence for the role of DNA replication in polyteny was obtained through the cytophotometric studies during the 1950s by Alfert (1954) and Swift (1962).

The development of polyteny has been studied most fully in larval salivary gland cells of *Drosophila* and *Chironomus*. In *D. melanogaster*, the cells are formed during the first 8–10 h of embryonic development. Cell divisions cease when invagination begins, at which time each pair of salivary glands consists of approximately 130 cells (Poulson, 1950; Rudkin, 1972). The gland rudiments consist of diploid cells. Eclosion coincides with the first endoreduplications. 4C and 8C nuclei are present in the gland 12 h later. By the time the first molt takes place, four or five cycles of DNA endoreduplication have occurred; by the second molt, another two or three cycles have taken place (Rudkin, 1973). The maximal DNA content in the gland cells of the prepupa of *D. melanogaster* is 512C; in glands of *D. virilis* the DNA content is 2048C (Rasch, 1970). All this tremendous growth of the genome occurs in just a few days, though all the while the number of cells remains constant.

An even greater degree of polyteny is achieved in the salivary gland of *Chironomus* which, in *Ch. thummi*, contains only 32–6 cells. During the 2 or 3 weeks of larval development, the nuclei of these cells undergo 11–12 endoreduplications, accumulating 4000–8000C of DNA. In salivary glands of larvae which have only just hatched there are already cells containing the 32C DNA content (Fig. 3.2). Just as in *Drosophila*, the DNA replication rate decreases in *Chironomus* as the larva grows.

Both in *Drosophila* and in *Chironomus*, DNA replication occurs during the period between molts and none occurs during the molt itself (Fig. 3.3). Consequently, the polytene cycle differs from the usual mitotic cycle in that it has fewer phases: mitosis is omitted and the period of DNA synthesis (S) alternates with the intersynthetic period (G) (S → G → S → G . . .).

In the prepupa, endoreduplications occur in only a few nuclei. The cells of the salivary gland degenerate soon before pupa formation. In the pupa, new cells develop and become polytenized; these function at this stage, for example, in synthesizing the cuticle of the foot pads. Some cells of adult tissues also appear, such as the ovarian nurse cells. The level of DNA in cells of Malpighian tubules of the adult fly corresponds to its level in the prepupa.

Studies of the regulation of this cyclical pattern of DNA synthesis, using inhibitors of transcription and translation, have shown that the beginning of replication is the most strictly programed (Darrow & Clever, 1970;

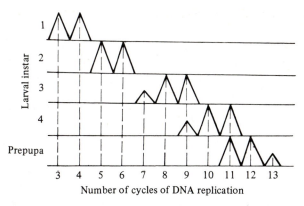

Fig. 3.2. Changes in the level of polyteny (the number of cycles of DNA replication) during the larval development of the salivary gland of *Chironomus thummi* (after Vlasova, Kiknadze & Sherudilo, 1972).

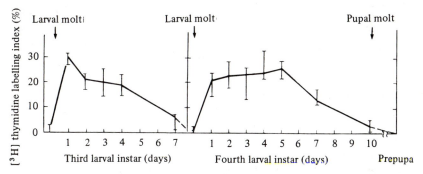

Fig. 3.3. [³H]thymidine labelling index (mean ± standard error) on different days during the development of the salivary gland of 3rd and 4th instar larvae of *Chironomus thummi* (after Vlasova & Kiknadze, 1975).

Rudkin, 1973; Vlasova & Kiknadze, 1975). Injection of actinomycin D or cycloheximide into a larva soon after a molt prevents DNA synthesis. New RNA and proteins are necessary for DNA synthesis to continue. Just as in other eucaryotes, reproduction of the genome in the salivary glands is not regulated by a trigger mechanism, but requires a constant flow of regulatory proteins. The blocking of DNA synthesis during a molt is an interesting phenomenon, the mechanism of which has not been elucidated. A slowing down of DNA synthesis is noticeable before a molt and it is surmised that replication is affected by the changing concentration of the hormones which regulate the molt (and possibly by other hormones, too).

A unique property of the polytene chromosomes of Diptera is the pairing of homologous chromosomes. Thus, in *D. melanogaster*, only five chromosomes, instead of the eight characteristic of the diploid set of this fly, are found in the cells of the salivary gland. However, this somatic pairing is a less regular process than meiotic conjugation. It is not characteristic of all Diptera and, in those species where pairing of homologues does occur, it is not observed in all tissues. Pairing does not occur to the same extent in all chromosomes. In *Calliphora erythrocephala* some homologs were observed to be completely paired, some were partially paired and others were not paired at all (Bier, 1959). In *Drosophila*, an absence of pairing is known to occur in some nuclei (Evgen'ev & Polianskaya, 1976). A recent investigation showed that the sites of pairing depend upon the extent of DNA homology between the homologous bands of the chromosomes (Riede & Renz, 1983).

A tendency for pairing of homologs was also observed in the non-polytene cells of *Drosophila*, as judged by the effect of mitotic recombination (Becker, 1976). Pairing of homologs was observed in both the polytene and the non-polytene mitotic nuclei of the plant *Bryonia dioica* (Barlow, 1975). The general significance of this pairing phenomenon is not clear, but it is obviously not characteristic of all polytene cells alone, nor of all those cells that are polytene.

The consequences of chromosome pairing, and especially the repeated multiplication of the chromonemata, is a colossal increase in the size of the polytene chromosomes. The increased width of the chromosomes is quite understandable, but the increase in their length is even more marked. In the salivary gland, during the final larval stages, the chromosomes are 100 times longer than metaphase chromosomes, either mitotic or meiotic. It is quite reasonable to assume that, during polytenization, not only does the number of chromonemata change but also the manner in which they are packed, thus affecting the shape of the chromosome itself.

The old data on heterochromatin under-replication in the polytene cells of Drosophila are now disputed. New cytophotometric research has revealed complete doubling of DNA in cells of salivary gland (Dennhöfer, 1981; Lamb, 1982). The DNA amounts fall into discrete classes (Fig. 3.4). But such total measurements may mean that changes in some part of chromatin are ignored. In fact, the amount of Hoechst 33258-bright heterochromatin was found to be similar in the polytene and diploid nuclei while the former had a much larger euchromatin content (Lakhotia, 1984). It was suggested that α-heterochromatin does not replicate during polytenization, which confirms earlier conclusions. But recently Spierer & Spierer (1984) have concluded that there is no large variation in polyteny among bands and interbands in isolated segments of polytenic DNA from the third instar and diploid DNA from embryos.

A slight under-replication is claimed to have been shown in the salivary gland of Sciaridae (Gambarini & Lara, 1974). However, many endocycles in the salivary gland of Sciaridae are complete in that all the DNA

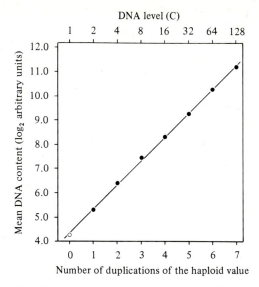

Fig. 3.4. Relationship between the number of replications and the mean DNA values in sperm (open circle) and both midgut and Malpighian tubule cells (closed circles) of *Drosophila melanogaster* (after Lamb, 1982).

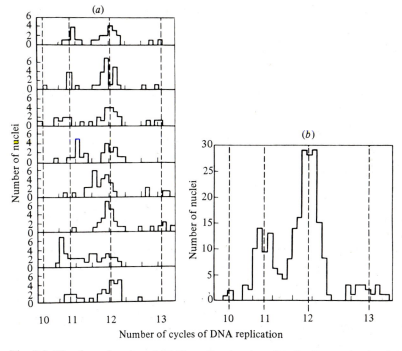

Fig. 3.5. Number of cycles of DNA replication in cells of salivary glands of 4th instar larvae of *Chironomus thummi*. (*a*) Nuclei from eight separate glands. (*b*) Summarized data from (*a*) (after Valeeva & Kiknadze, 1971).

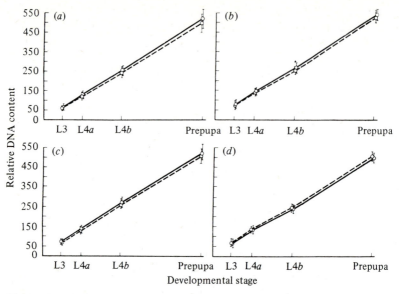

Fig. 3.6. DNA accumulation in heterochromatin (continuous lines) and euchromatin (broken lines) of chromosome I (*a* and *b*) and chromosome III (*c* and *d*) of larval salivary glands of *Chironomus melanotus* (*a* and *c*) and *Glyptotendipes barbipes* (*b* and *d*). L3, 3rd larval instar; L4*a* and *b*, the beginning and middle (respectively) of the 4th larval instar (after Walter, 1973).

is replicated. Under-replication has not been found in the ovarian cells, nor in the fat body cells of Sciaridae. Recently, some evidence of under-replication has been found in the polytene trichogen cells of *Lucilia cuprina* (Bedo, 1982). In these flies, the sex chromosome is much smaller than the autosomes, in the polytene cells; in diploid cells, by contrast, the chromosomes are all of a similar size.

In Chironomidae, only complete DNA replication occurs (Fig. 3.5 and 3.6). Studies of the quantity of satellite DNA in the salivary glands of five species of Chironomidae have shown good correspondence with the general DNA content in both the maximally polytene as well as in the diploid cells (Steinemann, 1978). A study has been made by hybridization *in situ* of the more active genes of the polytene nucleus of *Ch. tentans* (the ribosomal genes and the Balbiani rings). It was shown that, during polytenization, the rRNA genes, *BR-1* and *BR-2*, are duplicated as many times as is the entire genome. There is neither under-replication nor amplification (Hollenberg, 1976).

DNA amplification evidently takes place in some somatic cells of Diptera. Small rounded bodies, containing DNA in a ratio of 2:4:8 (Roberts, 1975) have been detected around the polytene chromosomes.

The neurons in the larvae of Diptera are one of the few examples of diploid cells, both in the number of chromosomes and the quantity of DNA. Formerly, polynemy was believed to occur in neuroblasts, since a DNA content of up to 8C was found in metaphase nuclei of the third larval instar (compared with 4C in the second instar; Gay *et al.*, 1970). When the measurements were repeated, employing a more precise cytophoto-metric method, no change in DNA content was demonstrated (De Flierdt, 1975). However, even though the quantity of DNA does not change, the metaphase chromosomes of the cells of the third larval instar are larger than those of the second instar.

What is the cause of the massive polytenization of the somatic tissues of dipteran larvae? As already mentioned, in addition to its occurrence in the cells of the salivary glands, polyteny is also characteristic of cells in the intestine, the fat body, the Malpighian tubules and a number of other tissues (Table 3.1). Although polytene chromosomes have long been usefully studied in investigations of gene function in the interphase nucleus, none of these interesting studies has told us why polyteny occurs nor whether polytene cells possess any advantages over ordinary diploid cells. Giant cell growth is evidently related to an extremely short period of tissue development. The entire larval development of the larva of *D. melanogaster* is accomplished in approximately 1 week, that of some Chironomidae in about 1 month. Let us recall that the only polytene tissue in mammals, the trophoblast, also functions (in mice or rats) for just a few days. It is difficult, though possible, to imagine such rapid tissue development being accomplished by mitotic cells. The polytene endore-duplication cycle takes 6–8 h (see Roberts, Whitten & Gilbert, 1974); a complete mitotic cycle in such differentiated cells would have lasted at least twice as long. Another important aspect of polytene growth is the compactness of the affected organ; usually it is formed by just a few cells whereas a similar diploid organ would consist of hundreds of thousands of cells.

The number of rounds of replication in the salivary gland of *Drosophila* larvae depends on the length of the development period. In wild-type *D. melanogaster*, DNA cytophotometry revealed seven or eight replications. In the *tu-h* mutant, where the larval period is longer, 9–10 replications occur (Rodman, 1967). In some mutants the length of the developmental period does not differ from that of the wild type; in these cases the number of replications does not change either. Rodman connects the number of cycles of DNA replication with the hormonal state of the larva. However, polytene cycles also occur in salivary gland tissue which has been transplanted to the body cavity of an adult fly (Hadorn, Ruch & Staub, 1964). In these circumstances, the lifespan of the larval cells increases two- or three-fold, and huge quantities of DNA accumulate in them. In some nuclei, the DNA content is 3.5 times greater than that normally charac-teristic of the prepupa. These observations also show that polytenization is probably part of the developmental program of salivary gland cells; DNA doubling occurs in these cells over a wide range of conditions.

Secretory proteins represent a considerable part (40–70%) of the proteins synthesized in the *Chironomus* salivary gland (Darrow & Clever, 1970; Kiknadze *et al.*, 1979). Synthesis of the secretion product begins during embryogenesis, when the chromosomes of the salivary gland are undergoing their first cycles of reduplication. A huge quantity of other proteins is also synthesized in the course of giant cell growth. According to data of Darrow & Clever, in the final (fourth) larval instar of *Ch. tentans* the salivary gland increases in weight fivefold, and secretory proteins comprise only 40% of this mass. According to the data of Kiknadze and coworkers, the proteins of the salivary gland of *Ch. thummi* increase approximately tenfold in quantity during the last instar. At that time, the cells undergo from one to three more endocycles. Thus, in Chironomidae (just as in mammals and other animals and plants) DNA synthesis coincides with a heightened synthesis of proteins which are not directly needed for proliferation.

In *Drosophila*, as in the Chironomidae, the salivary gland differentiates at an early stage of larval development. But the secretory granules in the cells of *Drosophila*, and the secretory proteins themselves, are not found in the salivary gland until approximately the middle of the third instar (Onishchenko, Mikichur & Maksimowsky, 1976). By that time, the nuclei have undergone several endocycles. Thus, polytene chromosomes form in the salivary gland before the secretion product is accumulated in the cells. The secretion product is released during pupation and is needed to fix the pupa to the substratum. In connection with the synthesis of secretory proteins, experiments with actinomycin D have shown that the corresponding mRNA is formed at 72–85 h of development and the first PAS-positive secretory granules (glycoproteins) are detected by 95 h (Kiknadze *et al.*, 1979). In *Calliphora erythrocephala*, post-transcriptional regulation of secretory-protein synthesis has been established (Griffith & Anderson, 1979); again, the formation of the secretion product occurs later than the synthesis of the relevant mRNA. In the fly *Dacus cucurbitae*, protein secretion begins during the third larval instar (after the maximum level of polyteny has been achieved). Synthesis of the secretion product and its accumulation coincides with polytenization, but not with its onset (Gopalan & Das, 1972).

An interesting example of the correlation between polytenization and the expression of a specific function is furnished by the cuticle-secreting cells of the foot pads of the meat fly *Sarcophaga bullata* (Roberts *et al.*, 1974). The pads are formed when the insect is in the puparium and the cuticle is synthesized by two giant polytene cells on each leg. The entire process of cell polytenization and of cuticle secretion takes approximately 12 days at 25 °C. During the first 4 days, 8–10 DNA endoreduplications take place. Then DNA synthesis ceases and, 5 days later, the cuticle is secreted. During the next 2 days, the cells undergo another 1–2 cycles of DNA replication. On day 11, the cells (which now have 4096C DNA) release the last cuticle layers; on day 12, degeneration of the cells begins, coinciding with the appearance of the imago. Thus, cuticle secretion occurs in the intersynthetic periods of the polytene cycles. The initiation of cuticle synthesis coincides in time with the completion of the first phase of polytenization.

As a result of successive polytene cycles, many larval cells of Diptera (and some other insects) attain huge sizes. This giant growth is accompanied by great synthetic activity of the cells. A noteworthy feature of this mode of growth is its extremely high rate. Just as in vertebrates, extending the period of differentiation of the larval tissues prolongs their overall period of endocycling. Polytenization of the cells is evidently regulated by direct genetic mechanisms, as proved by experiments involving salivary gland transplantation. Polyteny is a mode of intensive tissue growth; during evolution, tissues composed of a few cells with highly multiplied genome may have been preferred to tissues composed of many small diploid cells.

B Polyteny in the nurse cells (trophocytes) of the ovary

In the panoistic ovary of the primitive orders of insects (for example, Odonata, Orthoptera and Isoptera), the oocytes grow and synthesize all the substances and structures necessary for embryogenesis. In the more advanced orders (such as Coleoptera, Lepidoptera and Diptera), a meroistic ovary develops in which a large number of the potential germ cells does not transform into oocytes but develops as an auxiliary trophic apparatus (see Teflcr, 1975; Aizenshtadt, 1977). Such nurse cells, i.e. trophocytes, originate as abortive oogonia and are connected to the oocyte by cytoplasmic bridges. These bridges are formed by incomplete cell cleavage during multiplication of the oogonia. With regard to the spatial organization of the trophocytes, two types of meroistic ovaries can be discerned: telotrophic and polytrophic. In the former, all the trophocytes are localized at the apical end of the ovary and are bound to the oocyte by long cytoplasmic strands. In the polytrophic ovary, there are usually only a few trophocytes and these are in direct contact with the oocyte. The entire group of cells is surrounded by a follicle epithelium. A chain of these chambers forms the ovarian tube; the stage of maturity of the oocyte and trophocytes can be determined by their distance from the differentiation zone (at the apical end).

In the meroistic ovary, the oocyte receives a considerable amount of ribosomal (r-)RNA from the trophocytes (Bier, 1970; Mermod, Jacobs-Lorena & Crippa, 1977). After fertilization, this RNA, which has accumulated in an inactive form, determines the early events of embryogenesis. Yolk proteins are synthesized outside the gonad and then transported to the oocyte.

It has long been noted that the trophocytes of many insects are large cells. Cytophotometry of trophocytes has revealed an elevated DNA content (sometimes considerably so). The nuclei of the trophocytes of *Drosophila virilis* accumulate a DNA content of up to 4096C (King, 1970).

The structure of the trophocyte nucleus may change substantially as its DNA content increases. In the early stages of the development of *Drosophila* trophocytes, polytene chromosomes are visible in them, while diffuse chromatin is observed in the large nuclei of the more differentiated cells (Painter & Reindorp, 1939). In the meat fly, *Calliphora erythrocephala*, the trophocytes of the pupa (which have up to 16C DNA) also possess polytene nuclei. In the imago, the nuclei of the large trophocytes (the size of which corresponds to a 2048C and 4096C DNA content) fall into three classes (Bier, 1959). Some have diffuse chromatin, others contain bundles of chromatin threads, and, in the third type, there are numerous bundles with bands. Since growing trophocytes do not undergo mitosis, it is assumed that the number of chromonemata increases evenly in the nuclei with different morphologies – i.e. with recognizable and unrecognizable polytene chromosomes. The pattern of nuclear structure may depend on environmental conditions. For example, in the trophocytes of the meat fly, the number of nuclei with a polytene structure increases considerably if egg laying occurs at a low temperature (Bier, 1959). In some inbred lines of *Calliphora*, the polytene structure is retained through all stages of trophocyte development. In other lines, only a diffuse chromatin structure was observed (Ribbert & Bier, 1969).

Bier explains the disappearance of the polytene structure in developing trophocytes by the polyploidization of the polytene nucleus which entails the separation of the chromatids. It is more likely that the distances between the chromonemata of a single chromosome are increased. The result of such dispersal may be the disappearance of the polytene structure (see also Chapter 4). The appearance of the nucleus (polytene or diffuse) may depend on the concentration (the relative proximity) of the chromonemata, and possibly on the form of the chromosomes.

In nurse cells of the pond skaters *Gerris najas*, a strict doubling of DNA was observed from one cycle to the next (Won Chul-Choi & Nagl, 1977). In the same way, the DNA content in the nurse cells of the meroistic ovary of the bug *Oncopeltus fasciatus* increases in proportion to the diploid chromosome set (Cave, 1975). rRNA–DNA hybridization revealed the same number of ribosomal genes in somatic diploid cells, testicular and ovarian cells, in isolated trophocytes of *O. fasciatus*.

Conflicting data on the replication of the trophocyte genome in *Drosophila* have been reported. A comparison of the brain of *D. virilis* containing diploid cells, and the ovary with a maximum accumulation of

DNA in the trophocytes revealed a four-fold deficit in satellite DNA in the ovary (Renkawitz-Pohl & Kunz, 1975). A similar study of the ovary of *D. virilis* was published in the same year (Endow & Gall, 1975), in which the opposite effect was reported (satellite DNA was shown to be increased in the ovary relative to that in the brain). The under-replication of some rDNA cistrons has been recently demonstrated in the nuclei of polyploid nurse cells of *Calliphora erythrocephala* (Beckingham & Thompson, 1982). In the nuclei of trophocytes of the sciarid fly *Bradysia hydida*, the heterochromatin did not take up labelled thymidine (Sauaia & Alvez, 1969). It is not clear, however, whether its DNA synthesis is blocked, or whether it simply reproduces later than other forms of chromatin.

Nurse cells are typically associated with oogenesis, both in insects and also in some other invertebrates. An analysis of oocyte growth with different numbers of nurse cells led to the important conclusion of the existence of a link between polytenization and the intensity of functioning of this auxiliary apparatus (Aizenshtadt & Marshak, 1969). As already noted, in the polytrophic ovary of insects, usually only a few trophocytes are bound to each oocyte. In each egg chamber of *Drosophila*, there are 15 trophocytes; in the butterfly *Chryosopa perla* there are 7 trophocytes to each oocyte. All the trophocytes studied from polytrophic ovaries were highly polytene. In the annelid *Ophryotrocha puerilis*, the oocyte is linked with one nurse cell containing (according to cytophotometric results) 256C DNA (Ruthman, 1964). In the gonad of the snail-leech *Glossiphonia complanata*, the oocyte has approximately 2000 trophocytes (originally abortive oogonia) which are connected in a syncytium to each other and to the oocyte. According to cytophotometric data of Aizenshtadt & Marshak (1969), all the trophocytes of this leech are diploid. It is known that these cells actively synthesize rRNA, which is then transferred to the oocyte (Aizenshtadt, Brodsky & Gazaryan, 1967). The functions of snail leech trophocytes are, in principle, similar to those of insect trophocytes; and the snail leech ovary itself resembles a polytrophic ovary (except for the number of cells per oocyte). It may be supposed that the total number of cells, i.e. the real synthetic capacity of the nurse apparatus, determines the number of genome copies of the cells. Thus, the total tissue genome is regulated. The total quantity of DNA in the 15 trophocytes of *Drosophila* is actually greater than in the 2000 trophocytes of the leech. However, a comparison of the intensity of rRNA synthesis, in the course of oogenesis in *Drosophila*, with the number of available ribosomal genes has shown a direct relationship between the two (Mermod *et al.*, 1977). Therefore, all (or most) of the genes are still active after many endocycles. The number of nurse cells is probably programed during development, and depends on the number of oogonial divisions. The degree of genome multiplication in each nurse cell thus acts to regulate the oocytes' ultimate capacity for RNA synthesis.

Table 3.2. *The DNA content in the gonad cells of two species of insects and the duration, at 23 °C, of oogenesis and embryogenesis*

Type of ovary	Species of insect	DNA content (C)	Total quantity of DNA (pg)	Duration of oogenesis (days)	Duration of embryo-genesis[a] (h)
Panoistic	*Gryllus domesticus*	4×1^b	9	100	30
Meroistic	*Calliphora erythrocephala*	256×15^c	2928	6	1.5

After Ribbert & Bier, 1969
[a] Up to formation of the blastoderm.
[b] The single oocyte grows while the nucleus is in the G_2-phase.
[c] DNA in 15 trophocytes during the period of oocyte growth.

Trophocytes not only synthesize RNA for the oocyte but also provide some of its proteins (for example, the ribosomal proteins). Moreover, the trophocyte itself grows with every endocycle. The metabolic role of the trophocyte does not in essence differ from that of other polytenizing or polyploidizing cells.

The nurse cells of the ovary in insects and other animals therefore provide yet another example of a temporary nutritive tissue. Cells of some larval tissues of insects and of the mammalian trophoblast develop similarly toward the same end. As a rule, all these structures do not exist for long, sometimes only for a few days, attaining huge sizes by means of consistent (usually synchronous) polytene cycles. A splendid example of the advantageous possibilities of a highly polytene nutritive apparatus is found in the work by Ribbert & Bier (1969) and summarized in Table 3.2. Two systems are compared: one consisting of cells with highly polytene nuclei, which, although few in number, constitute the syncitial trophic apparatus, and the other consisting of numerous small diploid cells.

Also of interest is the relationship between cell proliferation and differentiation in the follicle cells of the cuttlefish *Rossia pacifica* (Aizenshtadt & Marshak, 1969). Unlike in other species studied, these follicle cells synthesize glycoproteins (components of the yolk). Mitosis in these cells ceases at the beginning of vitellogenesis. At this time, all the follicle cells are diploid. By the end of vitellogenesis, i.e. during the period of rapid synthesis of the secretory protein, the cells undergo two synchronous cycles of reduplication. By the time the yolk has accumulated, the follicle cell nuclei have an 8C DNA content and are approximately four times larger than the original diploid ones.

Trophic multicellular formations supply the oocyte with RNAs to

Table 3.3. *Mass of the genome in cells of the silk gland of* Bombyx mori *during the fifth larval instar*

	Region of the gland		
	Posterior	Middle	Anterior
Number of cells	520	255	350
Amount of DNA $(\mu g)^a$	119	160	3.4
Overall per nucleus	0.23	0.63	0.009
Number of endocycles	18–19	19–20	13

After Perdrix-Gillot, 1979.
[a] $2C = 1.06 \times 10^{-6} \mu g$.

support its giant growth. The growth can also arise through multi-nucleation of the oocyte itself. This possibility is realized in the marsupial frogs (Del Pino & Humphries, 1978). Acytokinetic oogonial divisions lead to a huge accumulation of nuclei. During the giant growth of the cell, the oocyte contains 1000–3000 nuclei. When growth is completed, the number of nuclei is reduced, and full grown oocytes have only one nucleus.

C The giant gland cells of the silkworm: are they polytene or polyploid?

The mature cells of the silk gland of *Bombyx mori* have, on average, a DNA content of c. 500 000C and as much as 2 000 000C (Rasch, 1974; Perdrix-Gillot, 1979). The cessation of mitosis and the growth of the cells and their nuclei are the first signs of silk gland differentiation. The number of cells in the gland then remains constant. In each of the three regions of the silk gland, the cells differ in size; this is determined by the number of endocycles they have undergone (Table 3.3). The formation of the silk gland, during which numerous endocycles occur, takes some 3 weeks to complete (Tazima, 1964).

The specialized silk proteins (fibroin and sericin) are synthesized in the posterior and middle regions of the gland. Towards the end of larval life, the cells of these regions (especially in the middle region) attain giant sizes – more than 3500 μm in length. The nuclei of these cells have a complicated arborescent form, with considerably increased surface areas.

DNA replication in the nuclei of the silk gland cells is complete. Neither amplification nor under-replication of any fraction of this DNA has been discovered. Evidence of this comes from cytophotometry of DNA in cells at different stages of development and from studies of DNA-reassociation kinetics and RNA–DNA hybridization (Gage, 1974a,b). The former study shows that the nuclei have exact multiples of the initial 2C value; the latter shows that the genomes of the giant are identical with the small diploid cells. The genome of the maximally differentiated gland cell contains up to 2×10^6 copies of the fibroin genes; this corresponds well with the number of haploid chromosome sets (Suzuki, Gage & Brown, 1972).

Mitosis ceases when growth of the gland cells begins. The conclusion that the nuclei are polytene would seem to be obvious. Nuclei of silk gland cells contain no apparatus for separating daughter chromatids and certain other observations have also been interpreted to indicate the polytene structure of the nucleus (Nakanishi, Kato & Utsumi, 1969). For example, a banding pattern was observed in bundles of threads in the nuclei of silk gland cells of fourth instar larvae. The number of bundles (about 60) corresponds closely to the diploid chromosome set of *B. mori*. Each bundle of this type was much longer and thicker than the metaphase chromosomes of the spermatogonia of the silkworm.

Certain doubt has been cast upon the observations of Nakanishi, by Perdrix-Gillot (1974, 1979). Two types of nuclei, with condensed or dispersed chromatin, have been noted in the embryonic and larval cells of the silk gland. [³H]thymidine is incorporated into only the dispersed nuclei. An important observation is that the number of chromatin bodies in the condensed nuclei increases from approximately 50 to several thousand though, by contrast, a single or minimally fragmented, but, nevertheless, unified, sex chromatin body is invariably seen at all stages of growth of the silk gland cells. These observations suggest the coexistence in one and the same nucleus of two modes of DNA reduplication. In the autosomes, endomitosis occurs, with the separation of the chromatids, while in the female sex chromosomes of the same nucleus, the chromatids double and remain together as though undergoing a polytene cycle.

In our opinion, polytene cycles may indeed occur in *all* the chromosomes of silk gland cells. True endomitosis (Geitler, 1939, 1953) has not been found in such cells. Alternation between the condensed and dispersed state is not restricted to the cells of the silk gland, but also occurs in other cells with a temporary or local polytene structure in the autosomes. Similar nuclei have been found in the trophoblast of mammals and here the dispersed nuclei also incorporate [³H]thymidine. It has been shown that, in the giant cells of the trophoblast, the number of chromocenters corresponds to the number of the diploid chromosome set. The chromatin bodies in the nuclei of *B. mori* are too small (up to 0.5 μm) to describe as chromocenters and there are too few of them to allow calculation of chromosome doubling. The only clear-cut chromosome marker is the sex chromatin body, and this body does not become divided up. The connected fragments of the sex chromatin may indicate that one or two polyploidizing mitoses occur before the polytene cycles.

Indirect evidence in favour of polyteny is the similarity of the kinetics of DNA synthesis in the nuclei of the silk gland cells (Fig. 3.7) with those in the obviously polytene nuclei of the salivary gland of *Drosophila*

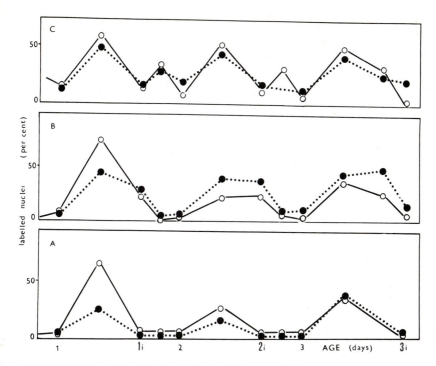

Fig. 3.7. [³H]thymidine labelling index in different regions (A, anterior; B, middle; C, posterior part) of the silk gland of *Bombyx mori* during the first three larval instars after labelling *in vivo* (solid lines) and *in vitro* (dotted lines). Each point indicates a mean value calculated from 2–5 individuals. The abscissa shows the age (in days) of each instar (1i, 2i, 3i) after the first molt (after Perdrix-Gillot, 1979).

(Darrow & Clever, 1970) and chironomids (Fig. 3.3). The peaks of DNA synthesis coincide with the intermolt periods while, during the molt, the rate of DNA synthesis diminishes.

A 'hidden' form of polyteny is noted in the nurse cells of the ovary. The neighboring cells may have typical polytene chromosomes with bands or just bunches of threads, or dispersed chromatin. At certain stages of development, the polytene pattern can be seen in all the nuclei. A case is known when polytene chromosomes were seen in relatively small nuclei as well as in the largest ones, but not in the nuclei of intermediate sizes.

Polyteny following polyploidy might well explain the fairly even distribution of fibroin and ribosomal genes in the nuclei of the gland cells of *B. mori* (Thomas & Brown, 1976). Cases of polytenization of a polyploid nucleus have been described in some plants (Nagl, 1970a,b). In these nuclei, the band structure did not appear but bands could be seen during polytenization of initially diploid cells.

As a result of the many replications in the silk gland, cells are formed with a colossal protein-synthesizing potential. According to calculations of Kafatos (1972), 4.5×10^5 molecules of fibroin are synthesized per minute per haploid genome, which is approximately the same rate of protein synthesis as in other gland cells studied. But, because of its degree of reduplication, the whole genome of the silk gland cell could theoretically be a million times more active. It would be interesting to determine the actual productivity of this giant cell.

As in other cases, when the number of cells in the tissue is constant, the degree of reduplication of the cell genome regulates the total genome content of the tissue, and consequently, the synthesizing activity of the tissue. In polyploid strains of *B. mori*, at the time of eclosion the caterpillars differ considerably from the weight of the diploid strains. Then the weight and size of the polyploid and diploid caterpillars evens out, and the total DNA content in their silk glands also becomes similar.

Thus, the study of the growth of giant cells in the silk gland of *B. mori* reveals again that polyteny may take various forms. In spite of its colossal scale, the growth of the cells in this case is accompanied neither by selective gene amplification, nor by any under-replication of the chromatin.

D The testicular wall epithelium of locusts and the imitation of endomitosis

The testicular follicle wall of locusts is one of the best known examples of a tissue which displays endomitosis. L. Geitler (1944) was the first to describe endomitosis as a mode of chromosomal division within the nuclear envelope, in the testicular epithelium of the locust *Psophys stridulus*. As a result of Geitler's observation, endomitosis came to be considered as the main, if not the only mode of polyploidization in somatic cells. We have already noted that endomitosis, in the form described by Geitler for some insects and plants, has not been observed in any polyploidizing tissue of mammals. Recently, the concept of endomitosis has come to be doubted, even in insects.

Kiknadze *et al.* (1975*a*) and Kiknadze & Istomina (1980) showed that the morphological pictures of endomitosis in the testicular wall of locust did not correspond to reproduction and division of the chromosomes. 'Endomitosis' was, in this case, one of the interphase states of the nucleus in which the chromatin bodies resembled mitotic figures within the nuclear membrane. These conclusions have been confirmed by studies of the testicular wall epithelium in another grasshopper genus, *Melanoplus* (Therman, Sarto & Stubblefield, 1983).

'Endomitoses', with the characteristic prophase or metaphase chromosome pattern, first appear in the epithelium of the testicle wall during the fourth larval instar; many only appear during the fifth instar or in adult locusts. Neither 'endomitoses' nor the neighboring nuclei incorporate

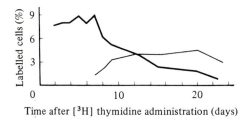

Fig. 3.8. Percentage of labelled cells in basal (thick line) and central (thin line) regions of the testicle wall of the locust *Schistocerca gregaria*, related to the time from [³H]thymidine incorporation (after Kiknadze *et al.*, 1975*a*).

[³H]thymidine, but they do take up [³H]uridine and synthesize proteins actively (see also Bakhtadze, 1981), as is characteristic of the interphase cell. Cytophotometry has established that many of the typical 'endomitotic' nuclei are diploid. This observation in itself proves that these cells could not be mitotic.

According to the data of Kiknadze, the true zone of proliferation of the testicular epithelial cells is spatially separated from the site of their transformation into 'endometaphases'. Proliferation is largely completed before 'endomitosis' occurs in the tissue.

In the third larval instar of the locust *Schistocerca gregaria*, small epithelial cells with dense nuclei are located near the outlet channels of the testis follicles. These cells both incorporate thymidine actively and display frequent mitoses. During the fourth instar, the 'endoprophases' appear at some distance from this region, near the center of the follicle. There are many mitoses near the channel at the base of the follicle, as before. Later on, mitoses are not found and [³H]thymidine is not incorporated, but the 'endomitotic' nuclear pattern persists.

The dynamics of the [³H]thymidine labelling during the transitional period (i.e. while there are still mitoses in the epithelium, and 'endomitoses' have already appeared; see Fig. 3.8) are interesting. Thymidine is only incorporated at the basal end of the section. The number of labelled nuclei remains high there for several days after the injection of [³H]thymidine (Fig. 3.8). During this period, there is not a single labelled nucleus in the zone of 'endomitoses'. They only appear here 7–8 days after the incorporation of thymidine into the cells of the basal zone. Thus, only when the cells have completed mitosis, do they migrate to the central zone of the follicular wall. Here endomitosis-like figures appear in the interphase nuclei. The reason for this chromatin condensation is not yet clear, but it does not involve reproduction of the genome. Thus, in the testicular wall, 'endomitoses' are separated from the chromosomal reproduction, both in space and in time.

To summarize, the epithelium of the locust testis contains 2C, 4C, and

8C cells. In the basal zone, polyploidizing mitoses are observed. Diploid and polyploid epithelial cells migrate to the central zone of the wall, where chromatin structures resembling endomitosis appear in these interphase cells.

The function of the epithelial cells of the testicular wall of insects is not clear. It is thought that these cells may play a role in spermatogenesis equivalent to that of Sertoli cells in the testes of vertebrates.

E Polyploidizing mitoses in the stomach epithelium of starfish

An original feature of the starfish stomach is its environment. During feeding, the esophagus and a considerable part of the stomach of the starfish are everted; they seize the prey and digest it outside the body. Contact with the sea bed, with the shells of molluscs and other hard objects damages the gastric tissues and, because of this, the tissue must be able to regenerate rapidly. Another peculiarity of the stomach, of no less interest for us, but which is characteristic of only a few species of starfish, is that cilia develop on the outward-facing surface of the epithelial cells. The beating of these cilia translocates fluid within the digestive system. In other starfish, there are special pump-like organs that create a flow of the fluids containing the food particles. Thus, in species with a ciliated epithelium special adaptations may be expected; these might include altered modes of cell growth. In the gastric mucosa of the starfish, two types of cells are found (Anderson, 1954). One type of cells has rounded, lightly staining nuclei with diffuse chromatin; the other cells have long, cigar-like, very densely staining nuclei (Fig. 3.9). The former cells are termed 'typical', the latter are called 'special', although what is special about their properties are not clear.

In the pyloric part of the stomach of the Far-Eastern starfish *Asterias amurensis*, there are mainly 'typical' cells; close to its oral tip there are more 'special' cells (Vorobyev & Leibson, 1976). The epithelium of the lower part of the stomach is formed almost entirely of 'special' cells. The surface of the stomach is uneven. The 'typical' cells are concentrated in troughs while, in the ridges, there are only 'special' cells. Cytophotometry of the DNA has revealed that the 'typical' cells are primarily diploid and tetraploid, and the 'special' cells are polyploid. DNA histograms at various levels of the stomach illustrate well the ploidy distribution of these cells (Fig. 3.10). The cells with the greatest DNA content are localized in the lower part of the stomach.

The very fact that the transition between 'typical' and 'special' populations of the cells is a smooth one – along the axis of the stomach and in the direction of the troughs and ridges – suggests that 'special' cells are formed from 'typical' ones. It may also indicate an increase in the level of ploidy as cells transform from 'typical' into 'special' cells. There are

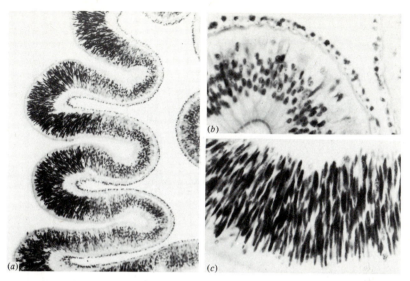

Fig. 3.9. The stomach epithelium of the starfish *Asterias amurensis*. (*a*) General view of the epithelium in which two types of cells may be discerned: (*b*) 'typical' cells with rounded, light nuclei and (*c*) 'special' cells with long, cigar-shaped nuclei. The polyploid 'special' cells are concentrated on the ridges of the gastric mucosa (photographs kindly provided by Dr V. A. Vorobyev).

transitions between the ploidy classes which are indicative of DNA synthesis. The migration of cells labelled with [³H]thymidine from the troughs to the ridges has been clearly shown (Vorobyev & Leibson, 1974, 1976).

Observations clearly point to a seasonal change in the cell populations. There are always cells of different ploidy – from 2C to 16C (Fig. 3.11) – in the epithelium, but in some months, especially in autumn, 32C cells appear. The percentage of 2C cells also changes cyclically, their peak frequency occurring when there are few cells of high ploidy. Evidently, the classes of 'special' cells derive from the polyploidization of 'typical' cells. As the starfish ages, the ratio of diploid to polyploid cells does not change (Vorobyev, 1972). During the year, a fairly stable ratio of the main ploidy classes is maintained; only the frequencies of the extreme classes change.

Seasonal changes in the number of mitoses correlate well with a model proposing an exchange of cell classes and populations of cells. Mitotic activity shows a double peak, in spring and late autumn (during summer and winter, mitotic activity is minimal). Mitoses were found mostly at the site of transition of the diploid 'typical' cells to the polyploid 'special' ones. In the lower part of the stomach, cell proliferation is mainly observed at the base of ridges (i.e. at the presumed site of 'typical' to 'special' cell transition).

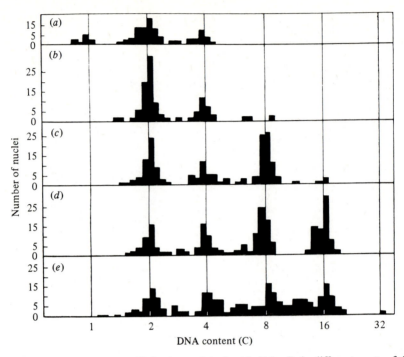

Fig. 3.10. DNA content (C) in the nuclei of epithelial cells in different parts of the stomach of *Asterias amurensis*. (*a*) Testicular cells, (*b*) cells of the pyloric stomach, (*c*) cells of the upper part of the cardiac stomach, (*d*) cells of the middle part of the cardiac stomach and (*e*) cells of the lower part of the cardiac stomach. The DNA contents of nuclei of cells removed from the testis (including sperm) serve as a reference. (After Vorobyev & Leibson, 1976.)

There are many polyploidizing mitoses in the epithelium of the stomach. Acytokinetic mitoses predominate and there are many binucleate cells among the 'special' cells. Other polyploidizing mitoses have also been noted. It is not clear whether mononucleate 4C cells are formed directly from diploid cells or whether the binucleate condition is intermediate in the formation of a mononucleate polyploid cell. Two mechanisms ($2C \rightarrow 4C$ and independently, $2C \rightarrow 2C \times 2$ or $2C \rightarrow 2C \times 2 \rightarrow 4C$) may operate here, as in various tissues of mammals.

There can be no doubt about the mitotic origin of the polyploid cells of the stomach epithelium in starfish. In addition to the results already described, we would like to emphasize the observation of polyploid metaphases. In studying cell transitions in starfish, it would be fruitful to apply the methods used in examining the mechanisms of polyploidization in mammalian cells.

Study of the dynamics of regeneration of the stomach promises to be

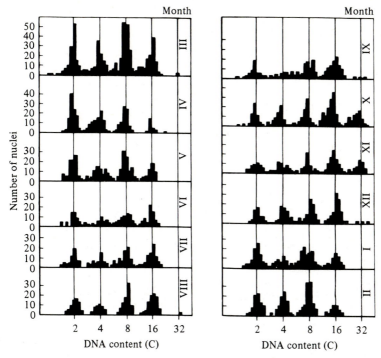

Fig. 3.11. Seasonal changes in the DNA content of epithelial cells in starfish stomach. Roman numerals indicate months of the year (I, January–XII, December). Specimens of *Asterias amurensis* were collected in Peter the Great Bay, near Vladivostok (figure kindly provided by Dr V. A. Vorobyev).

interesting. It is known that, after ablation of a considerable part of the stomach of starfish, the specific anatomical structure is restored within 3 or 4 weeks (Anderson, 1965). Which cells are responsible for this regeneration? Does further polyploidization of the epithelium occur, or are the ordinary classes of cells restored from diploid precursors?

A study of four species of the family Asteriidae has revealed similar anatomical and cytological characteristics in the stomach epithelium. In particular, the 'special' cells were always polyploid. There were, however, hardly any of these cells in the gastric epithelium of *Patiria pectifera*, a starfish from the family Asterinidae. The epithelial cells of the stomach of this starfish were mainly diploid. A noteworthy peculiarity of the digestive system of starfish of the family Asterinidae is the special pump-like organ, Tideman's sacs. In Asteriidae with polyploid 'special' cells, however, the flow of fluid in the alimentary canal is regulated solely by the cilia of the epithelial cells.

To understand the reasons for the polyploidization of the superficial ('special') cells, it is interesting to observe the change in the number of cilia with each polyploidization cycle. According to the data of Vorobyev, for 2–32C cells, the number of cilia forms a series $1:2:4:8:16$. Thus, the polyploidization of the stomach cells of the starfish family Asteriidae is a means of intensifying tissue function. The diploid cells with one cilium cannot, evidently, provide the stomach with the necessary circulatory function. Polyploidy in the stomach of Asteriidae is a necessary trait of ontogenesis. It is no accident that the polyploid cells are concentrated in the esophagus and on the ridges of the stomach, i.e. in the areas of most importance in the operation of the epithelium's circulatory function.

The poorly expressed (and functionally insignificant) polyploidy in the stomach of the Asterinidae, suggests that starfish of this family may be ancestral forms of the more modern family Asteriidae. In the Asteriidae, which possess no special pump-like organs, individuals with the most polyploid stomachs may have been selected by evolution.

F The giant polytene neurons of molluscs

Giant neurons are typical of two subclasses of Gastropoda, Pulmonata and Opisthobranchia, while in Prosobranchia the neurons are small. The ganglia of all molluscs are of one and the same type, and similar to those of other Protostomia: the bodies of the neurons are located on the periphery, and the single nerve cell process is directed into the core of the ganglion. In the ganglions of Pulmonata and Opisthobranchia, there are one or two neurons of huge size. For example, the body of the neuron R-2 in the sea slug *Aplysia californica* reaches 1 mm in size. The nuclei of such cells have a volume of more than 10^6 μm^3. In other Opisthobranchia, e.g. *Tritonia diomedia*, the neurons may be even bigger. These are the largest known somatic cells and can be discerned with the naked eye.

In Table 3.4 the results of cytophotometric analysis of DNA in giant neurons of molluscs are summarized. The smallest DNA content is 32C, which differs greatly from the diploid level characteristic of vertebrate nerve cells. The maximum values of 260 000C, characteristic of neurons R-2 or L-6 in *Aplysia* (Lasek & Dower, 1971), are among the most extraordinary examples of elevated DNA content in a single cell. Unfortunately, there are at the moment no precise data on the DNA content of the larger neurons of *Tritonia*.

The measurement of the DNA content in the neuron R-2 from *Aplysia californica* of various weights (0.5–600 g) and ages, revealed a gradual increase in genome size. The neurons from the youngest molluscs contain 2048C DNA (Coggeshall, Yaksta & Swartz, 1970), i.e. they have undergone 10 reduplications from the original diploid level, while in neurons of large (i.e. old) molluscs, there was 66 000–75 000C which corresponds to 15 replications. According to Lasek & Dower (1971) and Boer *et al.*

Table 3.4. *Cytophotometric data on the DNA content in the nuclei of giant neurons in molluscs*

Species	Maximum DNA content (C)	References
Aplysia californica	75 000	Coggeshall *et al.*, 1970
Aplysia californica	260 000	Lasek & Dower, 1971
Limnaea stagnalis	4 096	Boer *et al.*, 1977
Helix pomatia	500	Kuhlman, 1969
Planorbis corneus	300	Kuhlman, 1969
Triodopsis divesta	32	Cowden, 1972

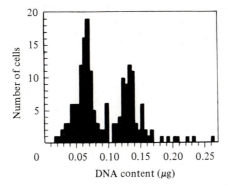

Fig. 3.12. DNA content in the nuclei of giant neurons of the mollusc *Aplysia californica*. The peaks in the histogram are centered at 0.067 and 0.131 μg, which correspond to 67000C and 131000C (the DNA content of a spermatozoon is 10^{-6} μg; after Lasek & Dower, 1971).

(1977), the DNA content in the nuclei of the largest neurons, where the nuclei reach a volume of 10^6 μm^3, is 130 000C and 260 000C. This corresponds well with 16–17 complete replications (131 072C and 262 144C); in absolute values this would be 0.131 and 0.262 μg DNA (Fig. 3.12). This compares with 10^{-6} μg in the spermatozoa of *Aplysia*. All researchers believe that DNA replication in the neurons of *Aplysia* is complete.

DNA synthesis in giant neurons of *Tritonia* has been studied in molluscs of different ages (Bezruchko *et al.*, 1969). In the comparatively small neurons of young molluscs which weigh 5–20 g, DNA synthesis was found in 95% of the nuclei after minimal incubation of the ganglion in a medium containing [^{3}H]labelled thymidine. In older animals, weighing 20–50 g, the labelling index averaged 10%; and in molluscs of 50–100 g, the labelling index was approximately 5%. Thus, neurons of the heavier molluscs rarely incorporated thymidine. Nevertheless, label is also sometimes found in neurons of the oldest of the *Tritonia* studied, which may weigh as much as 1 kg. Thus, the rate of DNA replication, which is high in young animals, decreases as the animals age. But, while the molluscs are growing, their nerve cells increase in size, and the extent of genome multiplication also increases. The mode of neuron growth in molluscs is, in principle, different from that in mammals. In the latter, the neurons also grow with the organism, but genome reduplication (simple divisions) ceases in the neuro-blasts (i.e. before growth of the axon begins). The size of the cells then increases, sometimes 100-fold, but retains a constant diploid genome.

The synthesis of DNA in giant neurons of molluscs is not accompanied by mitoses. The chromatin in the nucleus is distributed diffusively. It would be interesting to determine the number of chromocenters, and their sizes in nuclei with different quantities of DNA. It is, however, clear that the reduplications occur in the decondensed chromosomes but the chromatids do not then become condensed. Once again we are dealing with 'cryptic' polytene chromosomes.

Do the nucleus and cytoplasm of the giant neuron grow in proportion with the genome? In the garden snail *Helix pomatia*, whose neurons reach a comparatively low degree of polyteny – 512C (Kuhlman, 1969) – a strict correlation has been shown between the size of the nucleus and its DNA content (Narushevichus & Kviklite, 1968). It is not clear how the volume of the cytoplasm changes, nor whether the size of the nucleus doubles in all highly polytene cells in each endocycle nor what the correlation is after the 15–17 endoreduplications.

As the volume of a smooth spherical nucleus increases, its surface-area-to-volume ratio may decrease. Cell growth may impair both nuclear–cytoplasmic relations and also the cell contacts with the medium, since the relative amount of cell surface also becomes smaller. But this decline in surface-area-to-volume ratio probably does not happen. The giant nuclei of *Tritonia*, which seem oval when examined in the light microscope, possess numerous submicroscopic protruberances which penetrate deep into the cytoplasm. Cytoplasmic invaginations (enclosed by nuclear membrane) penetrate the nucleus, reaching the nucleoli (Borovyagin & Sakharov, 1968). Thus, the surface area of the nucleus of the giant neuron is much larger than may be deduced from its average diameter (see also Chapter 7B). In the super-large cells of various animals and plants, the nuclei occupy a considerable part of the cell; in *Aplysia* and *Tritonia*, for example, nuclei occupy up to 30% of the total cell volume (Lasek & Dower, 1971).

The specific glial environment facilitates the normalization of nuclear–cytoplasmic relations in the giant neuron. Thousands of glial cells surround each neuron. The glial cell outgrowths penetrate deep into the perikaryon, almost reaching the nuclear membrane. The neurons and gliocytes are separated from one another by two plasma membranes. Nevertheless, in molluscs, the components of the neuron and the surrounding auxiliary cells are in closer contact than they are in the diploid nervous system of vertebrates.

Some giant neurons probably have neurosecretory functions (Krause, 1960; Cowden, 1972; Manfredi Romanini, Fraschini & Bernocchi, 1972). Research on amino acid incorporation in the neurons of *Tritonia* has revealed more label in the big neurons than in the small ones. The relative intensity of protein synthesis in cells with different DNA content is not clear.

To conclude, let us consider an interesting aspect of cell giantism, which was briefly mentioned when describing the salivary gland of *Drosophila*. Each giant cell corresponds to a micro-organ. Whereas most of these cells are monofunctional organs, the gigantic neurons of molluscs may be polyfunctional, usually performing the functions inherent to a ganglion (nerve center) (Sakharov, 1965). The activities of the different axon branches of such a neuron may vary, and the nerve cell then has several outputs. In another state of nerve activity, the same branches are activated synchronously, and then the neuron operates like a single cell.

G Polyploid and polytene tissues of ascarids

Two peculiarities are usually noted in the development of ascarids – chromatin diminution in somatic cells and the constancy of cell number during tissue growth (the so-called eutelic growth).

Diminution, the elimination of some part of the chromosome material during cleavage of the ascarid egg, was first described by Boveri at the end of the nineteenth century. It has since been disputed many times. The results of quantitative studies of DNA, which ought to resolve the visual morphological observations, are contradictory. DNA content in germ-line and in somatic cells has been shown to differ. Cytophotometry of DNA during telophase of the third cleavage division of *Ascaris lumbricoides* reveals a difference of approximately 30%, in comparison to spermatozoa, taking into account the diploid chromosome set (Pasternak & Barrell, 1976). However, the number of ribosomal genes in spermatids and in the larval tissues of *A. lumbricoides* has been found to be identical (Tobler, Smith & Ursprung, 1972; Tobler, Zulanf & Kuhn, 1974). In another species (or subspecies) of ascarids, *A. suum*, a strict correlation between the DNA content in the various differentiated cells and the spermatozoa has been observed (Anisimov, 1973). In this work, an important point of DNA cytophotometry in ascarids was taken into consideration. It is known that the nematode spermatozoa have a so-called X-complex, consisting of several chromosomes (Wilson, 1925). An autosome + X (AX) spermatozoa differs from an AY spermatozoa in its size and more importantly, in chromosome number. In different species of ascarids the AX spermatozoa may

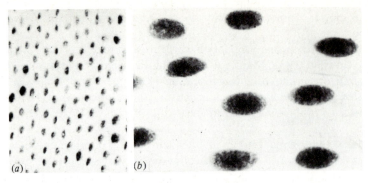

Fig. 3.13. The uterine epithelium of *Ascaris suum* (*a*) at the beginning and (*b*) at the end of its growth in worms weighing 0.35 and 8.00 g, respectively. Both photographs are at the same magnification (photographs kindly provided by Dr A. P. Anisimov).

have 5–9 chromosomes more than the AY ones. This may also be reflected in their DNA content. Thus, in *A. suum*, cytophotometry revealed differences of 37% between spermatozoa of two types. In analyzing the reasons for the discrepancies in the evaluation of the DNA content in germ-line and somatic cells, it should be taken into account that diminution may be expressed to varying extents in different species of ascarids. The correspondence between the DNA content in the somatic and germ-line cells has also been discovered in the free-living nematode *Panagrellus silusiae*, as well as in *A. suum* (Sin & Pasternak, 1970).

   The epithelia of the intestine and uterus of ascarids are convenient objects for cytochemical and, especially, for cytophotometric research since they consist of an almost pure single layer of prismatic epithelial cells (Fig. 3.13). Intestinal epithelial cells of *A. suum* are polyploid (Anisimov & Tokmakova, 1973). In the smallest of the ascarids studied (weighing up to 0.04 g) most cells are diploid, but 4C and occasional 8C and 16C cells are also found. In ascarids weighing 0.35 g, some cells are diploid, but the main mass of the epithelium consists of 4C and 8C cells. In worms of 1.75 g, there are 32C cells. This ploidy distribution – mostly 4C and 8C cells, and a few 16C and 32C cells – is retained in the later stages of ontogenesis (in worms weighing up to 8 g). At all stages of intestinal growth, there are binucleate cells and a few that are multinucleate. Thus, in early postnatal ascarid ontogenesis, complete (and practically synchronous) polyploidization of the intestinal epithelium occurs. Mitoses have been found in these epithelial cells, but among these there are many polyploidizing ones: acytokineses, metaphases with non-separation of the chromosomes (c-mitoses), and ana- and telophase anomalies. There are also completely normal mitotic figures, as well as multipolar mitoses and bimitoses.

Polyploid metaphase plates are also present in which a doubled DNA content can be measured. Consequently, the cells of the intestinal epithelium are genuinely polyploid and are not polytene.

An important observation is the increase in the number of cells in the intestinal epithelium as the ascarid ages (Anisimov & Usheva, 1973). The number of cells increases approximately 30-fold in worms weighing from 0.04 g to 7.78 g. These data refute the idea of total eutely, i.e. of a constant number of cells in all the tissues of ascarids. The most intense growth of the intestine occurs during the first half of postnatal life in ascarids. The mitotic index in this period is more than tenfold higher than in adult worms (mitoses still occur in adults, too). Since polyploidization is an early event in ontogenesis, the increase in the number of the cells in the population is the result of the multiplication of the polyploid cells. The average cell volume increases approximately sixfold during ontogenesis; this means that, besides polyploidization and the multiplication of the cells, the volume of cytoplasm also increases.

Another mode of growth is characteristic of the cells of the esophageal gland in ascarids. This gland is formed of just one dorsal cell and two subventral cells. In one of these cells in *A. lumbricoides*, some 30 000C DNA is found; a complete absence of mitoses during the growth of this cell was noted by Palanker & Swartz (1971). In *A. suum*, according to the data of Anisimov (1976), the quantity of DNA in the cells of the esophageal gland is 5–10 times greater. The volume of the gland cells grows continually as the gullet grows. In adult ascarids, the nuclei of these cells are so big that, after the Feulgen reaction, they are visible to the naked eye. Such cells have 130 000–260 000C of nuclear DNA (Table 3.5). The structure of the nuclei is diffuse. However, just as in the silk gland of the silkworm and in the trophoblast of mouse, nuclei in the esophageal gland are without doubt polytene.

It is more difficult to determine the mode of growth of the uterine epithelium in ascarids. In the first study of this tissue in *A. lumbricoides* (Floyd & Swartz, 1969) a large number of binucleate cells, the absence of mitotic figures, and 'strangulation' of the nuclei were noted. An increase in the average quantity of DNA in cells from different parts of the uterus was noted as the ascarid aged. In *A. suum*, approximately 90% of the epithelial cells of the uterus are binucleate; there are also multinucleate cells. During ontogenesis, a manifold synchronous reduplication of DNA occurs in the cells of the uterine epithelium (Fig. 3.14). In the histograms the quantities of DNA in individual nuclei can be seen; the figures must be doubled when applied to the cell since almost all the cells are binucleate. Just as in the intestine and the esophageal gland, the DNA reduplications are complete; neither amplification nor under-replication of the DNA can be detected. Mitoses were not found. Just as in the previously mentioned work, with *A. lumbricoides*, nuclear 'strangulations' were seen.

Table 3.5. *Changes in DNA content in esophageal gland cells during the growth of the ascarid A. suum*

Growth parameters of the ascarids		DNA content (C)				
Weight (g)	Length (mm)	Dorsal cell		Subventral cells		
0.05	55	66 400[a]	(65 536)[b]	8 400	(8 192)	8 430	(8 192)
0.05	60	64 600	(65 536)	7 790	(8 192)	8 300	(8 192)
0.15	95	136 000	(131 072)	17 600	(16 384)	18 300	(16 384)
0.80	160	137 000	(131 072)	18 300	(16 384)	21 100	(16 384)
3.50	270	135 000	(131 072)	34 300	(32 768)	35 300	(32 768)
3.50	260	149 000	(131 072)	35 400	(32 768)	36 600	(32 768)
7.70	380	268 000	(262 144)	70 000	(65 536)	65 500	(65 536)
10.00	400	284 000	(262 144)	123 000	(131 072)	75 500	(65 536)

After Anisimov, 1976.
[a] Measured value.
[b] The theoretically expected DNA content calculated according to the DNA value of the haploid standard.

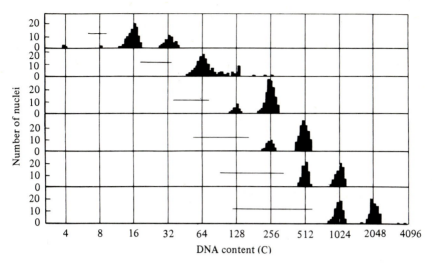

Fig. 3.14. DNA content (C) of nuclei in the uterine epithelium at different stages during the growth of *Ascaris suum*. The abscissa shows the theoretically expected DNA classes, calculated on the basis of the DNA content of AX spermatozoa. The horizontal bars to the left of each histogram show the relative length of the worm's body (after Anisimov, 1974).

According to the calculations of Anisimov (1974), the volume of the uterine epithelium grows proportionally to the growth of its cells. The number of its cells decreases significantly during ontogenesis. Binucleate cells are characteristic even of young animals. The quantity of DNA in the nuclei of binucleate cells is always equal. All the data contradict the amitotic division of the cells or only of the nuclei, which is discussed in the literature. Nuclear strangulation is, most likely, evidence of aberrant mitoses. Such mitoses may occur in the earliest stages of development, but these have not yet been studied. If acytokinetic, multipolar and other aberrant mitoses are found in early development, the cytological basis for the nuclear forms in the uterine epithelium will thereby have been elucidated.

Thus, the study of the epithelia of ascarids reveals two clear modes of cell growth. Polyploidy is found in the epithelium of the intestine. It does not differ in its mode of formation from the polyploidy in the intestine of the starfish (Chapter 3E) or in the organs of mammals. In the ascarid esophageal gland, polyteny has been found of the same form as the trophoblast of mammals, trophocytes of some insects, the gland cells of the silk worm, and in the giant neurons of molluscs. Various modes of cell growth may alternate during development of the ascarid uterine epithelium. This example is also interesting because here it is possible to compare genome multiplication with the functional activity of the cells. The epithelial cells of the ascarid uterus synthesize a protein from which the thick outer envelopes of the eggs are formed. A close correlation has been found between the number of eggs, and consequently the mass of their envelopes, and the polytenization of the epithelial cells (Anisimov, 1974).

Conclusion

A characteristic of invertebrate cell growth and development is polytenization of the genome. This mode of growth results in a far greater accumulation of DNA than does polyploidy. DNA contents as high as 64C are rarely achieved by means of polyploidy in either invertebrates or vertebrates, while hundreds and thousands of haploid DNA units often accumulate within a single nucleus through polyteny. Cell size increases proportionally. The record DNA content is approximately 1 000 000C in the gland cell of the silkworm. Some giant neurons of molluscs and the esophageal gland cells of ascarids have DNA contents of 100 000–300 000C.

Polyteny is usually associated with the specific structure of the interphase nucleus characteristic of *Drosophila* or *Chironomus*, in which multistranded chromosomes with bands are constantly visible and the apparent number of these huge chromosomes is reduced because of conjugation of homologs. However, giant nuclei with bundles of fibers and vague bands, or with diffuse interphase chromatin are found much more frequently in the polytene nuclei of animals and plants. These nuclei also pass through some endocycles but their multistrandedness is invisible. There are examples of banded polytene cells and cells with a diffuse structure being adjacent within a single tissue and possessing similar functions and the same DNA content. The frequency of nuclei with a visible or polytene structure can be varied according to the animals' (or plants') environmental conditions. The DNA and strand content does not change in these conditions.

Polytenization is a considerably more rapid mode of growth than division of the diploid non-polytene cells or polyploidization. The result of polytene growth is a compact organ which functions intensively, usually for a short period of time. Unlike the polyploidizing tissues, in some polytenizing ones the DNA content doubles repeatedly throughout the entire cell population.

Polyploidization of invertebrate cells, like that of vertebrate cells, is realized by means of incomplete mitosis. This process has been found in some starfish, locust and ascarid tissues.

DNA replication in the cells of invertebrates is, as a rule, complete. Underreplication is a rare phenomenon, which is characteristic neither of polyploidy nor polyteny. In isolated cases there is diminution of the chromatin.

4

Peculiarities of genome multiplication in plant cells

One of the differences between plant and animal cells is the exceptional variability in the mass of the diploid genome in plants. In angiosperms 2C DNA values may differ by 500-fold, varying from 0.5 pg in *Arabidopsis thaliana* to 255 pg in *Fritillaria assyriaca* (Bennett & Smith, 1976). The cells of many plants contain considerably more DNA than the cells of mammals. The reason for such a high DNA content is not clear (Chapter 1 B).

Polyploid and polytene cells are just as common in plant tissues as in animal ones. In Table 4.1 results of many authors are summarized, showing the maximum level of DNA in the cells of various tissues in different plant species. In the book by W. Nagl (1978), from which much of this information is derived, morphological observations of large nuclei in plant cells are also cited. Large nuclei have been found in oil cells, water-storing cells, some cells of fruits, and in hairs. There are many more data of this kind in addition to cytophotometric determinations of DNA content. Recently, Nagl (1981) listed examples of polytene nuclei in different plant tissues; these were mainly in suspensor, endosperm and antipodal cells.

The data in Table 4.1 characterize approximately the distribution of polyploid cells in the tissues of a single plant. In *Vicia faba* such cells have been found in six types of tissue; in *Bryonia dioica* they have been noted in four tissues; in *Cymbidium*, in seven. The ploidy of other tissues in the same plants has not been established. Species-specific differences may also exist. During a morphological study of 179 species of plants, cells with large nuclei were found in 140 species, but were absent in 39 species (Tschermak-Woess, 1956). In one species – *Helianthus tuberosus* – polyploid cells could not be detected by means of cytophotometry, in eight organs (Evans & Van't Hof, 1975). In another species of the same genus – *Helianthus annuus* – there are polyploid cells (Table 4.1). In some species (e.g. *H. tuberosus*), organs are purely diploid; while the same organ in another species (e.g. *Pisum sativum*) may contain polyploid cells (Fig. 4.1). A comparison of the properties of diploid and polyploid cells of the same organ may allow the reasons for cell polyploidization to be understood.

The increased quantity of DNA in plant cells as they develop may

113

Table 4.1. *Examples of multiplication of the DNA content in plant cells*

Organ	Tissue or cell group	Species	Maximum DNA content (C)
Root	Zone of elongation	*Zea mays*	16
		Hordeum vulgare	8
		Secale cereale	8
		Vicia faba	8
		Pisum sativum	8
	Metaxylem	*Allium cepa*	32
		Acorus calamus	32
		Zea mays	32
		Arisaema triphyllum	32
		Pisum sativum	32
	Cortex	*Cymbidium* sp.	16
	Exodermis	*Cymbidium* sp.	32
	Root hairs	*Cymbidium* sp.	64
Stem	Epidermis	*Vicia faba*	16
	Cortex	*Vicia faba*	16
	Parenchyma	*Cymbidium* sp.	8
Leaf	Epidermis	*Beta vulgaris*	32
Flower	Corolla hairs	*Bryonia dioica*	32
	Stamen filaments	*Bryonia dioica*	128
	Anther hairs	*Bryonia dioica*	256
	Tapetum	*Bryonia dioica*	32
		Solanum tuberosum	192×4
	Nectary	*Solanum tuberosum*	8
Embryo sac	Suspensor	*Phaseolus coccineus*	8192
		Phaseolus vulgare	2048
		Tropaeolum majus	2048
	Antipodal cells	*Scilla bifolia*	1024
		Triticum aestivum	256
Cotyledon	Parenchyma	*Pisum arvense*	16
		Vicia faba	16
		Gossypium hirsutum	16
Epicotyl		*Pisum sativum*	8
Hypocotyl		*Sinapis alba*	32
		Helianthus annuus	16
Protocorms		*Cymbidium* sp.	128

After Van Parijs & Vandendriesche, 1966; Brunori, 1971; Bennett, 1974; Pearson, Timmis & Ingle, 1974; Walbot & Dure, 1976; D'Amato, 1977; Nagl, 1978.

depend both on polyploidy (i.e. an increase in the number of chromosomes) and on polyteny (when only the number of chromonemata increases). Neither in their modes of formation nor, probably, in their functional significance do the polyploid cells of plants differ from such cells in animals. Let us examine several examples, to add to the material already

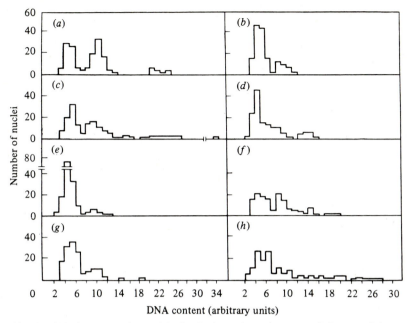

Fig. 4.1. DNA content in nuclei of cells in various organs and tissues of the pea
Pisum sativum. (*a*) Root, (*b*) petal, (*c*) stem, (*d*) pod, (*e*) leaf, (*f*) pistil, (*g*) sepal,
(*h*) stamen (after Evans & Van't Hof, 1975).

cited for animal cells. Most of these observations apply to the organs and
tissues of angiosperms.

The first example concerns root tissues. The root tip consists of cells that
comprise the cap, the quiescent center, the meristem, and the zone of
elongation. The meristem is the area of cell multiplication; mitosis is
largely confined to this zone, and to layers of cells in the root cap. The
number of cells in the meristem depends on the thickness of the root. The
meristem of very thin roots may extend for only ten cells, that of thick roots
may be several hundred cells long. The duration of the mitotic cycle in the
meristem depends on the mass of the diploid genome: the greater the
quantity of DNA, the longer the complete cycle and also the duration of
S-phase. First reported 21 years ago (Van't Hof & Sparrow, 1963), this
relationship has since been studied by many authors. Although there is a
wide range of values between species, there can be no doubt that there is
a general relationship between the DNA content and the duration of the
mitotic cycle (Fig. 4.2).

The duration of the mitotic cycle may also be influenced by the plant's
ecology. The amount of DNA in the diploid genome and the duration of
the cycle are often found to be greater in perennials than in annuals; and

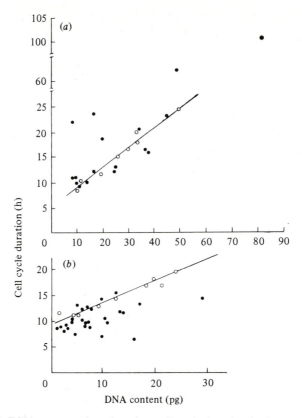

Fig. 4.2. DNA content plotted against cell cycle duration in the root meristem of different plant species. (*a*) Monocotyledons, (*b*) dicotyledons (after Ivanov, 1978, summarizing data from many authors; the open circles refer to species examined by Evans *et al.* (1972), the line is a regression drawn through their data points).

in annuals both are greater than in ephemerals (see Bennett, 1972). However, when species in individual families are analyzed, this dependence may not be evident (Ivanov, 1978). In the diploid and polyploid forms of the same species, the parameters of the cycle do not, as a rule, differ (see also chapter 7B).

Root cell differentiation begins in the meristem. The maximum number of mitoses occurs in the middle of the meristem. The number of mitoses diminishes as the cells approach the zone of elongation. This elongation phase is characteristic of plant cell development. Elongation is caused by the vacuolation and hydration of the cytoplasm with the concomitant accumulation of many substances, including proteins. During elongation, tissue-specific characteristics emerge. Elongation does not always block mitosis, and DNA synthesis occurs in some differentiating cells. More often

Table 4.2. *The number of endoreduplications (the percentage of mitoses with diplo- and quadruplochromosomes) in the cortical parenchyma of the root of* Bellevalia romana[a]

Thickness of the root	Number of mitoses	Type of mitosis (%)	
		$2 \times$ [b]	$4 \times$ and $8 \times$
Thin	130	39.2	60.8
Thick	1005	2.8	97.2
Very thick	2647	0.9	99.1

After D'Amato, 1952.
[a] Mitosis was stimulated by 2,4-dichlorophenoxyacetic acid.
[b] The number of chromatids in each metaphase chromosome.

than not, mitoses do not succeed DNA replication in the cells of the elongation zone. Division is also rare in the more distal cells of the meristem (e.g. the quiescent center).

As a result of DNA synthesis without subsequent mitosis, cells with a double (and sometimes quadruple) quantity of DNA accumulate just beyond the margin of the meristem and in the elongation zone (see, for example, Van Parijs & Vandendriessche, 1966; Brunori, 1971). Colchicine blocks mitosis, but does not prevent the accumulation of DNA and growth of the root cells (Barlow, 1977*a*). If mitosis is stimulated where it normally has ceased, double chromosomes (diplochromosomes) are discerned (Levan, 1939; Grafl, 1939; Gentcheff & Gustafsson, 1939). The appearance of such structures can only be caused by the occurrence of endoreduplication in the previous cycle. Another round of DNA synthesis leads to the emergence of four chromatids. Diplochromosomes, and even quadruplochromosomes, are common in the elongation zone of onion root after artificial stimulation of mitosis there (Levan, 1939; D'Amato, 1977). The thickness of the root not only corresponds to the size of the meristem, but also to the number of endoreduplications in the cortical parenchyma (Table 4.2).

Polyploid mitoses can also be induced in root tissues (Sinnott, 1960; Matthysse & Torrey, 1967; Libbenga & Torrey, 1973; Hervas, 1975). The further the cells are located from the root tip, i.e. the further they have progressed in their differentiation, the more tetra- and octaploid mitoses there may be. Thus, near the root tip of *Pisum sativum*, most of the mitoses (75%) are diploid, while 5 mm from the tip, most of the mitoses that can be induced are polyploid (Libbenga & Torrey, 1973). Thus, diploid cell multiplication in the apical regions of the meristem is succeeded by endoreduplications in its distal zone.

Interesting data reported by Barlow (1977*b*) on the doubling of DNA

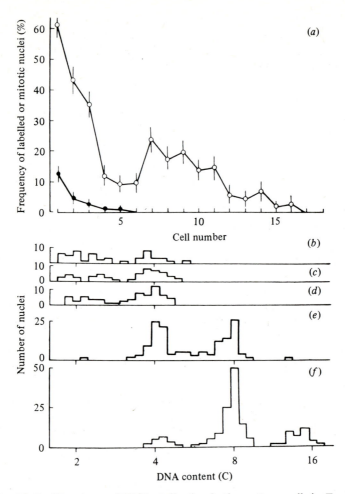

Fig. 4.3. Proliferation and DNA replication in the root cap cells in *Zea mays*. (*a*) Mean labelling index (open circles) and mean mitotic index (filled circles) in cells along the length of the columella of the root cap of *Zea mays*. Roots were fed with [³H]thymidine for 15 min prior to fixation. The abscissa indicates the position of the cells numbered from the cap meristem towards the tip of the cap. (*b–f*) DNA content of the nuclei in cells in different locations along the length of the cap columella. (*b–d*) DNA content in cells at the margin of the meristem (cells numbered three to five, respectively). (*e*) and (*f*) The nuclear DNA content in cells in the center of the cap and close to the tip of the cap, respectively (after Barlow, 1977*b*).

in the root cap cells of maize, support the conclusion of a role for endoreduplication in the growth of plant cells (Fig. 4.3). Mitoses were discerned only in the first five proximal rows of the root cap. There were no mitoses distal to this row, but the number of nuclei labelled with [³H]thymidine doubled or tripled. A cytophotometric study of DNA revealed 4C and 8C cells in the central part of the root cap, whereas in the more distal rows, there were 8C and 16C mononucleate cells.

Endoreduplication is characteristic of leaf tissues, too (see, for instance, Cionini *et al.*, 1983). This process is especially typical of the leaves of succulents; moreover, environmental stress may cause the number of polyploid cells to increase in these leaves (D'Amato, 1977). Thus, diplo- and quadruplochromosomes have been found after stimulation of mitosis in the leaf mesophyll in *Kalanchöe blossfeldiana*, which were grown in long-day conditions. In plants grown in short-day conditions, the cells underwent more endoreduplications, the chromosomes possessing 16 or even 32 chromatids. A similar process was observed in *Bryophyllum crenatum*. In succulent leaves, which developed in response to a saline environment, 64C DNA was detected in the nuclei; in non-succulent leaves, the nuclei contained 16C DNA (Catarino, 1965).

The quantitative changes in DNA in embryonic organs of plants are rather more considerable than in mature tissues. A particularly large amount of DNA accumulates in the suspensor cells, where polytene chromosomes are found. The suspensor moves the embryo into the central part of the embryo sac. At this time, endosperm (the source of the nutritive substances for the embryo) is formed in the embryo sac. Giant cells are frequently found both in the suspensor and in the endosperm. The largest cells are usually located in the basal part of the suspensor. In species with a small suspensor, the cells in the endosperm are especially large (D'Amato, 1977; Nagl, 1978).

The greatest number of studies has been made on the bean *Phaseolus*, whose suspensor is formed of 100–150 cells. Near the embryo, the suspensor cells are small. In young embryos there are diploid cells in this zone (although most of the other cells have 4–64C DNA). Giant cells are found in the basal part of the suspensor, even in young embryos. During the final stages of suspensor development, nuclei with 2048C, 4096C, and 8192C of DNA are discerned in such cells in an approximately equal proportion (Fig. 4.4).

A polytene structure has repeatedly been described in the nuclei of giant cells in the suspensor of *Phaseolus* (see, for example, Nagl, 1970*a*; Brady & Clutter, 1974). Polytenization of the chromosomes does not take the same pattern as it does in the cells of the salivary gland of Diptera, but rather it shows similarities with the mammalian trophoblast and the gland cells of the silkworm. Bundles of chromatin threads and chromocenters can be seen. The bands are indistinct, even in the most condensed chromosomes. In many nuclei there are no bundles, but only lumps of chromatin are visible. Nagl discerned some 20 large chromocenters in nuclei containing 2048C DNA and possessing condensed chromosomes. In the diploid set of *P. coccineus*, there are 22 chromosomes. Consequently, the giant nuclei with the condensed chromosomes are indeed polytene, and not polyploid; unlike in the polytene nuclei of Diptera, conjugation of the

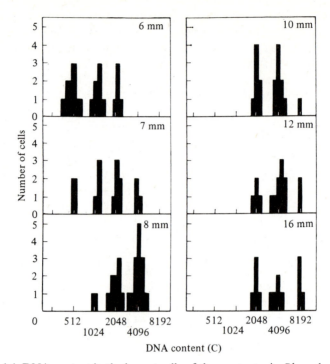

Fig. 4.4. DNA content in the largest cells of the suspensor in *Phaseolus coccineus* seeds measuring 6–16 mm (after Brady, 1973).

homologs does not take place in the nuclei of the bean suspensor. Nagl also noted that, in nuclei where bundles of chromonemata could not be discerned, the chromocenters were more numerous than in nuclei with the same quantity of DNA (2048C), but with condensed chromosomes. Such nuclei had on average, 44, 108, or 214 chromocenters, while the sizes of the chromocenters were inversely proportional to their number. At the beginning of the development of the suspensor, Nagl found polyploid mitotic figures: 4n, 8n, or 16n. This author assumes that the bundles of threads, i.e. the condensed chromosomes,are visible in cells which changed over to the polytene cycle directly after completing mitosis, i.e. the cells remain with a diploid genome. Therefore the number of chromocenters (centromeres?) in such cells corresponds to 2n. The diffuse structure of the nucleus was characteristic of polytene cells as well, but was most evident in those where polyploidizing (restitution, according to Nagl) mitoses had taken place before the endocycles. The number of chromocenters is doubled once, twice or thrice, depending on the number of such mitoses and, consequently, the nucleus houses a 4n, 8n, or 16n chromosome set.

The distribution of the polyploid set of chromosomes within the nucleus determines its diffuse structure, in spite of the polytenization of the

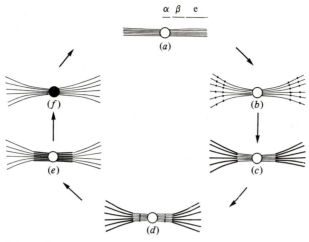

Fig. 4.5. Schematic representation of chromatin changes during the replication cycle in the polytene chromosomes of *Phaseolus coccineus*. (*a*) Condensed, nonreplicating chromosome showing distribution of the centromeric (α-) and adjacent (β-) heterochromatin and the euchromatin (e); (*b*) the initial stage of replication, chromatin strands unwinding; (*c*) euchromatin replicating; (*d*) euchromatin completing replication and β-heterochromatin beginning to replicate; (*e*) β-heterochromatin completing replication; (*f*) α-heterochromatin replicating (after Brady & Clutter, 1974).

chromosomes. The even distribution of the [³H]thymidine label in many such polytene nuclei (Avanzi, Cionini & D'Amato, 1970) is quite under-standable. Another model has been put forward to explain the diffuse structure of the polytene nucleus. Autoradiographic data (Brady & Clutter, 1974) suggest that in plants (unlike in Diptera) the distance between chromonemata increases during replication (Fig. 4.5). Banding is not evident because of the looser arrangement of the replicating chromomeres.

Thus, polyteny in plants is of the 'cryptic' type, which is also characteristic of many animal cells. Just as in these cases, the condensed or diffuse structure of the chromosomes in the suspensor depends on environmental conditions. In unfavorable conditions (cool temperatures, an exceptionally long day, after the effect of certain growth inhibitors) the extent of chromosome condensation may change (Nagl, 1978). Nuclei with the looser distribution of chromonemata in the diploid polytene chromosome set are probably more frequent than nuclei with a polyploid set of polytene chromosomes, although the second case is common in the suspensor.

Endosperm, the source of nutrition for the embryo, is yet another example of a cell type showing genome multiplication. Endosperm cells are initially triploid in angiosperms. In the endosperm of the palmyra palm

Borassus flabellifer, in addition to 3n cells, 6n, 12n, 24n, 48n and 96n cells have also been found (Stephen, 1974). Here polyploidy is the result of incomplete mitosis. Usually a metaphase block is observed: the nuclear membrane disperses, the chromosomes become spiralized, but there is no spindle. Stephen (1974) writes that there is no endomitosis in palm endosperm. In maize endosperm, triploid mitoses occur at the outset of its development. Then the mitoses cease, but the cells continue to grow and bundles of chromatin threads begin to be observed in some nuclei (Stephen, 1973), indicating the probable onset of polytenization. In some nuclei the polyploid chromosome set may also be polytenized. Nor has endomitosis been demonstrated in the earlier morphological studies of the endosperm of various species of angiosperms. A typical mode of cell polyploidization was the inhibition of mitosis in the anaphase (Enzenberg, 1961). The discovery of especially large nuclei in the endosperm of many plants (Tschermak-Woess, 1967) makes this organ an interesting subject of study in terms of polyploidy and polyteny.

Three stages can be distinguished in the development of the cotyledon of leguminous plants: cell division, cell growth (characterized by the accumulation of considerable quantities of substances, especially proteins), and then dehydration. While the cells are increasing in number, they remain diploid. Then, during the period of cell growth, the nuclei go through two cycles of endoreduplication without mitosis, and their DNA content increases fourfold (Manteuffel *et al.*, 1976). The quantity of RNA and proteins increases in proportion to the size of these cells.

The anther tapetum provides an example of the binucleate and multinucleate variants of polyploidy in angiosperms (D'Amato, 1977). Thus, during the development of the potato tapetum, numerous transformations in the diploid nucleus are possible (Fig. 4.6). The cells complete three mitoses. Some of them remain mononucleate, and considerably increase their DNA content as a result of polyploidizing mitoses. Most of the cells undergo acytokinetic mitosis and become binucleate. During the subsequent bimitosis, the daughter chromosome sets fuse in each mitotic figure but the cell remains binucleate with a higher level of ploidy. During bimitoses of other cells, their nuclei divide completely, and tetranucleate cells are formed. Bimitosis may be repeated, and binucleate and tetranucleate cells with a higher degree of ploidy are formed. A rare variant is the fusion of some of the daughter nuclei, whereby trinucleate cells with nuclei possessing different ploidy are formed.

An interesting example of the development of polyploidy is found in galls, growths of plant tissue which occur around some animal parasites, such as the larvae of insects of the family Cecidomyiidae (Diptera) and some nematodes. Judging from the size of their nuclei and chromocenters, the cells become polyploid or polytene. The growth of these cells is greater the closer the cell is to the chamber with the parasite (Fig. 4.7). At present,

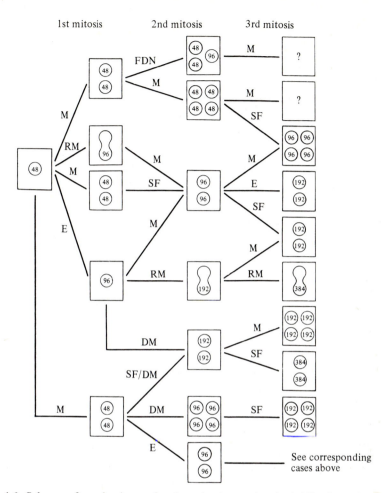

Fig. 4.6. Scheme of mechanisms of polynucleation and polyploidization of the anther tapetum cells of the potato *Solanum tuberosum* (2n = 48). M, Mitosis; DM, mitosis with diplochromosomes; RM, restitution mitosis; E, endomitosis; FDN, fusion of daughter nuclei; SF, spindle fusion; SF/DM, spindle fusion in diplochromosome mitosis. The numbers in the circles indicate the number of chromosomes in the nucleus (D'Amato, 1977).

the largest cells have been found in the galls induced by the midge *Mayetiola poae* on stems of the grass *Poa nemoralis*. Compared with the diploid cells of the normal leaf, the size of the central cells of this gall corresponds to a ploidy level of 4096C (Hesse, 1969). In the nuclei of some giant cells, threads with signs of bands can be seen, i.e. these are highly polytene cells whose nuclei possess a similar structure, and evidently a similar degree of DNA accumulation, to those of the suspensor cells in some plants. In other

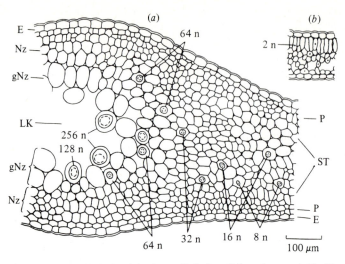

Fig. 4.7. Ploidy gradient in (*a*) the gall induced by the cynipid *Neuroterus numismalis* on leaves of the oak *Quercus robur*, and (*b*) a transverse section of a normal leaf. E, epidermis; P, gall parenchyma; ST, starch-containing cells; Nz, nutritive cells; gNz, giant protein-containing nutritive cells; LK, chamber which houses the larva. Cells of various ploidy are indicated (after Hesse, 1968).

galls, multinucleate cells are formed and acytokinetic and other poly-ploidizing mitoses are seen (Hesse, 1971). Proliferation, polyploidization and polytenization are stimulated by substances released by the parasite. The existence of growth factors in gall tissue has been proved in experiments in which an extract of gall was added to seedlings in sterile culture. A 'gall' tissue was formed, owing to the effect of this extract (Tandon, Vyas & Arya, 1976).

In one species of plant – *Bryonia dioica* – not only has condensation of the chromosomes during their polytenization been described, but also conjugation of the homologs (Barlow, 1975). At the moment, this is the only clear instance of this type in somatic tissues apart from those described in the larval tissues of Diptera. Nuclei with condensed chromosomes have been found in large basal cells (with a volume of up to $10^6 \mu m^3$) of hairs on the anther of *B. dioica*. They contain large amounts of DNA; as much as 512 pg, which correspond to 256C, has been found. Bands are invisible, as in most polytene chromosomes, except in the chromosomes of Diptera.

The possible deterioration of function in huge nuclei, due to the lack of correspondence between their surface area and their volume, has been discussed in respect of both giant plant cells and animal cells (Bennett, 1974). It has already been mentioned how giant cells in animals overcome such difficulties by increasing the structural complexity of the nuclear and cell surface (Chapter 7B).

Up till now, we have noted the complete DNA replication in the tissues of various plant organs. This has been shown both in cytophotometric works and also in studies of DNA-reassociation kinetics. Thus, not only has a doubled DNA content been shown, but also there is a proportional change in the various fractions of DNA, in the cotyledons of the broad bean *Vicia faba* (Millerd & Whitfeld, 1973) and of the cotton plant *Gossypium hirsutum* (Walbot & Dure, 1976). However, the DNA composition of different tissues seems to vary more in plants than in animals. But the significance of the observed variation is not always clear.

It has been shown that satellite DNA is contained in large quantities in the genome of the meristems (in particular, in the root meristem) and, to a much lesser extent, in differentiated tissues. The heterochromatin content in the nuclei also changes. In some cells, the heterochromatin may be under-replicated; in others it may be over-replicated (Pearson, Timmis & Ingle, 1974). In the orchid *Cymbidium*, differential replication of AT- and GC-rich heterochromatin has been discovered (Schweizer & Nagl, 1976; Nagl, 1977), a phenomenon that is still unknown in other plant cells and also in animals.

Judging by the clear doubling of the DNA content in the suspensor cells of the bean (Fig. 4.4), the conclusion may be drawn that complete reduplication occurs during each of the 10 to 12 endocycles. However, incomplete reduplication of the ribosomal genes and over-replication of the satellite DNA have been revealed by studies of fractionated DNA (Lima-de-Faria *et al.*, 1975). The overall effect of these opposed processes is a balanced, strictly multiple 2n quantity of DNA. In the related species, *P. vulgare*, no variation of any kind has been found in the content of satellite DNA (Ingle & Timmis, 1975).

In the suspensor of the nasturtium *Tropaeolum majus*, chromatin threads could be seen in most of the nuclei, i.e. condensed polytene chromosomes, with large areas of heterochromatin. In other cells, there was little heterochromatin (Nagl, Peschke & van Gyseghem, 1976). The conclusion is drawn that there is incomplete replication of heterochromatin in cells of the second type. It has, however, been determined by DNA hybridization and an analysis of the DNA on a CsCl gradient that there was the same quantity of satellite DNA in the cells of the suspensor as in the leaves of the nasturtium (Deumling & Nagl, 1978). The suspensor cells are perhaps heterogeneous, but under-replication may not be the only reason for this.

In the hypocotyl of *Sinapis alba*, Feulgen–DNA cytofluorometry and some biochemical data are interpreted as an indication that over-replication of DNA has occurred in this region (Capesius & Stöhr, 1975). In the epicotyl of *Pisum sativum*, under-replication of certain repeated sequences was revealed; this was compensated a few days later by over-replication (Van Oostveldt & Van Parijs, 1976). Neither of the effects exceeded 15% of the total DNA content. The more considerable changes previously found in epicotyl of this species (Broekhaert & Van Parijs, 1975) turned out to be the result of bacterial contamination.

Conclusion

Examples of polyploidy and polyteny are as numerous in plants as in animals. Polytene cells with huge DNA contents are characteristic first and foremost of rapidly growing nutritional organs (such as the suspensor and the endosperm). The analogy with some nutritional organs of animals, such as the mammalian trophoblast, seems obvious. The rate, mode of formation and structure of the nuclei are similar in all these tissues. Two or three rounds of DNA replication are usually realized in plant cells without any intervening mitoses. In animal cells, endoreduplicated chromosomes are seldom found, but when they are, it is mainly in cells raised *in vitro* or in tumour tissue. Polyploidizing mitoses are observed in differentiating plant tissues. An interesting example of genome multiplication in plants occurs in gall tissue.

The polytene nuclei of plant cells, like those of most animal cells, differ in structure from the polytene nuclei of dipteran cells. Bundles of chromatin threads with vague bands or a diffuse chromatin structure are typical of the nuclei that have accumulated huge quantities of DNA without undergoing subsequent mitosis.

In plants, just as in animals, endoreduplications may be connected with an increase of functional activity. Where environmental conditions are unfavourable, for example, when day length is artificially changed or the plants are exposed to salt, the number of endocycles increases.

Cases of differential DNA replication have been described more often in plants than in animals.

5

Genome multiplication in protozoans

Polyploidization in unicellular organisms is a phenomenon that has long been known. Detailed reviews on the karyology of protozoans and, in particular, on polyploidy and other changes in the genome are due to Poljansky (1965), Raikov (1972, 1982), Ammermann (1973), Grell (1973), Raikov & Ammermann (1976). This chapter is based on these reviews.

It is essential for us to note the analogy between somatic polyploidy in Metazoa and polyploidy in Protozoa, the mechanisms of polyploidization and, its converse de-polyploidization, in their life cycle, and also variants in the scheme of DNA replication in the protozoan genome.

Two types of polyploidy, generative and temporary, can be distinguished in protozoans. Generative polyploidy (that is, polyploidy inherited through the germ line) is similar in protozoans, angiosperms, and some metazoans: a polyploid race of organisms is formed from a diploid ancestor. In Protozoa, this occurs through incomplete mitosis, when the replicated chromosomes remain in the undivided nucleus. If it is viable and competitive, such a polyploid individual may form a polyploid clone. If the polyploid clone possesses some advantages over the diploid ancestral clone, further differentiation of the line may be the basis for speciation. One can assess the likelihood of such a process from the number of polyploid species in some genera of Protozoa (just as in Metazoa and plants). Fairly stable polyploid strains of Protozoa can be obtained under laboratory conditions.

Examples of generative polyploidy have been found among flagellates of the genus *Holomastigotoides*. Species of this genus have only two chromosomes. Strains of some species with n = 4 and n = 8 are, however, known. Generative polyploidy is evidently specific to ciliates of the genus *Paramecium*. Thus, the micronuclei of different races of *P. bursaria* are of different sizes and stain with different intensity after Feulgen's reaction for DNA. There is a varying number of chromosomes in the first prophase of meiosis (Fig. 5.1). According to cytophotometric data, the DNA content of micronuclei of different races of *P. bursaria* fall into the following ratio: 1:2:3:4. In *P. caudatum*, the ratio of the quantity of DNA in the different races is 1:2:4:6:8.

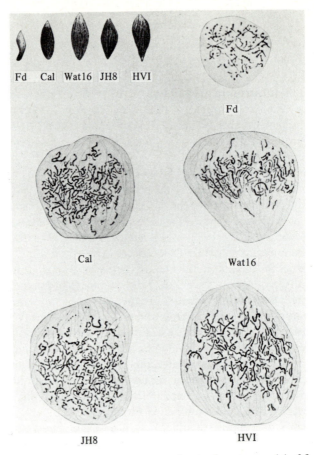

Fig. 5.1. Variation of the chromosome number in the micronuclei of five races of *Paramecium bursaria* during pregamic division. In the upper left-hand corner are drawings of the vegetative resting micronuclei of the same races (after Chen, 1940).

Temporary polyploidization occurs during the life cycle of some Protozoa. Divisions of the nucleus may occur immediately after replication of the genome, and the resulting small daughter nuclei contain a number of complete genomes. Consequently, the initial polyploid nuclei must also be polygenomic. It should further be noted that genome segregation may start during polyploidization. This type of polyploidy has been called 'cyclic' by Raikov. The alternation of diploid, polyploid, and segregated stages is specific to radiolarians. The macronuclei of some ciliates are also polyploid. Polyploidization of macronuclei in ciliates is irreversible, unlike that in radiolarians. At the beginning of the protozoan life cycle the macronucleus is formed anew from the diploid derivative of the synkaryon. This, initially diploid macronucleus, is then polyploidized. This type of polyploidy in Protozoa is similar to somatic polyploidy in Metazoa.

Temporary, or cyclic, polyploidy has been thoroughly studied in radiolarians, especially in *Aulacantha scolymantha*. For us, the point of interest is that repeated genome multiplication occurs without subsequent mitosis. Nevertheless, during reproduction, whole genomes can be singled out. Synaptonemal complexes are found in the nucleus of *Aulacantha*. There are grounds for believing that conjugation of homologous chromosomes occurs in each replication cycle. In this way, preparation for genome segregation takes place during polyploidization of the nucleus.

Genome multiplication in radiolarians produces composite chromosomes, which are linked up end to end. Each chain of chromosomes is a whole genome. This arrangement of the polyploid (or, to be more correct, polygenomic) nucleus influences the subsequent correct segregation of the genomes; the nucleus, during the act of division, may look as though it is undergoing amitosis. The process of genome segregation begins with the destruction of the envelope of the polygenomic nucleus. Groups of composite chromosomes can be discerned, around each of which new membranes are formed. The cell becomes multinucleate. After this, the cytoplasm segregates into small multinucleate units. Nuclear division continues within these cells. Multipolar mitoses can be seen; these are evidence of continuing de-polyploidization. The final stage is the formation of the mononucleate zoospores.

An example of cyclic polyploidy has been observed in *Pyrsonympha*, an intestinal parasite of termites. Synaptonemal complexes are observed in the polyploid nuclei of these flagellates, just as in radiolarians. Once again, this structure is not part of a meiotic process, but is rather a component in somatic reduction, facilitating the pairing of the homologous chromosomes before they separate into daughter nuclei. Small, dispersive forms of the parasite are formed as a result of this segregation.

An interesting variant of polyploidy in unicellular organisms is found in the multinucleate flagellate *Opalina ranarum*, which lives in the intestines of the frog. Multinuclearity is, in itself, indicative of cell polyploidy. In some nuclei of opalines, both diploid as well as polyploid chromosome sets can be observed (Fig. 5.2). It is noteworthy that the nuclei of opalines enter into the mitotic cycle and begin mitosis asynchronously. This rare exception among multinucleate cells goes against the idea that a cytoplasmic factor triggers the initiation of mitosis (Prescott, 1982). It is possible that the unicellular Protozoa differ from cells in multicellular organisms in some of the mechanisms of cell cycle regulation.

The DNA content of the macronucleus of many infusoria is considerably higher than that of the micronucleus. The macronucleus, a 'vegetative'

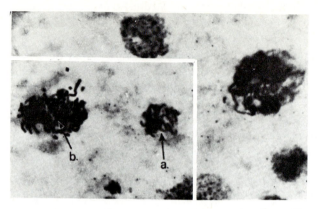

Fig. 5.2. Mitoses of different ploidy (a and b) in the multinucleate ciliate *Opalina ranarum*. Interphase nuclei are shown near the mitotic figures (Kaczanowski, 1968).

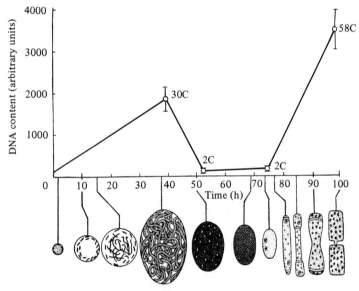

Fig. 5.3. Changes in DNA content (C values indicated in the graph) and the rearrangement of nuclear and chromatin structure (drawn below the graph) during development of the macronucleus in *Stylonychia mytilus* (after Ammermann *et al.*, 1974).

nucleus, functions throughout the life cycle of infusoria. During the cycle, the genome of the macronucleus may undergo complicated transformations caused by polyploidization or polytenization. In the course of these transformations (or following them) changes may occur in the number of genes – some genes are amplified, others are destroyed. Even in the latter case, the absolute DNA content in the macronucleus may be so great (because of the preceding replication and amplification) that, when there is no data apart from that derived by cytophotometry, these nuclei are considered to be polyploid. Such observations are being re-examined.

Thus, by the end of the development of the macronucleus only a small part of the genome may remain. The degree of multiplication of this part may be huge, since the other genes have either not been reproduced or may even have been destroyed. This complex process of macronucleus development, including temporary polytenization of the nucleus and chromosome diminution, is typical of *Stylonychia mytilus* (Fig. 5.3).

In the developing macronuclear anlage of *Stylonychia*, the chromosomes are at first condensed and distributed around the periphery of the nucleus. At this stage, the DNA content is already four times higher than in the initial anlage. At the next stage of development, approximately one third of the chromosomes decondense and move to the center of the nucleus; the rest (the compact chromosomes which remain at the nuclear periphery) are then destroyed. The DNA content then increases further and ribbon-shaped polytene chromosomes are formed (Fig. 5.4). DNA replication in the polytene chromosomes of *Stylonychia* does not occur evenly; many more replication cycles take place in some bands than in others. After this, the chromosomes fragment into individual bands. As a result, thousands of chromomeres can be seen. At this stage the DNA content decreases sharply, falling almost to the diploid level. During this second gene diminution, more than 90% of the DNA in the polytene chromosomes disintegrates; the anlage decreases in size and RNA synthesis begins (the polytene chromosomes were not actively transcribed at the earlier stage). The synthesis of DNA starts again, its quantity growing 32-fold in *Stylonychia*, compared with the preceding minimal level.

Thus, genome multiplication occurs during development of the macronucleus in *Stylonychia*. Polytenization of the chromosomes is also observed, but these processes are not analogous to gene reproduction and polyteny in metazoans. All in all, some 14 cycles of DNA replication take place during the development of the macronucleus but, in the final count, only a negligible part of the genome (less than 2% of the original starting material) is reproduced; the remaining genes are destroyed. According to the latest data, the degree of multiplication of the remaining fragments of the genome is 15 000-fold (Steinbrück *et al.*, 1981). Polytenization of the chromosomes (after the first diminution) is a temporary phenomenon, which is essentially a preparation for the second diminution. Unlike the polytene chromosomes of metazoans, the polytene chromosomes of *Stylonychia* and other Hypotrichida are transcriptionally inactive, synthesizing hardly any RNA.

In *Stylonychia* and other Hypotrichida, the switching over of part of the genome to the performance of somatic functions brings about the

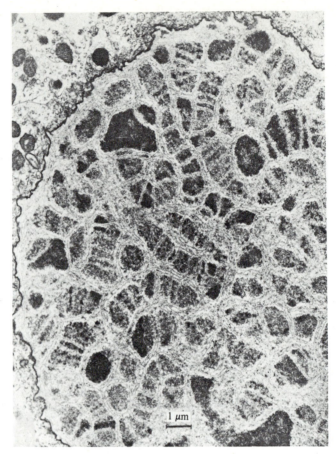

Fig. 5.4. Electron micrograph of the macronucleus of *Stylonychia* showing banded polytene chromosomes (after Kloetzel, 1970).

destruction of all remaining genes. In practically all other protozoans, as well as in metazoans and plants, genes which are not needed for the function of the fully differentiated cell are, as far as we know, not destroyed. The active functioning of a small number of genes in differentiated cells is achieved by inactivation of the 'superfluous' genes. But such idle (or sometimes minimally active) genes are preserved in the genome and may, in principle, be reactivated. Amplification of the active genes is a phenomenon known in metazoans, but it is rare and not involved in the process of differentiation (Chapter 6F).

   The macronucleus may be diploid throughout the vegetative period of the cycle, for example, in some lower ciliates of the order Karyorelictida.

Table 5.1. *The mean ratio of the DNA content in the macronucleus and the presynthetic micronucleus of different ciliates*

Species	$DNA_{macronucleus}/DNA_{micronucleus}$
Dileptus anser	
Single macronucleus	6
Macronuclei in total	1000–2500
Colpoda steinii	8
Chilodonella uncinata	32
Nassula ornata	15
Nassulipsis elegans	125
Tetrahymena pyriformis	21–5
Tetrahymena patula	375–450
Paramecium primaurelia	430
Paramecium tetraurelia	467
Paramecium calkinsi	280
Paramecium caudatum	55–80
Ichthyophthirius multifilius	
(mature trophont)	6300
Metaradiophrya varians	290
Metaradiophyra lumbrici	660
Spirostomum ambiguum	6575
Bursaria truncatella	2620
Euplotes eurystomus	200

After Raikov, 1982.

But, in most ciliates, the DNA content of the macronucleus exceeds the 2C level (Table 5.1). Frequently, the macronuclei are indeed polyploid. In lower infusorians, polyploid macronuclei have been found in representatives of the orders Colpodida, Synhymenida, Nassulida. In free-living Heterotrichida (*Bursaria, Stentor, Spirostomum*), the chromosomes are observed to double several times without apparent subsequent reduction. The chromosomes of the parasitic ciliate *Nyctotherus cordiformis* evidently also polytenize without diminution; the polytene chromosomes of these infusorians are not fragmented. In some other ciliates studied (Gymnostomatia, Suctoria and Peritrichia), DNA replication in the macronucleus is also not accompanied by subsequent diminution (at any rate, not by any substantial diminution).

The macronuclei of higher ciliates (Heterotrichida) are huge: these are the biggest somatic nuclei. Thus, in *Spirostomum ambiguum* or *Bursaria truncatella* the macronucleus may reach 2 mm in length. The surface area of such an enormous nucleus is further expanded by protruberances and invaginations. In *Helicoprodon gigas* the nodules of its beaded macronucleus contain cavities (Fig. 5.5) filled with cytoplasm with numerous

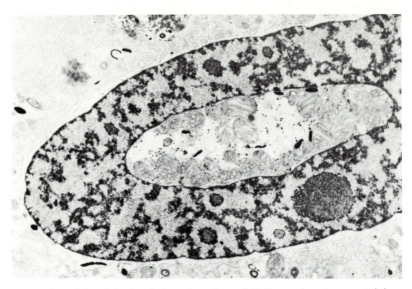

Fig. 5.5 A nodule of the bead-shaped nucleus of *Helicoprodon gigas* containing a cavity filled with cytoplasmic material (electron micrograph kindly provided by Dr I. B. Raikov).

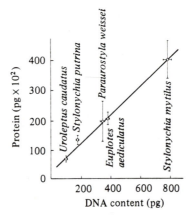

Fig. 5.6. Relationship between the mean cellular protein content of some ciliates and the mean DNA content of their macronuclei. Error bars about each point indicate a standard deviation (after Ammermann & Muenz, 1982).

mitochondria. Such expansion of the surface for nuclear–cytoplasmic contact is also characteristic of other large Protozoa, as well as of the larger polyploid and polytene cells in metazoans.

It is interesting that the quantity of DNA in the macronucleus (but not that of the micronucleus) can be correlated with the overall protein content in ciliates of different species (Fig. 5.6). Such a connection may not be fortuitous. In ciliates, the two types of nuclei have separate functions: control of the cellular metabolism resides in the macronucleus, while the micronucleus is concerned with the inheritance of the genetic material from generation to generation. In cells of higher animals and plants, where nuclear functions are not split in this way, the connection between the genome mass and the protein content may be disguised since a single nucleus carries out these different functions, not just metabolic ones.

The correlation between macronuclear DNA content and cell mass is only observed when mean values are considered (see also Berger, 1982). As Berger noted, DNA content and cell mass are both highly variable, even within the life cycle of ciliates of a single species. It is thought that DNA content, the rate of protein synthesis and cell mass are co-ordinately regulated and that this leads to the maintenance of an optimum gene concentration.

With reference to the cytophotometric data of Table 5.1, it should be noted that a direct evaluation of the level of ploidy in relation to DNA content in the macro- and micronucleus may be incorrect, and not only because of complex transformations of the genome (similar to those described in *Stylonychia*). The micronucleus may not be diploid, but polyploid. Such micronuclei have been found in some species of the genus *Paramecium*. Thus, in *P. bursaria*, a comparison of the DNA contents of the macronucleus and micronucleus (assuming that the latter is diploid), reveals 14–40n in the macronucleus. From the rate of DNA renaturation in the macronucleus of certain ciliates, 560n is found. Polyploid micronuclei are also characteristic of *P. caudatum* and *P. putrinum*. But, in *P. aurelia*, *P. woodruffi* and *P. calkinsi*, the micronuclei are diploid. Polyploid micronuclei are rarely found outside the genus *Paramecium*.

Multiplication of the number of nucleoli in the macronucleus of many ciliates requires special study. In metazoans, a large number of nucleoli is usually a sign of polyploidy. However, as a result of the amplification of ribosomal genes in the oocytes of some animals, for example, the number of nucleoli increases to many hundreds. In ciliates, the number of macronuclear nucleoli varies. In some species, there are few nucleoli or just one. In one of the latter cases, however, it has been shown that a single nucleolus may be composite since under the electron microscope it appears to be an aggregate of several nucleoli.

It is interesting that in three species of *Tetrahymena* and in *Paramecium tetraurelia*, several classes of macronuclei can be distinguished with a ratio of DNA content of $1:2:4:8$. The DNA content in the micronuclei within these individuals is constant, however.

Descriptions of ciliate macronuclear division are pertinent to our review of the properties of polyploid cells. Division resembles amitosis; the nucleus stretches and constricts, forming two parts. Approximately equal fragments are usually formed but, in some species, unequal division (or even budding) of the macronucleus is observed. In ciliates which reproduce by means of cysts, DNA synthesis does not directly precede the division of the macronucleus; as a result of their particular mode of division, the quantity of DNA in the daughter nuclei may decrease tenfold. Classical papers described division of the macronucleus as an example of amitosis. An in-depth study of this process, using electron microscopy has, however, revealed that the chromatin unites in a regular manner and the microtubules become orientated. It has become clear that the individual chromatin aggregates may contain euploid quantities of DNA. For example, according to cytophotometric data, each of the four or eight chromatin aggregates of the divided macronucleus of *Colpoda steinii* contains as much DNA as the presynthetic micronucleus (Frenkel, 1978). The special arrangement of the chromatin in ciliates has been described (Lipps, Gruissem & Prescott, 1982). Thus, divisions of the macronucleus, even those that appear to be unequal, may be a variant of genome segregation. The precise mechanism differs from that observed in radiolarians, but it also leads to euploidy in the daughter nuclei. Such a division is more likely to be cryptic mitosis than direct division of the nucleus.

In Protozoa, in contrast to Metazoa, there is no general or universal form of mitosis. The forms vary and the division process does not always resemble mitosis. However, the essence of mitosis is always retained in that nuclear division leads to the equal distribution of the reproduced chromatids between the daughter nuclei.

The universal form of mitosis, customary in cytology, is characteristic of some autotrophic flagellates, of the sarcodinids (for example, *Amoeba*), and of some gregarines. In these Protozoa (as in metazoan cells) the nuclear envelope is disrupted during mitosis, the nucleoli disperse, and a spindle apparatus is formed whose microtubules are of two types – free and connected with the kinetochores of the chromosomes.

In flagellates of the orders of Volvocida and Chloromonadida, the nuclear envelope is only ruptured at the spindle poles. A mitotic spindle of the usual type is found inside the nucleus, and centrioles are found in the cytoplasm. The nucleolus usually breaks down at division.

Completely enclosed, intranuclear, mitosis is typical of many Protozoa (haemosporidians, trypanosomes, foraminiferans, radiolarians) and also of some fungi and yeasts. In these cases, the spindle-organizing centers are weakly bound to the nuclear envelope and the mitotic apparatus is located in the nucleoplasm. The spindle itself is not at first bipolar, and often remains assymetrical until division is completed. Another version of intranuclear mitosis is observed in the micronuclei of ciliates, and sometimes in opalinids and heliozoans. The connection between the spindle and the nuclear envelope here is even weaker, the spindle is organized in the nucleoplasm and has a bipolar, symmetrical form.

Closed, extranuclear mitosis has been observed in flagellates of the order of Trichomonadida and Hypermastigida. In this case, the nuclear envelope is also retained, but the spindle is located outside the nucleus; the chromosomal spindle microtubules are fixed to the nuclear envelope and are attached to the kinetochores of the chromosomes, or the kinetochores are situated in the nuclear pores.

It is assumed that all the 'atypical' forms of mitosis are evolutionary precursors of the 'universal' mitosis found in metazoans. It cannot, however, be ruled out that the unique forms of mitosis in Protozoa are the result of parallel development of different modes of equivalent genome division. It is possible that, in metazoans, it is neither the most complicated nor the most perfect mode that has become fixed, but the one which occurred in the ancestral metazoans.

The mitotic cycle in Protozoa is a distinctive one. In its essence (namely DNA replication), the cycle is the same as in Metazoa, but its structure can vary considerably. The interphase periods customary in the mitotic cycle of Metazoa – G_1, S and G_2 – are clearly expressed in flagellates, some trypanosomes, and euglenids. In the micronuclei of ciliates, the S-phase sometimes occurs in the middle of interphase and the G_1 and G_2 periods can then be discerned. Often, DNA synthesis occurs at the start of interphase, immediately after mitosis. In some ciliates (for example, *Paramecium bursaria*) DNA synthesis takes place immediately before mitosis, in which case G_1 lasts for a long time and there is no G_2 phase. The structure of the cycle in the micronucleus and macronucleus in a single species is identical, but DNA synthesis in the micronucleus lasts for a fraction of the time necessary for DNA replication in the macronucleus.

In *Amoeba proteus*, there is no G_1-period: DNA synthesis begins immediately after mitosis and the main mass of DNA is synthesized in the first few hours following mitosis. Then, after an interval, the remaining fraction, chiefly of nucleolar DNA, is replicated. Consequently, at this time, ribosomal genes are replicated or amplified. The omission of the G_1-period is also characteristic of some other amoebae. However, in *Naegleria gruberi*, DNA synthesis does not start immediately after mitosis, and the G_1-period is the same length as the S-phase.

Conclusion

Though the mitotic cycle, mitosis and the form of genome multiplication are all fairly distinctive, Protozoa have all the main features of eucaryotic cells, including the ability to polyploidize their own nucleus. The significance of polyploidy in the cell cycle of Protozoa may be different from that in metazoan cells where polyploidy is not a standard feature either.

One of the unique features of some ciliates is their extraordinary mode of differentiation involving the destruction of 'unnecessary' genes. It is not clear whether this is an evolutionarily successful mode of differentiation or not. Some traces of this process can be discerned in ascarids. This mode of ciliate development leads to a huge accumulation of DNA by means of amplification of the few remaining genes; these and other changes in DNA content (which are unusual in metazoans) were formerly considered to be polyploidization. Re-evaluation of these data is required to reveal genuine polyploidization, which does take place in some protozoans.

Unlike in metazoans, genome multiplication in protozoans may only be the prelude to genome diminution, rather than the intensification of cell functions. The latter process begins only after subsequent reproduction of the remaining genes. However, true activation of cell function by polyploidization occurs in many protozoan species.

Another interesting peculiarity of the unicellular polyploid genome is its regular reductional segregation. In metazoan cells, somatic reduction is a rare phenomenon.

Cells with two very different nuclei (nuclear duality) are a specific feature of ciliates. The different functions of the nuclei make it possible to study the relationship between the quality and quantity of the active genes and the metabolic patterns of the cell.

The evolution of mitosis, the mitotic apparatus, and the whole of the mitotic cycle as well as its incomplete versions can be studied in protozoans taking as examples the range of variants of these processes which these organisms display.

II

Modes of and reasons for genome multiplication

The stem cells of all animal and plant tissues are invariably diploid. During early stages of tissue development cells proliferate, but at the terminal stages of differentiation the cells often become polyploid or polytene. There must be a mechanism for the change-over from routine cell multiplication to one in which the genome alone multiplies; in other words: why does a diploid cell become polyploid and how are polyploid cells formed?

Polyploidization may be important for the fulfilment of specific functions in tissues and organs. What, therefore, are the advantages and the shortcomings of the polyploid as compared to the diploid genome in the performance of these functions? What is the biological significance of genome multiplication?

Comprehensive answers to these questions are not yet available, having been posed only recently. In this section of this volume, we shall generalize the information presented in the first part, cite findings relevant to, and discuss current hypotheses that bear on, these questions.

6

Mechanisms for changing the number of genomes

In eucaryotes, the only mechanism of cell reproduction is the mitotic cycle, and the act of division itself is mitosis. The main event in the mitotic cycle is the reproduction and doubling of the number of chromosomes. The purpose of mitosis is the equal distribution of the now-doubled chromosomes (chromatids) between the daughter cells. The nuclear envelope disintegrates during mitosis and a division spindle is formed. In some protozoans the spindle is formed inside the nuclear envelope (Chapter 5), though in many others a more normal type of mitosis is also found.

The mitotic cycle does not always conclude with cell division. Two main forms of incomplete cycles and, correspondingly, two modes of genome multiplication, are known (Fig. 6.1) and are discussed in this chapter.

1. The mitotic process is interrupted at some stage and subsequent phases are omitted. The cell does not divide and a polyploid mononucleate or a bi- or multinucleate cell results. Because the daughter chromatids disjoin, this polyploidizing mitosis leads to the formation of cells with double the number of chromosomes, often with spatially separated chromosome sets.

2. The entire process of mitosis is omitted, and the cell is blocked in the G_2-phase. As a result, the cell nucleus contains doubled chromonemata (chromatids) which, after several endocycles, give rise to polytene chromosomes.

Endomitosis has been described in insects, some other invertebrates, and in plants. It is the division of the chromosomes within the nucleus, without disruption of the nuclear envelope, and without spindle formation. Endomitosis may be regarded as a variant of an incomplete mitotic cycle. Polyploid cells may also be formed as a result of the fusion of two or more diploid cells.

Reduction of the number of chromosome sets, gene amplification, gene underreplication and diminution of the DNA are also discussed in this chapter as they are events associated with polyploidization.

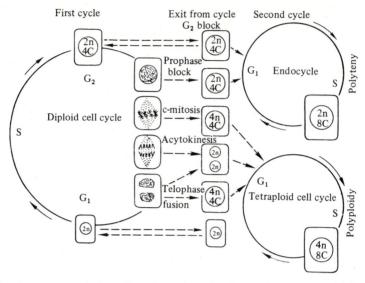

Fig. 6.1. Forms of the cell cycle and mechanisms of genome multiplication. C, haploid level of DNA; n, haploid chromosome set (after Brodsky & Uryvaeva, 1970).

A Polyploidizing mitosis

Deviations from the course of mitosis which lead to polyploidy were noted long ago (Wilson, 1925; Mazia, 1961). Lothar Geitler, the originator of the concept of endomitosis, presented pictures of incomplete mitosis – 'Restitutionskernbildung' – in his monograph of 1934. Subsequently, 'restitution mitosis' was noted in plant tissues by other Viennese cytologists (e.g. Nagl, 1962; Tschermak-Woess, 1971). This process, which it would be more correct to call 'polyploidizing mitosis', has attracted increasing attention in the last 10 years, following its recognition in mammalian cells. It has now also become possible to evaluate comparable data on cell polyploidy from the tissues of invertebrates and plants.

The rapid accumulation of information on cell polyploidy in mammals during the 1950s and 1960s was unfortunately not accompanied by studies of the corresponding modes of polyploidization. In most works it was considered sufficient to designate polyploidy as 'endomitotic' and illustrations of purported endomitoses were usually not published. Yet, many of the descriptions correspond to variants of polyploidizing mitosis rather than to a true endomitosis. These variants include acytokinetic mitosis with separation of the nuclei, mitosis of binucleate cells (bimitosis), and mitosis with subsequent fusion, or non-separation of the daughter nuclei.

In animal and plant tissues, the variant of mitosis which is most

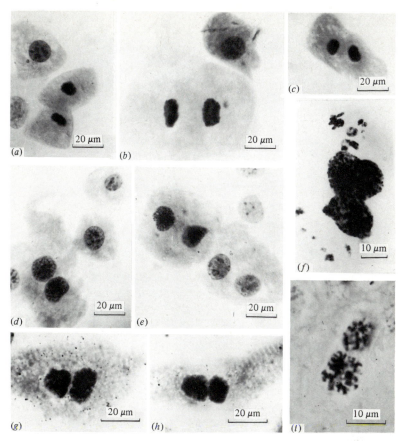

Fig. 6.2. Acytokinetic mitoses and newly formed binucleate cells illustrated in cells isolated from mouse liver (*a–f*), heart (*g,h*) and trophoblast (*i*). (*a*) Normal telophase for comparison; (*b,c*) acytokinetic telophases of different ploidy; (*d,e*) binucleate, newly formed hepatocytes with a post-mitotic nuclear structure; (*f*) typical picture of a nucleus formed by means of telophase nuclear fusion in regenerating mouse liver, pretreated with an alkylating drug (showing multiple micronuclei around the nucleus); (*g, h, i*) late telophases without cytokinesis.

frequently observed is one in which events are as usual up to the separation of the daughter nuclei but which then stops: i.e. cytokinesis does not occur (Fig. 6.2). In diploid cells, the result of this acytokinetic mitosis is a binucleate cell with two diploid nuclei (2n × 2). In mammals, such binucleate cells are the predominant component in liver, myocardium, pigment epithelium of the retina and sympathetic ganglia; they are also often found in the fibroblasts of loose connective tissue, in mesothelium, and in the uterine decidua. This list, though far from complete, is sufficient to give an idea of the extent of acytokinetic mitosis in mammalian tissues. In

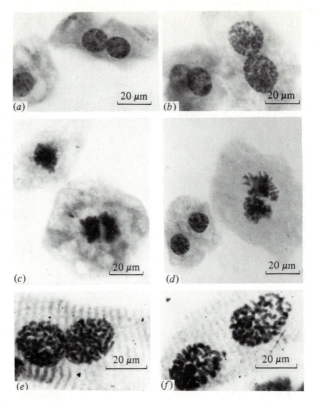

Fig. 6.3. Pictures of bimitoses in isolated cells of mouse heart and regenerating liver. (*a* and *b*) Biprophases in hepatocytes; (*e* and *f*) biprophases in cardiomycetes; and (*c* and *d*) bimetaphases of hepatocytes.

invertebrates, acytokinetic mitosis and binucleation have been described in the gastric epithelium of starfish and in the intestine of ascarids. They are also typical of the tissues of plants (see Chapters of Part I). Acytokinetic mitosis often occurs as a spontaneous phenomenon in cell culture, where it is responsible for the accumulation of binucleate cells. It appears to reflect a certain nutritional deficiency of the culture (Pera & Schwarzacher, 1968). It is interesting that the initial phase of ageing of a diploid culture is also accompanied by the accumulation of binucleate cells, while mononucleate polyploid cells appear later (Kaji & Matsuo, 1979).

The inclusion of just one acytokinetic mitosis and the resulting binucleation in a sequence of mitoses is sufficient for mononucleate cells of high ploidy to be obtained in subsequent cycles. In the next mitotic cycle, the nuclei of the binucleate cell usually synthesize DNA and enter prophase simultaneously (Fig. 6.3). The further course and result of such bimitoses may differ depending on the cell type and also on their form and size.

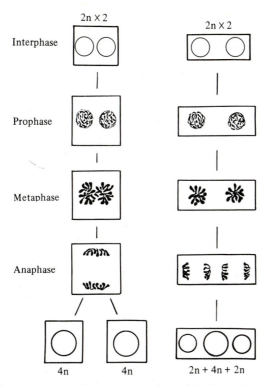

Fig. 6.4. Rearrangement of nuclear material and the formation of polyploid nuclei by bimitosis (after Nadal & Zajdela, 1966; Giménez-Martín *et al.*, 1968).

Frequently, two normal division-organizing centers are formed. The two spindles, as well as the two sets of chromosomes, may then fuse during pro-metaphase, forming a single metaphase plate which nevertheless exhibits certain features suggesting the isolation of the parental genomes (Fig. 6.3). Anaphase and telophase occur normally, and nuclei of the next level of ploidy are formed. This mode of polyploidization was first suggested for hepatocytes by Beams & King (1942) and Wilson & Leduc (1948) and then confirmed experimentally (Nadal & Zajdela, 1966).

In cells with an elongated shape, the two mitotic figures may be coaxial and after the anaphase separation, the adjacent group of chromosomes may fuse and form a tetraploid nucleus (Fig. 6.4). Trinucleate cardio-myocytes are formed in this manner (Gräbner & Pfitzer, 1974). Bimitosis with the mitotic spindles at right angles to each other and other modes of multipolarity, also result in multinucleate cells (Giménez-Martín *et al.*, 1968; Hervás, López-Sáez & Giménez-Martín, 1982). The caffeine-induced binucleate cells of onion roots, undergo mitosis with the non-fusing coaxial division figures (González-Fernández, López-Sáez & Giménez-Martín,

1966). This parallel bimitosis gives rise to binucleate daughter cells similar to the parental type. It is interesting that this variant of bimitosis which reproduces the parental type of binucleate cell is typical neither of all animal tissues nor of cells in culture. For example, during regeneration of mouse liver, bi-anaphases are never observed, while bi-metaphases (in which the chromosomes of the two nuclei fuse) are found with a frequency proportional to that of the binucleate interphase cells (Gerhard, Schultze & Maurer, 1973). In a culture of murine myeloma cells tetraploid clones were formed also by the fusion of metaphase nuclei (Moriwaki & Imai, 1978). It has been suggested that artificial binucleation by means of cytokinetic arrest, with subsequent fusion of daughter nuclei, could be used as a method of obtaining tetraploid cell lines (Hoehn, Sprague & Martin, 1973; Dickerman & Goldman, 1974).

It is important to emphasize that any redistribution of the nuclear material of binucleate cells, during their transformation into a mononucleate or multinucleate condition, is achieved solely by means of mitosis. It is known that the cell classes in the non-growing liver of rodents are stable as long as mitotic activity is not stimulated (see Chapter 2A). The fusion of interphase nuclei appears to be impossible. Even in multinucleate, caffeine-induced plant cells, nuclear fusion does not occur during interphase (Hervas *et al.*, 1982). A delayed cytokinesis may, however, occur in such cells (Röper, 1976).

Mononucleate polyploid cells may also be formed by the fusion of the daughter chromosomal complexes during telophase. It is this process that corresponds to Geitler's 'Restitutionskernbildung' and has often been observed in plant tissues and in invertebrates (Nagl, 1978). Telophase nuclear fusion was noted during observation of living HeLa cells (Moorhead & Hsu, 1956). Although the fusion of telophase nuclei is a dynamic process, typical pictures may be found in fixed preparations of myocardium, regenerating liver, or megakaryocytes (see for example Fig. 6.2).

The reversion of cells to interphase from mitosis is also possible when it stops at the early stages of prophase and metaphase. It is obvious that some kind of abnormality in the centriolar and microtubular apparatus is responsible for this anomalous transition. But, since they do not possess any remarkable morphological features, they cannot easily be discovered. However, pro-metaphase arrest and the consequent accumulation of mitotic figures and of the resulting tetraploid cells were induced in a culture of animal cells by means of diazepam. This agent inhibits the separation of centrioles (Andersson *et al.*, 1981). The well known colchicine mitosis (c-mitosis) is a model of mitosis blocked in metaphase. The agents blocking cytokinesis *in vitro* – cytochalasin for animal cells (Carter, 1967) and caffeine for plants (González-Fernández *et al.*, 1966) – allow the binucleation process to be observed in detail.

Hsu & Moorhead (1956) described incomplete mitoses in tumor cells and referred to the stages in the arrest of mitotic events as pro-, meta-, ana- and teloreduplication, respectively. It is obvious that the types of polyploidizing mitosis just mentioned correspond to these forms. However, the terminology used by Hsu & Moorhead is not entirely convenient because of its excessively rigid definitions. From a practical point of view, it is usually difficult to ascertain the exact type of incomplete mitosis, except in the case of acytokinetic mitosis. We use the term 'polyploidizing mitosis' to contrast the mitotic mode of polyploidization with other modes of formation of an increased genome, primarily those relating to the polytene cycle and cell fusion. In the latter cases, the cells do not enter into mitosis at all.

As a rule, polyploidizing mitosis results in the formation of true polyploid cells with separate daughter chromosomes. There is also the important case where disjunction of chromatids occurs in abnormal mitosis without the spindle apparatus as, for example, in c-mitosis.

Although the sister nuclei of binucleate and multinucleate cells usually enter DNA synthesis and mitosis simultaneously, there are exceptions (Mazia 1961). Examination of the behavior of artificially induced binucleate cells in culture may indicate some reasons for mitotic asynchrony. It has been shown cytophotometrically that sister nuclei in caffeine-induced binucleate cells of pea roots frequently differ in their protein content. In spite of a common cytoplasm, the nucleus with the greater protein content may begin DNA synthesis first (Armstrong & Davidson, 1982). The asynchrony of the nuclei in binucleate cells induced with cytochalasin may be due to a certain structural autonomy of the cytoplasm, at any rate during the first mitotic cycle (Ghosh, Paweletz & Ghosh, 1978). Another explanation was put forward from observations on the multinucleate cells of onion root meristems induced by agents causing multipolarity and acytokinesis (Hervas *et al.*, 1982). Some aneuploid nuclei in such cells lose their replicative ability since they evidently lack the chromosomes on which the formation of a hypothetical intranuclear replication factor depends.

In Table 6.1, some examples are cited of the natural transformation of cell classes, with the help of polyploidizing mitosis, that occurs in the cells of mammals, invertebrates and plants. The succession of transformation of cell classes cited in the table is not always well founded but, nevertheless, in all the examples polyploidy is without doubt the outcome of an incomplete mitosis. Other examples of polyploidizing mitoses are also presented in various other chapters of the book.

What are the mechanisms of the abortive course of mitosis in normal development? Altmann (1966) suggested the hypothesis of the gradual blocking of mitosis in differentiated tissues. According to this model, acytokinesis is the least severe step, while a block at metaphase represents more severe damage to the mitotic apparatus. Klinge (1968, 1970, 1973) regards disruption of the formation of the division spindle as an important

Table 6.1. *Examples of naturally occurring polyploidizing mitoses*

Transformations of cell classes[a]	Cell type	References
$2n \to 2n \times 2 \to 4n \to 4n \times 2 \to 8n \to \ldots$	Mouse hepatocytes	Brodsky & Uryvaeva, 1977
$2n \to 2n \times 2 \to 4n \to 4n \times 2 \to 8n \to \ldots$	Rat hepatocytes	Nadal & Zajdela, 1966
$2n + 2n \to 4n + 4n$; $2n \to 2n \times 2 \nearrow 4n \to 8n$	Rat melanocytes	Marshak, 1974
$2n \to 2n \times 2 \to 2n \times 4$; $\searrow 4C \to 8C$	Mouse cardiomyocytes	Brodsky *et al.*, 1980a
$2n \to 2n \times 2 = 4n \to 4n \times 2 = 8n \to 16n \to 32n$	Megakaryocytes *in vitro*	Kinosita, *et al.*, 1959
$2n \to 2n \times 2 \to 4n \to 4n \times 2 \to 8n \to 16C \to 32C$	Trophoblast tertiary cells	Zybina & Grishchenko, 1970
$2n \to 2n \times 2 \to 4n$	Kidney cells *in vitro*	Pera & Schwarzacher, 1968
$4n \to 8n \to 16n$	Epithelium of mosquito intestine	Grell, 1946a
$2n \to 2n \times 2 \to 4n \to 4n \times 2 \to 8n$	Epithelium of starfish stomach	Vorobyev, 1977b
$2n \to 2n \times 2 \to 4n \to 4n \times 2 \to 8n$	Epithelium of ascarid intestine	Anisimov & Tokmakova, 1973
$3n \to 6n \to 12n \to 24n$	Endosperm of palmyra palm	Stephen, 1974
$3n \to 6n \to 12C \to 24C$	Maize endosperm	Stephen, 1973
$2n \to 2n \times 2 \to 4n \times 2 \to 8n \times 2$	Antipodal cells of kingcup	Grafl, 1941
$2n \to 2n \times 2 \ldots \ldots 16n \times 2$; $\searrow 4n \ldots \ldots 16n$	Potato tapetum	D'Amato, 1977
$2n \to 4n \to 8n \to 16n$	Cells of pea root *in vitro*	Landgren, 1976
$2n \to 2n \times 2 \to 4n$	Murine myeloma *in vitro*	Moriwaki & Imai, 1978

[a] n, haploid chromosome set per nucleus; C, haploid quantity of DNA per nucleus; – – – an undefined path of polyploidization.

cause of polyploidization in cardiomyocytes and hepatocytes; he takes his examples from pathological material from the human heart and liver. Spindle abnormalities during polyploidization have also been noted in some plant cells (Landgren, 1976).

A number of authors have suggested that the defect in the spindle is caused by a block in the reproduction and distribution of the centrioles (Altmann, 1966; Nadal, 1970; Brodsky & Uryvaeva, 1974). This suggestion has not been confirmed. It is known that, in cycling polyploid cells, the number of centrioles corresponds to the level of ploidy (see Chapter 7B). Thus, disturbance of the centriolar apparatus cannot be responsible for acytokinetic mitosis in populations of hepatocytes, megakaryocytes, and probably in other cases where there is a regular polyploidization of cells during growth and differentiation.

Recent studies of delayed cytokinesis in cells *in vitro* point to the adhesion of cells to the substrate and, consequently, the state of the cytoskeleton and plasma membrane as being important for cytokinesis (Orly & Sato, 1979; Ben-Ze'ev & Raz, 1981). It is known, for instance, that 3T3-4E cells can become multinucleate in a liquid medium (Ishii, 1980), but when such cells are transferred to a solid substrate they divide and give rise to mononucleate daughter cells. Cytokinesis is therefore induced by cell spreading; without it, the final act of mitosis is omitted. The conclusion to be drawn is that cytokinesis is an anchorage-dependent process. It remains unclear to what extent this phenomenon *in vitro* is related to the mechanisms of polyploidizing mitosis *in vivo*, and whether surface-membrane changes occur during the course of the extensive binucleation which takes place in normal development.

B Endomitosis

'Als Endomitose wird die im natürlichen Ablauf der Entwicklung eines Organismus erfolgende Zweiteilung der Chromosomen bzw. ihrer Äquivalente im Zellkern ohne Bildung einer Spindel und ohne Teilung des Kerns bezeichnet' (Geitler, 1953). Thus, according to L. Geitler, endomitosis is the division of chromosomes inside the nuclear membrane, without the formation of a spindle and without nuclear fission. Endomitosis was believed to be a mode of formation of polyploid, as well as polytene, nuclei. The discovery of the mitotic cycle showed that this view was incorrect since it then became clear that mitosis was unnecessary for the reproduction of chromosomes, and that polyteny resulted from the non-separation of sister chromatids. Nevertheless, the initial idea was not rejected. Endomitosis was used to explain cases where neither mitosis nor the chromosomes were visible, and where nuclear growth and doubling of the DNA content were the only observable processes. However, this gave rise to a discrepancy between the term and the observed results. Therefore, besides the visually distinguishable endomitosis, cryptic (masked) endomitosis was suggested.

Endomitosis was first seen and described by Geitler (1939) in the pond skater, *Gerris lateralis*, and was later called the 'insect-type' endomitosis. Cryptic endomitosis was considered to be typical of angiosperms. Such an endomitosis, in which changes in the chromosomes could not be visualized, was called the 'angiosperm-type' (see also Tschermak-Woess, 1971, 1973), although there are also numerous examples of nuclear growth in the absence of mitosis (i.e. cryptic endomitosis) in animals.

The insect-type of endomitosis was immediately accepted in the literature, almost as the only mode of polyploidization of animal cells. The difficulty in accepting the angiosperm type of endomitosis was caused by its vagueness with respect to chromosome division. Attempts were made to overcome this lack of clarity by turning to the concept of the condensation cycle of chromosomes. Long before endomitosis was proposed, E. Heitz (1928) had put forward the idea of a condensation–decondensation cycle of the chromatin. This cycle was especially expressed in plant cells. Heitz called the least condensed state (when the nucleus containing chromocenters in the interphase becomes maximally homogenous in appearance after staining) the Z-phase (from 'Zerstäubungsstadium', dispersion phase) and referred to it, without any particular evidence, as very early prophase. Z-phase was also found in plant endomitosis and hence it was thought to mark the prophase of endomitosis. This endoprophase differed from the prophase of ordinary mitosis in that the chromosomes were dispersed and by the fact that the chromosomes divided during the endoprophase. The latter suggestion obviously cannot be checked, the former is incorrect.

By cytophotometry and autoradiography, Barlow (1976, 1978a) has shown that, contrary to the previously mentioned hypothesis, the least condensed state of the chromatin (the Z-phase) coincides with the S-phase of the mitotic cycle and not with early prophase (endoprophase). Z-phase is separated from prophase by the G_2 (Fig. 6.5). Thus, the Z-phase cannot be used as a marker of either mitosis or endomitosis in general. Heterochromatin is usually replicated in a decondensed state, although (according to Barlow, 1978a) the processes of DNA synthesis and chromatin decondensation are regulated independently. In many angiosperms, heterochromatin replication and (accordingly) the Z-phase are characteristic of the end of S-phase, but in some angiosperms it occurs at the beginning of S-phase. Coincidence of the Z-phase with the period of DNA replication was also shown by Cavallini, Cionini & D'Amato (1981). In the root apices of seedlings, nuclei with a typical dispersed structure are found more frequently than prophase nuclei, suggesting that Z-phase cannot be simply an early prophase. DNA cytophotometry and the incorporation of [³H]thymidine shows that the dispersion phase corresponds to the second half of S-phase. Z-phase is rarely seen in the nuclei

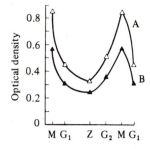

Fig. 6.5. Cyclical changes in the optical density (OD) of Feulgen-stained chromatin in diploid nuclei of the root tips of *Bryonia dioica* during the mitotic cell cycle. The points on the two curves A and B refer to the OD value above which 10% and 50% of the total chromatin area lies. M, Mitosis; G_1, pre-DNA-synthetic phase; Z, dispersion phase ('Zerstäubungsstadium'); G_2, post-DNA-synthetic phase (after Barlow, 1978*a*).

of animal cells but, when it is, chromatin decondensation also coincides with the S-phase and not with mitotic prophase. Thus, in the tropho-blast, [³H]thymidine is incorporated into nuclei with the most dispersed chromatin (see Chapter 2G).

Thus, the Z-phase of the angiosperm endomitosis represents nothing other than S-phase.* In cells where nuclear growth is not accompanied by mitosis (not even by its initial manifestations) there is no mechanism for the division of the chromatids. To use 'mitosis', as in the term 'endomitosis', to describe this phenomenon does not seem appropriate. The views of Barlow and D'Amato on the need to revise the concept of angiosperm endomitosis are indisputable. D'Amato suggests that the term 'angiosperm endomitosis' should be replaced by 'chromosome endoreduplication', and rejects the usual synonyms 'masked endomitosis', 'cryptoendomitosis' and even the very term 'endomitosis', when speaking of polytenization.

The insect-type of endomitosis, which has been discovered not only in insects but also in various invertebrates, has so far been investigated using cell kinetic methods in only one system. Kiknadze and coworkers have examined DNA content and synthesis in the grasshopper testicular epithelium (Chapter 3D), one of the classical subjects for morphological description of endomitosis. The clear pictures of endoprophases and endometaphases have proven to represent an interphase state of the chromatin, i.e. an imitation of endomitosis. DNA synthesis, as well as both normal and polyploidizing mitosis occur in another zone of the testicular epithelium and, what is more important, separated in time from 'endo-mitoses'. One week after the incorporation of [³H]thymidine into the cells of the basal zone and the completion of their mitoses, labelled cells were

* For the latest data on the interrelationships between the Z-phase and DNA synthesis, see Barlow, P. W. (1984). *Caryologia*, **37**, 167–76.

recorded in the central zone of the testicular wall where 'endomitoses' are seen. Intensive RNA transcription and protein synthesis are typical of these 'endomitotic' nuclei and are also characteristic of interphase nuclei. These data have been confirmed by recent autoradiographic studies of endomitosis-like nuclei in septal cells of the testicular follicles of the grasshopper *Melanoplus* (Therman *et al.*, 1983). Endomitosis-like nuclei were also discerned in interphasic glandular cells of the mollusc *Succinea putris* (Anisimov, 1984). These nuclei contained chromosome-like structures and incorporated [³H]uridine intensively.

These interesting and important observations throw doubt, therefore, on the concept of insect-type endomitosis. Moreover, they indicate a method of ascertaining the extent to which the appearance of endomitosis-like nuclei corresponds to the expectations of an endomitotic process, which should lead to multiplication of the chromosome set. The endomitotic nuclei, which correspond morphologically to Geitler's classical definitions (1939, 1953), are also described in the cells of trophoblast in various mammals (Zybina, 1961), and in human trophoblast tumors (Sarto, Stubblefield & Therman, 1982; Therman *et al.*, 1983). The giant cells of the trophoblast are known to form by a series of consecutive endocycles and are highly polytene (see Chapter 3D). Typical of them is a tendency to rearrange their chromocenter structure to aggregate and disaggregate their chromatin bodies. The function of these rearrangements of the nuclear structure and their relation to the process of genome multiplication remain unclear. In the normal trophoblast, the endomitosis-like nuclei do not incorporate [³H]thymidine during 3-hour-long incubation. However, typically endomitotic nuclei in the samples taken from human placenta affected with trophoblastic decrease – hydatiform moles – were frequently labelled (Therman *et al.*, 1983).

In the first part of this volume, many examples of polyploidization of various cells in animals and plants were cited. Whenever the dynamics of polyploidization have been studied by a quantitative cytochemical method, and mitoses observed objectively, the polyploid cells have been found to be of mitotic and not of *endo*mitotic origin. It is noteworthy that in mammals, where somatic polyploidy occurs no less frequently than in insects, endomitoses have never been described. Incomplete, ordinary mitoses are constantly noted however. Acytokinetic mitoses, the fusion of anaphase and telophase groups of chromosomes, c-mitoses, and multipolar mitoses are consistently observed in some tissues during normal ontogenesis and during various regeneration processes. The arrest of normal mitosis is also characteristic of some tissues in both invertebrate animals and in plants.

We describe here some instances in which endomitosis has (incorrectly) been thought to occur. One of our first papers on somatic polyploidy (Zhinkin, Brodsky & Lebedeva, 1961) was entitled: 'Mitosis, endomitosis and amitosis in the cells . . . ' In fact, only mitoses were actually seen. Endomitosis and amitosis were felt to have obviously occurred. Cells that were polyploid in their DNA content were found and this was sufficient for it to be concluded that endomitosis had taken place. Binucleate cells were observed and, at the time, these were considered to be evidence of amitosis.

The conclusions on nuclear polyploidization did not always correspond to the morphological observations of cell growth. Thus, the considerable changes in nuclear size in cells of *Cucurbita pepo* developed in the absence of mitosis, led to the assumption that they were undergoing endomitotic polyploidization from 2n to 256n. But the published illustrations of such cells (Kuster, 1956, Fig. 133), however, show the same number of chromocenters in the 2n and 256n cells, i.e. nuclear size increases but the number of chromocenters does not increase, as would be expected if polyploidization were occurring. The chromocenters grow in size, however, as occurs in some polytene cells (for example, in the trophoblast or in the *Phaseolus* suspensor).

Thus, it should be noted that even though the terms 'endomitosis' and 'endomitotic polyploidy' have been widely used, these processes have not been demonstrated to occur in the way that these terms would lead us to expect. To summarize: Geitler (1953) discerned the following variants:

1. Endomitosis can be distinguished visually. According to Geitler's definition, the nucleus appears to be in metaphase while retaining its membrane; a spindle is not formed.
2. Endomitosis is not visible. Its occurrence is surmised from the existence of a nucleus with giant interphase chromosomes.
3. The nucleus grows, but mitotic figures cannot be seen and chromosomes cannot be distinguished. Endomitosis is only a supposition.

Only the first-mentioned phenomenon (1) can be considered to be a variant of mitosis. The imprecision of attaching the concept of 'mitosis' to polyteny (point 2) is obvious. Nevertheless, the dictum 'polyteny by means of endomitosis' is used in some publications without qualification. But it is much more common to write of 'endomitotic polyploidy' without connecting it with the changes in the chromosomes that are actually observed. To apply this term, it has been deemed sufficient to show that there is a doubling of the DNA content, or simply that there is an increase in nuclear size. This is not only terminologically inaccurate, but it also reveals an incomprehensible confidence in the universality of endomitosis; this unquestioning attitude has held back the study of the true mechanism of cell polyploidization.

Endomitosis is not the main mode of cell polyploidization, nor is it even very widespread. It may be possible in the near future to ask 'does endomitosis exist?' rather than 'how common is endomitosis?'. But, the present is not the right time to ask this question. Only one example of the

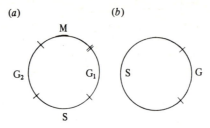

Fig. 6.6. The cycle (*a*) of a polyploid cell and (*b*) of a polytene cell (after Pearson, 1974).

insect-type of endomitosis (in cells of the wall of the grasshopper testis) has been studied. This does not mean that the concept of endomitosis will not be substantiated by cell kinetic analysis in other systems, where its morphological manifestations can be seen.

Major deviations from the course of ordinary mitosis can be seen in protozoans: the complete or partial retention of the nuclear envelope, an intranuclear spindle, and so forth (see Chapter 5). These images resemble neither endomitosis nor polyploidizing mitosis; rather, they reflect the phylogenetic evolution of mitosis.

C Endocycle and polyteny

Polyteny, the doubling of the number of chromonemata of the interphase chromosomes without their subsequent spiralization and division, is a common mode of genome growth in the cells of plants and invertebrate animals. Polytene chromosomes are formed as a result of recurrent blockage of mitosis. It is not clear in which part of the G_2-period the block occurs, or even whether a true G_2-period exists. Mitosis is omitted completely during the sequence of reduplications, and the structure of the nuclear cycle therefore includes only two periods, the DNA synthetic and the intersynthetic one (Fig. 6.6). It is understandable that the expression 'endomitotic polytenization' is still widespread in the literature, it is nevertheless a regrettable misunderstanding. In recent reviews this rather imprecise term has ceased to be used (Brodsky & Uryvaeva, 1970, 1977; Pearson, 1974; D'Amato, 1977). From the structure of the interphase nucleus, three types of polytene cells can be discerned.

The most well known polytene cells are those of the larval tissues of *Drosophila*, Chironomidae and some other Diptera (see Chapter 3A and also Ashburner, 1970; Berendes, 1973). In such cells, especially in the salivary glands, long chromosomes with a characteristic banded structure are constantly found. Another peculiarity of the polytene cells of Diptera is the pairing of homologous chromosomes, which produces a haploid (or near-haploid) chromosome number. Besides the Diptera, somatic pairing has been described in only a few other types of cells. Even in insect species, the pairing of homologous chromosomes is not a general phenomenon and examples of typical polyteny are known in endopolyploid nuclei with a diploid chromosome number (Pearson, 1974). Thus the classical type of polytene cells of the larval tissues of some Diptera is evidently rather a rare case.

Other, less striking, morphological features of polyteny are found more often. In many types of cells, bundles of threads can be seen after many DNA endoreproductions without any intervening mitoses. The bundles are not found in all nuclei and are usually seen only in some parts of the nucleus. Only in some chromosomes and then, as a rule, in only small areas, can bands be discerned. It is more usual that bands cannot be seen. Typical examples of such cells are the following: the giant cells of the trophoblast, in mammals; the nurse cells of many insects; the glandular cells of the silk worm; some plant cells; and the macronuclei of certain ciliates. According to the number of chromocenters (i.e. the pericentromeric regions) and, sometimes, according to the number of bundles (i.e. chromosomes), it may be judged whether there is a diploid chromosome set in this type of polytene nucleus. The quantity of DNA in such nuclei reaches thousands (sometimes many thousands) of haploid units. The number of chromonemata is also multiplied proportionally and amplification does not occur in the majority of cases. Consequently, these nuclei are obviously polytene, but their structure differs from that of the usual type of *Drosophila* polytene nucleus.

The morphological expression of polyteny may differ from cell to cell. In mammalian trophoblasts, insect nurse cells and in other tissues mentioned above, some nuclei contain long aggregates of chromatin threads, while others, with the same quantity of DNA, contain only few or no bundles simply having chromocenters. In the latter case, the nucleus is referred to as having a 'diffuse' structure. This type of nucleus, is not usually called polytene. The extent of polytenization can be judged from their DNA content. There are two possibilities for such nuclei: either they have doubled amounts of DNA and are really polytene, or they are polyploid, the mechanism of division of the chromosomes remaining to be explained.

The indicators that this is masked (cryptic) polyteny are not always easy to see but, when they are, the results speak for polyteny and not polyploidy. One of these indicators is the number of chromocenters. This number can be compared in nuclei of the diploid precursor cells and in the nuclei with high DNA content. Cases are known where the number of chromocenters remains diploid, thus proving that the nucleus is polytene; in others the number of chromocenters increases, indicating polyploidy (Fig. 6.7). Doubling of chromocenters is an indicator of the complete replication of the chromosome set. So, the chromocenter number doubles in the cells that

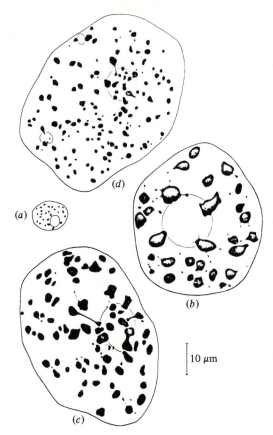

Fig. 6.7. Interphase nuclei of cells in the corolla of *Cucurbita pepo*. (*a*) Diploid nucleus; (*b–d*) nuclei which have enlarged 62 times compared with (*a*). The number of chromocenters in (*b*) is the same as in (*a*), while it is doubled in (*c*) and quadrupled in (*d*) (after Tschermak-Woess & Hasitschka, 1953).

form crystals of calcium oxalate in the orchid *Vanilla planifolia* (Kausch & Horner, 1984). This doubling series coincides with nuclear growth up to octaploid level. But the ensuing nuclear growth is not accompanied by chromocenter multiplication; it cannot be suggested that there is further polyploidization.

Another indicator of polyteny is the sex chromatin body (Barr, 1959). The sex chromatin body (or Barr body) is not present in all nuclei, even in those from females, and its structure is not always easily distinguished from other chromatin (see, for example, Lee & Yunis, 1971). Barr bodies may include not only the pycnotic X-chromosome, but also fragments of autosomes; their structure varies in different animals and even in different tissues of a single animal (Sieger, Pera & Schwarzacher, 1970). All the same,

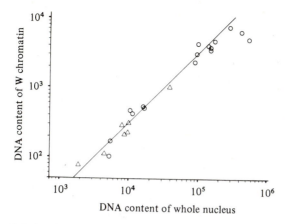

Fig. 6.8. Relationship between the whole nuclear DNA (abscissa) and the female-specific W heterochromatin DNA (in arbitrary units, in each case) in polyploid cells of wing imaginal discs (triangles) and in the cells of the silk gland (circles) of the moth, *Ephestia kühniella* (after Traut & Scholz, 1978).

if only one sex chromatin body is found in a cell and the DNA amount in it doubles in proportion to the total DNA content, such an observation strongly suggests polyteny. Precisely such a variant was found in the giant cells of the trophoblast (Fig. 2.33) and in the wing imaginal discs and silk gland cells of a moth (Fig. 6.8). In female fibroblasts *in vitro*, a doubled sex chromatin body was observed in interphase, and diplochromosomes were found in the mitotic figures (Schwarzacher, 1968). In polyploid cells where the chromosome sets separate, the number of the sex chromatin bodies is usually doubled. Thus, in epithelial cells of the human amnion, one sex chromatin body is most frequently observed in diploid cells, two in tetraploid cells and four in octaploid cells. The total quantity of DNA in the sex chromatin bodies of an individual 8C cell corresponds to the DNA content of the bodies of eight diploid cells (Klinger & Schwarzacher, 1960). In tetraploid rat hepatocytes, two sex chromatin bodies were found, while in diploid hepatocytes there is one body (Ohno, Kaplan & Kinosita, 1959).

Nucleoli serve as yet another, though less reliable, indicator of nuclear polyteny and polyploidy. The number and sizes of the nucleoli depend on the intensity of RNA transcription and on the rate of ribosome transport from the nucleus to the cytoplasm. However, fewer and larger nucleoli are observed in cells where masked polyteny is suspected, compared to those seen in polyploid cells with the same quantity of DNA.

Masked polyteny is apparently common in plants. In animals, it is known to occur in the giant neurons of molluscs and the cells of the esophageal gland in ascarids. In the trophoblast and other cell types mentioned earlier, the indicators of polyteny are sometimes more clear, although, besides the nuclei containing condensed chromosomes (and sometimes with bands), nuclei with the same DNA content can have a diffuse structure.

Why should there be such substantial differences in the structure of nuclei that are genetically and functionally similar? It is hardly likely that the polytene nucleus is a morphologically exceptional form. Although on the whole they do not differ from one another with respect to genome multiplication, polytene and polyploid cells are both the result of various changes in the mitotic cycle; these changes are probably under different genetic controls (Chapter 8). Other peculiarities are probably secondary, although distinctive, features of the polytene cell – namely chromosome pairing, puffing, and DNA under-replication. These interesting properties are characteristic of only a few cell types. Many typical polytene cells, with banded chromosomes, show no signs of chromosome pairing or under-replication. In any case, it is not these properties that determine the nature of polyteny or the structural differences between polytene nuclei. At present one can only guess at the probable reasons for the condensed or diffused state of the chromatin in polytene nuclei.

It is known that in newly hatched larvae of *Chironomus thummi* the cell nuclei of the salivary gland have a diffuse structure (Kiknadze *et al.*, 1975*b*). However, these nuclei already contain as much as 32C DNA. Two or three endocycles later, these nuclei become typically polytene. There can be no doubt that, earlier, they were polytene too. It is natural to suggest that a minimum number of chromonemata and, more importantly, a particular distribution density, are required for the visualization of polytene chromosomes. Many observations indicate this; we shall cite the most illustrative of these.

In the nuclei of the giant cells of the trophoblast, bundles of chromatin threads are found only after four endoreduplications. However, unlike *Chironomus*, the morphology of the polytene nuclei of the trophoblast is very variable, even after seven or eight DNA replications. A well ordered polytene structure is mainly discerned in the region of the nucleolar organizer; this is where there is the greatest concentration of chromatin threads, and only here are bands and dense heterochromatin areas visible. The chromatin at the nucleolus-forming regions becomes thicker with further DNA replication. This is additional evidence for the occurrence of polyteny, and not polyploidy, in the trophoblast.

In other systems, features of polytene structure are also characteristic solely of dense areas of the chromosomes. This has been noted in the trophocytes of *Calliphora erythrocephala* (Bier, 1957, 1959), in silk gland cells of the silkworm (Nakanishi *et al.*, 1969), and in cells of the salivary glands of the mosquito *Phryne cincta* (Wolf & Sokoloff, 1976). The authors of the last-mentioned study point out that the external appearance of the polytene chromosomes is determined solely by the concentration of the chromonemata. This dependence is also illustrated in the description of plant polytene cells (Fig. 4.5).

Many observations of the effect of the cellular environment on the external manifestation of polyteny indicate the small biological significance of the banded structure itself. Thus, in the X-chromosome of *Phryne*, the development of the bands depends on the moisture and salinity of the environment though the extent of nuclear growth does not change in the different media. The incorporation of [³H]uridine by chromosomes with different structures is also constant (Wolf & Sokoloff, 1976). Consequently, the banded structure is not essential either for growth or for nuclear functioning. Further, it was shown that the larval cells of the mordellid fly, *Aphiochaeta*, which had been raised on a diet rich in amino acids, had chromosomes with definite bands; only single threads could be seen in the same cell nuclei where the larvae had been subjected to amino acid deprivation (Barigozzi & Semenza, 1952). The banded chromosome pattern appears in the salivary gland cells of *Harmandia laevi* (Diptera, Cecidomyiidae) only in late larvae just prior to the degeneration of these cells, while the nuclei of other larvae have a diffuse reticulate structure (Zhimulev & Lychov, 1972).

In the nurse cells of the meat fly *Calliphora*, the number of nuclei with a visible polytene structure increased considerably if egg-laying occurred at a low temperature (Bier, 1959). In a recent study of adult ovarian nurse cells in mutants of *Drosophila*, polytene nuclei may be seen in many (but not in all) cells in females reared at a low temperature (King *et al.*, 1981). The DNA content is the same in the mutant nuclei with either polytene or diffuse chromosomes, and also in the nuclei of wild-type nurse cells (where a polytene structure is never found).

Thus, even in its classical manifestations, the external appearance of the polytene nucleus in dipteran larvae depends on many conditions which do not affect polyteny itself. Temperature, lighting, moisture and salinity also have an effect on the condensation of the polytene chromosomes in plant cells.

Transition from polyteny to polyploidy has been suggested in some works. When studying the salivary glands of the gall-producing larvae of the fly *Dasyneura crataegi* (Diptera, Cecidomyiidae), Matuszewski (1965), and later Henderson (1967), found nuclei whose structure varied from markedly polytene to diffuse. Just as in many other cases, there were nuclei with an intermediate structure, exhibiting a friable distribution of the chromonemata. Henderson particularly noted that in most cases the chromosomes do not split into individual chromatids, but only into elementary ('oligotene') threads attached to the centromeric heterochromatin. There was no suggestion that the homologous chromosomes were unpaired. Thus, it cannot be concluded that these nuclei had become polyploid. According to Bier's observations (1957, 1960) in the nurse cells of *Calliphora*, a polytene structure is found mainly in small nuclei with a DNA content of up to 16C and, later, in the largest nuclei; a diffuse structure is characteristic of intermediate stages of development. Bier's observation is an important one: the thickness of the 'secondary' polytene chromosomes in the older cells is much greater than in the small, young cells. This makes it highly unlikely that the chromatids separate, that polyploidization of the polytene nucleus occurs in the nuclei of the intermediate stages, and that this is later followed by polytenization. It is more likely that the polyteny is constant throughout nurse-cell development, but that the degree of chromatin condensation varies with developmental stage.

One can imagine that real polyploidization, i.e. separation of chromatids in a polytene nucleus, should be accompanied by the appearance of hundreds or even thousands of chromocenters, bodies of sex chromatin, and nucleoli.

The converse process – polytenization of a polyploid nucleus – is well established. A particularly fine example is described by Nagl (1970*a*). He revealed that different morphologies of polytene nuclei in the cells of the bean suspensor were determined by whether the original nucleus was polyploid or not. If it was polyploid, then a diffuse polytene nucleus resulted. This result (polyteny after polyploidy) is probably observed in some other plant cells (Fig. 6.7) and in the cells of the silk gland of *Bombyx mori* (Chapter 3C).

Polytenization is the result of a repeated blocking of mitosis after DNA replication. Cases are known where endoreduplication occurs just once or twice. The fate of these cells varies. After the first endocycle, cells of a so-called G_2-population are formed; these are ready for mitosis, even after a lengthy period. Their chromosomes are indistinguishable from ordinary premitotic ones; it is impossible to tell the mitosis of a previously G_2-arrested cell from ordinary mitosis. An example of this G_2-population is cited in the section on erythropoiesis in the case of induced anemia in pigeons (Chapter 2I).

If the cells do not enter mitosis after the first DNA replication, but undergo a second endocycle and then enter mitosis, paired mitotic chromosomes (diplochromosomes) are seen at metaphase (Bell, 1974; Mittwoch, Lele & Webster, 1965; Herreros & Giannelli, 1967). Three consecutive cycles of chromosome replication are needed for quadruplo-chromosomes to emerge. The sister chromosomes of the diplochromosomes are replicated synchronously in a pattern similar to the normal chromosomes (Lau, 1983).

Hence, the well known phenomenon of endoreduplication of chromosomes (Levan & Hauschka, 1953) is a result of endocycles, but can only be registered retrospectively by a mitosis. Chromosome endoreduplication is an abnormal phenomenon. Diplochromosomes appear spontaneously in tumor tissues and in cells *in vitro* (Levan & Hauschka, 1953; Schwarzacher & Schnedl, 1965, 1966; Takanari & Izutsu, 1981). Endoreduplication may be induced artificially in plant and animal cells by a number of chemical and physical treatments (Rizzoni & Palitti, 1973; Sutou & Arai, 1975; Evans & Verville, 1978; Speit, Mehnert & Vogel, 1984).

Polyploidizing mitosis and the endocycle are the main modes of doubling the genome in somatic cells. Repeated endocycles lead to polyteny, a widespread mode of cell growth in plants and invertebrate animals. In vertebrates, only one example of polytenization is known so far – that of the giant cells of the mammalian trophoblast. Morphological manifestations of polyteny vary even in a single cell lineage and in cells with the same DNA content and function. At present, it is still not clear whether genetic and physiological differences are concealed by the different morphologies of the polytene nucleus.

D Fusion of cells *in vivo*

Cell fusion is an important component of natural biological processes including fertilization of the ovum and formation of skeletal muscles. Interest in this phenomenon has greatly increased of late, owing to the study of its applications *in vitro*; these include the introduction of new genetic information and the consequent change in the cell's genetic make-up.

Of special interest are the cases of induction of new phenotypes by cell fusion and multinucleation; these changes do not, in all probability, result from genotypic changes. Thus, multinucleation of transformed murine fibroblasts results in the partial normalization of their morphological characteristics, such as spreading on the substrate, and contact between actin filaments and the substrate. These multinucleate cells were induced by cell fusion, using polyethylene glycol, or by acytokinesis using cytochalasin B (Bershadsky, Gelfant & Vasiliev, 1981). In another study, cultured human keratinocytes were transformed by infection with the oncogenic virus SV-40. The resulting cell fusion and attendant multinucleation resulted in increased cornification (Steinberg & Defendi, 1982).

We shall not discuss here the extensive literature on induced cell fusion *in vitro*. The experiments mentioned above interest us as models from which certain functional explanations of the general phenomenon of multinucleation may be inferred. We shall examine only those cases of cell fusion which occur normally *in vivo*. The fused cells may be considered analogous to mononucleate polyploid ones, since their genetic material is also increased. Besides, the subsequent cycling of these cells and the mitotic union of the nuclei may provide a clone of mononucleate polyploids or result in multiple polyploid nuclei within the symplasm, as occurs in mitotic multinucleate cells (see chapter 6A). This phenomenon is described in the fat body of some dipterans (Wigglesworth, 1966).

During mammalian development *in vivo*, cell fusion is a fairly rare phenomenon. Where it does occur, it is regularly associated with differentiation of a specific type of tissue. In addition to the skeletal muscle fibres already mentioned, the differentiated cells of the bone and connective tissue (the osteoclasts and chondroclasts) arise by fusion of monocyte-like stem cells. After fusion, the nuclei of these tissues lose the ability to synthesize DNA or undergo mitosis (Fischman, 1979). These cells of the macrophage family degrade mineralized tissue. The multinucleate cells exhibit greater resorptive properties than the mononucleate macrophages, as was demonstrated by their ability to bind and degrade [^{14}C]labelled bone particles *in vitro* (Fallon, Teitelbaum & Kahn, 1983). Cell fusion is involved in the formation of the giant superficial cells of the epithelium of urinary bladder in some mammalian species (Chapter 2E). The syncytiotrophoblast (which is part of the human placenta and synthesizes hormones and transports many substances between the maternal and the fetal blood) is formed by fusion of a large number of mononucleate cells from the cytotrophoblast. Like the other tissues mentioned, the nuclei of syncytiotrophoblasts are incapable of DNA replication (Fischman, 1979).

The giant 'inflammatory' cells, which are formed during aseptic inflammation in various tissues, are more likely to be a manifestation of pathological rather than of normal development. They are a mixed group of functionally different types of cells, among which may be phagocytes and producers of collagen. They are formed by fusion of precursor monocytes, which circulate in the blood; these, in their turn, originate from the hemopoietic stem cell (Khrushchov, Lange & Satdykova, 1978). After cell fusion, the nuclei remain capable of DNA replication and mitosis and, up to terminal differentiation, their number may be increased by both mitotic division and further cell fusion (Fig. 6.9). Langhan's cells are similar in their genesis and function to these giant inflammatory cells. They are typical of chronic infections in human beings such as tuberculosis, lues and others, and characterized by the nuclei lying around the cell periphery (Fischman, 1979). The examples cited above are the only known cases in mammalian cell types where cell fusion occurs regularly.

Plant cells are capable of cell-wall breakdown and cell fusion, too. For example, infection of broad bean roots with a nematode results in the formation of specialized giant cells. It is thought that syncytia grow by a process involving cell fusion and synchronous acytokinetic mitosis (Bird, 1973).

It was formerly thought that giant mononucleate and binucleate cells of uterine decidua and the polyploid cells of the trophoblast are formed by fusion. Two groups of authors reliably showed that cell fusion does not take place in these tissues. Allophenic mice, chimeric for two electrophoretic

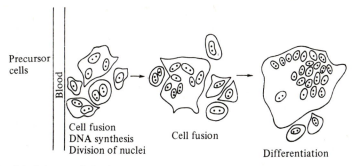

Fig. 6.9. Scheme of the formation of giant foreign body cells at a focus of aseptic inflammation (after Khrushchov, Lange & Satdykova, 1978).

variants of the enzyme glucosephosphate isomerase, did not reveal the heteropolymeric forms of the enzyme in either decidual tissue (Ansell *et al.*, 1974) or cytotrophoblast (Chapman *et al.*, 1972). The heteropolymer should be formed in heterokaryons. It is interesting that the syncytio-trophoblast in murine placenta did not exhibit heteropolymeric variants of the enzyme either. Cell fusion has been shown conclusively to occur in this tissue. The results suggest that redistribution and mixing of the nuclear material or nuclear products in fused cells does not occur in this syncytial tissue (Gearhart & Mintz, 1972).

Some authors distinguish two variants of cell fusion (Martin, 1972; Wheatley, 1972). One type of cell fusion involves adjacent (usually non-sister) cells. The second occurs as the immediate sequel to mitosis. The daughter cells in such cases are united immediately after division or after some stage of incomplete cytokinesis, the so-called refusion (Wheatley, 1972). It is interesting that the result of treatment of cells with cytochalasin B corresponds precisely to refusion. Observations using time-lapse cine-matography demonstrate that the normal cleavage furrow develops and the two daughter cells separate, but remain for some time connected by a narrow cytoplasmic bridge. The daughter cells soon reunite and form a binucleate cell (Krishan & Ray-Chandhuri, 1969). 'Refusion' was also observed in living HeLa cells, but without addition of any drug (Moorhead & Hsu, 1956).

It is obvious that these two variants of cell fusion are fundamentally different and have different consequences. Refusion is the result of a mitotic defect; it is essentially acytokinesis. The fusion of adjacent cells is more likely to be a specific event in differentiation (at least the examples mentioned earlier) and may be preceded by specific changes in the plasma membrane which increase its fluidity.

Genetically identical sister cells are united by refusion. Fusion of adjacent cells may involve cells which are genetically different although, usually, they belong to one and the same tissue type. Thus, during mitosis within the giant inflammatory cells, chromosomal abnormalities are frequently discovered (Mariano & Spector, 1974; Dreher *et al.*, 1978) suggesting that the nuclei of the cells contributing to fusion are genetically defective. These authors believe that the union of nuclei in a single cytoplasm may result in a genetic complementation between the defective genomes.

Recently, some authors have given great significance to the exchange of genetic information which is believed possible in a single nucleus created from the nuclei of fused cells (Hayflick, 1981). It happens that spontaneous fusion occurs fairly often *in vitro* and it has been suggested that it also sometimes occurs *in vivo* as an accidental phenomenon caused by viruses or chemical carcinogens. The hybrids thereby formed may acquire immortality, thanks to genetic recombination, and start a clone of cancerous cells. If this does in fact occur, cell fusion must be such a rapid process that it leaves no trace in fixed preparations. At the moment, there are no suitably sensitive methods for detecting accidental cell-fusion events *in vivo*.

E DNA and chromosome reduction in polyploid cells

In most of the cases of polyploidy and polyteny studied, the quantity of DNA and the number of chromonemata (chromatids) double many times, and then remain constant throughout the rest of the life of the cell. Thus any polyploid mammalian cell is the result of duplication of all the chromosomes of the diploid set, from a single doubling (4n) up to five doublings (64n). Polyploid cells preserve a complete chromosome set, right up to the death of the cell. Illustrative examples of gigantic growth of the genome are the cells of the esophageal gland of ascarids or of the silk gland of *Bombyx mori*. In the former case, the nucleus accomplishes 16 or 17 doublings and, in the latter, 18 or 19 doublings. The complete doubling of all the DNA is shown by cytophotometric data and the kinetics of DNA reassociation (see Chapter 3C). However, a few examples of reduction of the quantity of DNA, the number of chromonemata in the chromosome, and even the number of chromosome sets in the polyploid cell (i.e. its depolyploidization), are known. Chromatin diminution will be discussed in section F of this chapter.

The most vivid example of reduction divisions in the somatic nucleus is segregation of the genomes in the macronucleus of ciliates (see also Chapter 5). The polygenomic nature of the macronucleus in some ciliates has been shown by DNA-reassociation kinetics. It is also suggested by the regeneration of the macronucleus from a small fragment of it (a 5–10% fragment). Consequently, even a small part of the macronucleus should contain at least one complete chromosome set. The macronucleus divides, frequently into two, but sometimes into many fragments. It is interesting that complete sets of chromosomes can be found in the fragments. Fragmentation of the macronucleus is seemingly different from mitosis. However, it is becoming clear that such a division is not an aberration. In the fragmenting macronucleus, orientated microtubules have been discovered. In some species of ciliates, bunches of microtubules are located in the cytoplasm in contact with the nuclear membrane. During division the chromatin of the macronucleus is arranged in a definite manner (Lipps *et al.*, 1982) and special structures of reduction division, similar to the synaptonemal complex of meiosis, have been discovered (Raikov, 1982).

There are very few examples of the reduction of the number of chromosomes in metazoan cells. The case described by Grell (1946*a,b*) for the cells of the ileum of the mosquito *Culex pipiens* is similar to genome segregation in protozoans. At the start of pupal metamorphosis, cells which have reached a ploidy level of 32–64n undergo reduction divisions; the homologous chromosomes evidently pair and then divide. As a result of a series of divisions without any intervening genome duplication, the cells increase considerably in number and diminish in size. The result is that 2n, 4n and 8n cells are found in the adult ileum.

A similar mechanism of somatic reduction is thought to operate in polyploid strains of buckwheat. Diploid seeds and plants appeared among experimentally produced tetraploid forms (Sakharov, 1946). Reduction divisions have been claimed to occur in some somatic plant cells (Huskins, 1948). However, mitotic anomalies, caused by indole-acetic acid and other reagents with which the roots had been treated, may have been their cause.

Genome segregation may be the result of multipolar mitosis. For example, in primary cultures of kidney cells of the field vole *Microtus agrestis*, complete chromosome sets (the smallest set being haploid) move to each pole of three- or four-polar mitoses (Pera & Schwarzacher, 1969; Pera & Rainer, 1973). In rare cases, a multipolar mitosis is completed with the division of the cytoplasm resulting in reduction of the parental genome. Haploid cells have not been found; i.e. haploid nuclei are viable only as part of the multinucleate cell. Triploid daughter cells were capable of entering into a new cycle while they behaved like diploid cells in S-phase and during mitosis. Segregation of 1n, 2n and 3n genomes as a result of multipolar mitosis has also been found in a culture of the kidney cells of the Rhesus monkey (Rizzoni, Palitti & Perticone, 1974). Here, the viability of the segregating cells (which could enter into bipolar mitosis) was also noted. The mechanism of genome segregation is not clear.

A two-step reduction division was observed in kangaroo rat cells *in vitro* (Brenner *et al.*, 1977). Time-lapse photography of living tetraploid cells showed that chromosomes with double chromatids separate to the poles at the beginning of bipolar division; then two separate metaphase plates form and a four-poled mitosis occurs. The outcome of this very rare type of division was either four cells, or a tetranucleate cell, or two mononucleate and a single binucleate cell. In all cases, the genomes were complete. *In vitro* selective forces then determined the viability of the resulting cells, i.e. which of them were able to enter into ordinary bipolar mitosis.

Multipolar mitosis may be the mechanism of genome segregation in hybrid cells (see, for example, Teplitz, Gustafson & Pellett, 1968). As the result of cell fusion, a tetraploid cell with the genomes of the two diploid parent cells is formed. In the subsequent four-polar mitosis, diploid cells may result, carrying a chromosome set from each parent.

It is not clear whether segregation of the genomes occurs during multipolar mitoses *in vivo*. Numerous observations of a similar phenomenon *in vitro* have, in principle, indicated that somatic genome reduction is possible. The multipolarity of mitosis and the release of complete genomes may be caused by the behavior of the centriole. G. E. Onishchenko (see Chapter 7B, p. 187) showed that at least one centriole is needed for the movement of one genome. Premature maturation of the daughter centrioles and their independent separation may be the cause of multipolar mitosis. At the poles of the mitotic cell, double and single centrioles have both been observed. In the latter case, a haploid chromosome set may be caused to separate from the general metaphase plate.

The separation of the genomes ('Genomsonderung') has been described in rat liver (Gläss, 1957). During metaphase, both euploid as well as other groups of chromosomes which had separated from one another, were observed. The work was justly criticized for the poor method of sample preparation and for inadequate analysis of results (Hsu, 1961; Schwarzacher & Klinger, 1963). Nevertheless, separate genomes are often seen in metaphases, although this phenomenon is not connected with segregation but, on the contrary, with the union of the genome of a binucleate cell during mitosis (see Chapter 6A).

A special case of segregation of the genome is the fragmentation of the giant nuclei of the trophoblast (Zybina & Rumyantsev, 1980). In the terminal stage of their development, the nuclei divide into a series of small fragments which frequently contain complete chromosome sets. The mechanism of segregation is unclear. The nuclear membrane and the contractile structures of the cytoplasm have been reported to play a role. The entire tissue eventually degenerates, but nuclear fragmentation occurs before this takes place.

A decrease in DNA content and in the number of chromatids may occur during division of endoreduplicated cells. Prolonged arrest in the G_2-period results in persistence of doubled DNA content and chromatid number. If the cell undergoes a further cycle of DNA synthesis and finally enters mitosis, diplochromosomes will be seen in the metaphase. Mitosis of a cell with diplochromosomes (8C) gives rise to two tetraploid cells and is thus a form of reduction division. A reduction in the number of chromosomes and in DNA content in a polyploid nucleus may then occur

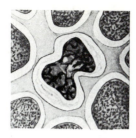 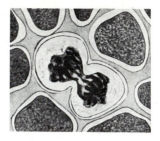

Fig. 6.10. Pseudoamitosis in the corneal epithelium of *Salamandra maculosa* larva following X-irradiation (after Geitler, 1934).

as a result of amitosis. The possibility of direct, amitotic nuclear division was heatedly discussed until the 1960s (Bucher, 1959, 1971). The main source of doubt with regard to the widespread nature (or even the actual occurrence) of this process stems from many observations of anomalous mitoses, which give rise to the controversial 'irregular' shaped nucleus. Thus, anaphase- and telophase-chromosome bridges, which persist in interphase as dumbbell-like figures, are formed during division of a nucleus with chromosome aberrations. This is probably the origin of the dumbbell-like nuclei in the liver, which are rare in the undisturbed organ but are common during regeneration. As is known, a high level of chromosomal aberrations is characteristic of hepatocytes of some mammalian species (see Uryvaeva, 1981).

Dumbbell-like nuclei are characteristic of the myocardium in mammals. Until quite recently, binucleate and multinucleate cells were thought to be a manifestation of amitosis. The main argument in favor of this interpretation is the low level of mitotic activity of the myocytes in the postnatal period. Present-day research has revealed the mitotic origin of the traditional 'amitotic' cell forms, which made other, extremely doubtful, interpretations unnecessary (see Chapter 2B; and also Pfitzer, 1980).

Multipolar mitoses without cytokinesis are responsible for the lobulate nuclei of megakaryocytes (Goyanes-Villaescusa, 1969). Because of their form, megakaryocytes were earlier thought to be multinucleate cells of amitotic origin. It is now understood that it is the result of events occurring during the final mitoses.

Colchicine treatment of onion roots revealed the correlation between abnormal groupings of chromosomes in cells with a disorganized spindle and the amitosis-like figures in the daughter interphase nuclei (Sidorov & Sokolov, 1965). Lobed nuclei (rather than the normal spherical ones) result from the irregular movement of the chromosomes. In the first c-mitosis, an irregular grouping of the chromosomes becomes evident at the end of metaphase. In second and subsequent c-mitoses, the groups of chromosomes can be seen right from the beginning of spiralization. Every distribution pattern of the metaphase chromosomes corresponds to a particular irregular nuclear form (dumbbell-like, lobed or ring-shaped). The amitosis-like figures are therefore the result of irregular mitosis and not amitosis ('pseudoamitosis' – Fig. 6.10).

Pseudoamitosis may also be the result of deformation of a nucleus by cytoplasmic secretions, by fat or by other inclusions.

Some doubt is cast upon observations of amitosis *in vitro*. It could be an illusion caused by one nucleus of a binucleate cell overlying the other (Půža, 1969).

F Special cases of changes in nuclear DNA content

In the concept of the relative constancy of DNA (Chapter 1A) the term 'relative' is not to be ignored. This means that, although the DNA content strictly corresponds to the genome in the majority of cases, exceptions are known. We should immediately note that polyploidy and most of the examples of polyteny do not break the rule of relative constancy of DNA, i.e. the stable content of DNA in a definite (initially diploid) chromosome set (Boivin, Vendrely & Vendrely, 1948). The doubling of such a set does not change the DNA ratio: the 4C or 4096C DNA content correspond equally to their own 4n or 4096n chromosome set (just as the 2C quantity corresponds to the 2n set).

There are, however, cases in which the nuclear DNA content is inconsistent with the 2^n series, i.e. intermediate between 2C and 4C, 4C and 8C and so on. We do not wish to consider cases of aneuploidy, for example, where changes in the number of chromosomes in the somatic cell are caused by errors in mitosis or by damage to the interphase chromosomes. Aneuploidy is a common phenomenon, but it bears no relation to normal developmental processes. Nor shall we discuss the numerous sources of error in evaluating the DNA quantity in the cells. Mistakes in cytophotometry and biochemistry have often led to erroneous conclusions about anomalies in the DNA content during ontogenesis and during changes in the functional state of the cells. Suffice it to refer to works in which such conclusions have been corrected for various cell types: for spermatozoa (Bahr & Wied, 1966; Gledhill *et al.* 1966), adrenal gland (Cohn & Van Duijn, 1967; Arold & Sandritter, 1967), blastomeres (Brodsky *et al.*, 1971*a*), embryonic nervous tissue (Brodsky *et al.*, 1971*b*). One of the latest examples correcting earlier methodological errors of DNA determination is the digital cytophotometric analysis of Feulgen staining in euchromatic and heterochromatic areas of chromosomes from the root meristem of *Trillium kamschatcense* at different temperatures by Grif (1980). It was repeatedly noted in early work that the quantity of DNA in plant cells decreased at low temperatures. Grif's analysis of the distribution of the optical densities in different regions of the chromosomes has led to the conclusion that the extent of DNA spiralization in the heterochromatic regions is temperature-dependent and that the quantity of DNA actually remains constant.

There is one interesting, but as yet incomplete, study in which it is not clear whether functionally conditioned quantitative changes in DNA occur or not. DNA cytophotometry, after Feulgen's reaction and biochemical analysis with diphenylamine, have revealed changes in the DNA content in human leucocytes in the course of a 24 h period (Fontaine & Swartz, 1972). Control analyses with nucleases revealed that the diphenylamine reaction was not absolutely specific to DNA. An error of 15–30% was found, similar to the changes observed over the 24 h period. The Feulgen reaction was considerably more reliable, but physiological rhythms of various cell properties during the course of a day might cause a non-standard Feulgen reaction to occur. The authors justly point out the desirability of checking one's data by means of ultraviolet cytophotometry. This method, like any other, is not perfect but it is based on different principles from the Feulgen reaction. Therefore parallel changes in DNA content, assessed by two such different techniques, is highly unlikely if these changes are methodological.

Well known exceptions to the rule of the relative constancy of DNA are the following: gene amplification, under-replication, chromatin diminution and chromosome elimination.

Amplification* differs from gene repetition, firstly in that repeated sequences are a widespread phenomenon typical of practically all cells (see Chapter 1B), while amplification has reliably been found in only a limited number of systems. Secondly, repeated DNA sequences are frequently arranged linearly in the chromosome (in tandem); amplified genes are, by contrast, often entirely separated from the chromosomes in the form of extrachromosomal elements. Repeated sequences have the same composition in different tissues of an organism and do not, as a rule, change during ontogenesis; amplified genes occur only in some cells; it is also only a temporary phenomenon in the life of a cell.

Amplification has been clearly demonstrated in oocytes (reviews Adrian, 1971; Tobler, 1975). At a certain stage in the growth of the oocyte, the number of ribosomal genes multiplies. This is manifest in the increased number of nucleoli (sometimes up to hundreds and even thousands). Amplification of ribosomal genes has been found in the oocytes of nematodes, molluscs, crustaceans, insects, echinoderms, fish, amphibians, birds, and mammals.

DNA amplification has been reliably shown in the macronuclei of some

---

*Exhaustive information on the present state of the problem as well as a historical review are represented in a recent book *Gene Amplification* (Schimke, 1982).

ciliates (see Chapter 5). Sometimes amplification is so great that these cases have, in the past, been interpreted as polyploidization of the nucleus. In actual fact, although some genes are multiplied, most are destroyed.

There is some information on DNA amplification in metazoan somatic cells. Since these studies are interesting but require confirmation, we shall mention the main ones.

A number of observations apply to the embryonic tissues. Thus, a reversible change in the number of moderately repetitive sequences at some stages in the differentiation of cartilage has been discovered (Strom & Dorfman, 1976). The genes for ribosomal and transfer RNA form part of these sequences. These genes are known not to determine cell differentiation. The functions of other moderately repetitive sequences are unclear. Cytophotometric studies of embryonic tissues of *Triturus vulgaris* also sometimes reveal small quantitative changes in nuclear DNA (see, for example, Lohmann, 1972). In such works, it is not always clear whether the change in the total DNA content is related to the proportion of proliferating, DNA-synthesizing cells.

An increased quantity of ribosomal DNA was determined by rRNA–DNA hybridization in a regenerating lens as compared to a normal lens. Increased nucleolar size may also point to the amplification of the rDNA, and also to intensified RNA synthesis in the regenerating lens (Collins, 1972).

Hyperdiploid DNA quantities have been determined by different methods in Purkinje cells of the cerebellum (Chapter 2H). The suggestion that changes in the perinucleolar chromatin are a result of gene amplification is, at the moment, only a working hypothesis requiring further evidence. Another indication of amplification of the ribosomal genes is the numerous nucleoli in the trophocytes in the ovaries of some craneflies (Tipulidae). The trophocytes of other insects have one, or few, nucleoli (Chapter 3B).

Cases of amplification of AT-rich DNA have been described in the somatic cells of the orchid *Cymbidium* (Schweizer & Nagl, 1976; Nagl, 1977). The changes mainly apply to satellite DNA, the function of which is still unknown. The satellite DNA content may vary in different plant tissues and manifests as a variable heterochromatin mass. Recently, chromatin loss during differentiation of orchid tissue has been observed by light and electron microscopy (Nagl, 1983).

Amplification of unique genes has been revealed in cells of the ovarian follicle of *Drosophila*; these cells synthesize proteins for the egg membranes (Spradling & Mahowald, 1980). The synthesis of these chorion proteins occurs in just 5 h; their specific mRNAs are synthesized in 1–2 h. Before that, the follicular cells are polyploidized to a 16C level. The synthesis of DNA also continues after polyploidization, right up to the time of chorion-protein-gene expression, but only these genes are reproduced. It has been shown that the genes for the chorion proteins are amplified more than tenfold.

During the last 5 years the understanding of gene amplification has rapidly progressed due to the study of the genetic mechanisms of the building-up of drug resistance in cultured mammalian cells and in human tumor cells. As was first shown by Schimke *et al.* 1978, the resistance to methotrexate in cultured murine cells resulted from the manifold increased content of the methotrexate target enzyme, dihydrofolate reductase. Molecular hybridization of [³H]DNA complementary to dihydrofolate-reductase-specific mRNA with cellular DNA showed that enzyme-coding sequences are increased in a resistant cell line by several hundredfold (Dolnick *et al.*, 1979). Characteristic chromosome abnormalities (hypertrophied chromosomal segments called homogeneously staining regions, and small paired bodies called double minute chromosomes) accompanied the conversion of cells to a resistant phenotype. Amplified genes were located in these regions. Subsequently, selective duplication of specific genes was revealed in many other cases of the resistance of cultured cells to cell-killing agents as well as resistance developed to cancer chemotherapeutic drugs in patients.

It is known that malignant transformation may be the result of increased expression of a cellular oncogene (Chapter 7D). One of the ways of increasing the dosage of such gene products is by gene amplification. Recently oncogene amplification has been revealed in cultured human myeloid leukemia cells and primary leukemic cells from the same individuals (Collins & Croudine, 1982; Dalla Favera, Wong-Staal & Gallo, 1982).

It has been established that some genes (especially informational, tissue-specific genes) are not amplified. Thus, genes for serum albumin, ovalbumin, globin, fibroin, and crystallins occur as single copies in the haploid chromosome set and the number of such genes does not change in the corresponding terminally differentiated cell. For example, the number of globin genes in erythroid cells is the same under conditions of acute anemia as it is in normal cells (Chapter 2I). The number of fibroin genes per diploid set is the same in all stages of development of the silk gland of the silkworm (Chapter 3C). But recently, Zimmer & Schwartz (1982) published evidence of amplification of the actin gene in postfusion myoblasts in vitro. The DNA species accumulated for about 100 h and then totally degraded within 24 h after reaching its greater concentration.

At the moment, well founded data on gene amplification are restricted to only a few systems. However, it may be assumed that this mode of intensification of the cellular function is more common than appears today. Low levels of gene copying are not to be ruled out. If, moreover, the effect is reversible (i.e. characteristic only of brief stages of development and then not in all tissues), amplification may be undetected by modern methods.

Under-replication is the incomplete reproduction (non-doubling) of DNA during the polytene cycles of some cells. Under-replication of part of the heterochromatin (consisting mainly of satellite DNA) is characteristic of the cells of some larval organs of Drosophilidae and Sciaridae (Chapter 3A). Recently, under-replication of the intron-containing rRNA cistrons has been discovered in the polyploid nuclei of the nurse cell of *Calliphora erythrocephala*; intron-free cistrons do not differ in their DNA content from those of embryonic DNA (Beckingham & Thompson, 1982). Some signs of under-replication were noted in the sex chromosomes of the pupal trichogen cells in the fly *Lucilia cuprina* (Bedo, 1982). The X and Y chromosomes in diploid cells were of similar size to the autosomes while, in the polytene cells, the sex chromosomes were proportionally much smaller due to under-replication. However, there was selective replication of non-C-banding material (following Giemsa staining) and brightly fluorescing material (following quinacrine staining) in these chromosomes. Some differential replication also occurred in the autosomes.

Under-replication should not be thought of as a general characteristic of polyteny. In the salivary gland of *Chironomus*, and also according to recent data on *Drosophila* (Dennhöfer, 1982; Spierer & Spierer, 1984), the DNA is completely replicated. DNA is also completely synthesized in the giant polytene cells of mammalian trophoblast, in the suspensor of plants, in the esophageal gland of ascarids, in the nerve cells of molluscs, and in the silk gland of *Bombyx mori*. Some of these cells undergo up to 15–20 endocycles, which should provide opportunity to detect under-replication.

An interesting variant of DNA changes is under-replication of one part of the genome and hyper-replication of another part. Such cases are described in the development of a single plant tissue (see Chapter 4) and in the macronuclei of some ciliates.

Diminution of the chromatin and *elimination* of chromosomes are different processes which are sometimes confused. From DNA analysis it is difficult to judge whether chromosome fragments have been destroyed or whole chromosomes have been removed. The elimination of some chromosomes was described in Cecidomyiidae, Sciaridae and Hironomidae; the diminution, i.e. excision of some heterochromatic chromosome segments, was revealed in *Ascaris* and copepods (for references see Beerman, 1977). The number of eliminated chromosomes varies considerably. Extreme examples of the chromosome elimination were observed in Cecidomyiidae. Only 6 or 8 chromosomes remain in somatic cells of *Wachtliella persicariae* as compared with 40 chromosomes in the zygote of this gallfly (Geyer-Duszyńska, 1959). Fewer than 10 species have provided evidence of diminution (Sager & Kitchin, 1975; Kovaleva & Raikov, 1978). In metazoans, the removal of part of the DNA occurs most often in blastomeres (after the first cleavage division). Sometimes, a considerable part of the DNA (up to 80%) is removed.

The function of diminution has not been determined. It has been suggested that, in somatic cells, only those genes are removed which are required for gametogenesis (Neifakh & Timofeyeva, 1977). But, of late, it has been shown that during chromatin diminution in *Ascaris suum*, only the satellite DNA is eliminated from the chromosomes (Roth & Moritz, 1981) and that the entire germ line genome is retained in both larval and adult cells. These data, and the fact that the DNA content is stable in all somatic cells of ascarids, point to the irrelevance of diminution to differentiation in metazoans. By contrast, in a few protozoans (for example, *Hypotrichida*) large-scale genetic diminution is a means of forming the definitive macronucleus (see Chapter 5). But it is an exceptional example.

Thus, the occurrence of quantities of DNA inconsistent with a strict doubling series has reliably been shown. As a result of the changes that bring this about, the proportionality between the number of the chromosome sets and the DNA content in the nucleus is disrupted. It should, however, be noted that in cases of amplification, or DNA under-replication, the basic diploid level and the composition of the DNA do not change. In the case of amplification, it is only supplemented by copies of a certain fraction of DNA, usually rDNA. Where under-replication occurs, the changes relate to the polytene nucleus and usually it begins after several cycles of reduplication have been completed. Therefore, in this case, the initial diploid genome set does not change. Substantial quantitative and qualitative changes in the genome take place during the destruction and diminution of part of the chromatin or the removal of whole chromosomes.

It may also be concluded that the above-mentioned changes in the DNA content are characteristic of a limited number of systems. At the moment, there are no grounds for considering these changes as a common feature of cell differentiation. In most of the well known cases, ontogenesis is accompanied by a stable DNA composition and content. In addition to the results of cytophotometry and data from molecular studies, the evidence for this fundamental stability comes from experiments involving transplantation of nuclei from somatic cells to the cytoplasm of an egg cell; the result is the development of an embryo. Furthermore, a whole plant may be obtained from a single somatic cell.

Conclusion

In response to the question of how a polyploid cell is formed from the diploid cell, it is necessary, first and foremost, to distinguish two forms of cells with a doubled genome: polyploid cells (with a reduplicated set of chromosomes) and polytene cells (in which only the number of chromonemata has been reduplicated but the chromatids have not separated). In both cases, the genetic material doubles, and so in this sense polyploidy and polyteny are similar. Therefore we bring both of these phenomena under the heading 'genome multiplication'. In both cases, too, the reduplicated genome is formed as a result of the premature withdrawal of the cell from the mitotic cycle; the time at which the cell leaves the cycle differs. A block immediately after DNA replication is responsible for the persistent doubling of the number of chromonemata (polytenization). By contrast, because a mitotic apparatus is needed for polyploidization, polyploid cells are formed by blocking mitosis at metaphase or later. Accordingly, the lengths of the cell cycles may be different. In polyploidizing mitoses, as compared with ordinary complete mitoses, only the final stages of the cycle are omitted. In the polytene (endo-) cycle the whole of mitosis is omitted and, possibly, the G_2-period as well. Definite differences in mechanism between polyploidy and polyteny, as well as in their consequences and developmental controls, force us to pay attention to the delimitation of these modes of genome multiplication through study of actual examples.

Practically all the examples of genome replication in the cells of animals and plants cited in this book have as their basis either polyploidizing mitosis or a G_2-block.

The morphological expression of polyteny may differ, even in cells of a single type with the same DNA content. The typical polytene thread-and-band pattern or the diffuse chromatin in nuclei with multiple chromonemata are probably not the only criteria for recognizing polytenization. Elucidation of the reasons for and functional significance of masked (cryptic) polyteny is an interesting task. The concept of 'endomitosis', and its derivatives 'angiosperm endomitosis' and 'masked endomitosis', can certainly not be applied to polyteny.

It is not clear how common endomitosis (the division of the chromosomes within the nuclear membrane) is in general. Further research on polyploidizing mitosis may possibly refine the term 'endomitotic polyploidization', so widely in use today. Another form of polyploidization, cell fusion, which gives rise to binucleate and multinucleate cells and symplasms, is characteristic of a limited number of tissues *in vivo*.

Chromatin under-replication and especially chromatin diminution seem to reflect the quest during evolution for a mechanism suppressing gene activity. Heterochromatinization has been adopted as an optimal path; other paths have probably also been tried, some of them have been preserved in a few species. The main modes for increasing gene activity, other than those directly influencing transcription, are polyploidization and polytenization. But in one case of giant cell growth, that of the oocytes, where polyploidy cannot be used, another mode of heightening ribosomal gene activity (amplification) has emerged and been adopted. This path may also have been used in some other cells, when production of a large quantity of proteins is needed within a short time, and where specific mRNA can not be accumulated beforehand.

7

The biological significance of polyploidy and polyteny

The biological function of genome multiplication has been discussed for at least 20 years now. But only recently have the early intuitive ideas been replaced by hypotheses which can be tested experimentally. Some of the benefits and drawbacks of genome multiplication for all growth and function have been advanced, and the significance of polyploid- and polytene-cell formation, as well as the possible reasons for and mechanisms of genome multiplication, have been discussed (Brodsky & Uryvaeva, 1977). Some recent publications (Gahan, 1977; Nakanishi & Fujita, 1977; Barlow, 1978*b*; Nagl, 1978; Uryvaeva, 1981) have also drawn attention to important but hitherto ignored properties of polyploid cells.

A The relationship between differentiation and genome multiplication

Polyploidization characterizes the development of many tissues. Differentiated cells – for example, hepatocytes, cardiomyocytes and melanocytes of mammals, the intestinal cells of ascarids or starfish are often polyploid. High-ploid and polytene cells in plants and some invertebrates, and also the giant cells of the trophoblast and megakaryocytes are found in the terminal stage of cell differentiation. Curtailment of the mitotic cycle and the corresponding onset of genome multiplication within a cell occur in parallel with differentiation.

Whether differentiation depends on polyploidization has long been discussed in the literature. The suggestion has been made that polyploidization is a means by which cells acquire a certain differentiated structure and function.

This relationship may, in fact, occur in some (and possibly all) populations of polytene cells, the differentiated function evidently deriving from polyteny. By contrast, differentiation of polyploidizing cells, in the cases studied, is not caused by polyploidy.

Differentiation of homologous tissues in related species of plants may take place with or without polyploidy (Evans & Van't Hof, 1975; Broekaert & Van Parijs 1978). Decondensation of chromosomes, an increase in nuclear volume and a decrease in the volume of the nucleoli

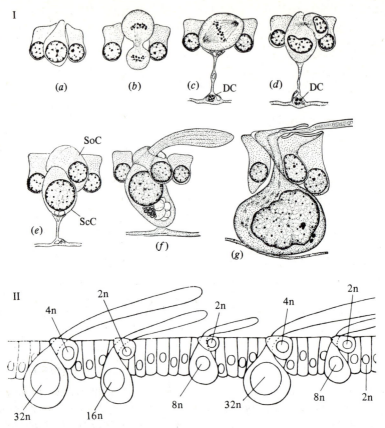

Fig. 7.1. Polyploidization of cells during scale formation in the wing of *Ephestia kühniella*. I. *a*, Primary scale stem cells; *b*, first asymmetric mitosis; *c–d*, second asymmetric mitosis; *e–g*, scale-forming cell (ScC), socket-forming cell (SoC), and degenerating cell (DC). II. Diagram of the epidermis showing nuclear ploidy class distribution. Uppermost and longest scales, 32n; intermediate scales, 16n; deeper, smaller scales, 8n; scale socket cells, 4n and 2n; epidermal cells, 2n (after Kühn, 1971).

(which are typical markers of nuclear differentiation), occur equally at the diploid and the polyploid level (Brossard, 1978).

The role of polyploidy in the stomach in some starfish was noted earlier (Chapter 3E). Here only polyploid cells with a large number of cilia are capable of carrying out this function. However, cilia also characterize diploid cells; but polyploidy makes this feature more pronounced.

A vivid example of the unique contribution of large polyploid (or polytene) cells to the structure of an organ is cited by Kühn (1971). In the butterfly *Ephestia kühniella*, the size and location of the wing scales (outgrowths of the epidermal cells) correspond to the ploidy of these cells

(Fig. 7.1). The uppermost, longest scales are formed by large cells (nuclear size and chromocenter number correspond to 32n). Small scales are formed by 8n cells, and the intermediate-sized ones by 16n cells. The socket cells differ correspondingly; one such cell surrounds the base of each scale. Tetraploid socket cells connect with the uppermost scales and the diploid cells with those of the inner layers. Polyploidy acts as a useful and necessary factor in scale- and socket-cell development. These elements may, however, be formed by cells of different ploidy. During its formation, the precursor of the scale cell undergoes two asymmetric mitoses (one of the daughter cells of the first division dies, one daughter of the second division gives rise to the socket cell). Then the cells change in form, and grow without mitosis; this leads to their specialization.

Species differences in the pattern of polyploidization of liver parenchymal cells were mentioned earlier (Table 2.2). Even among rodents, there are species that differ greatly in this feature. Whereas the liver of mice or rats consists mainly of polyploid cells, in guinea pig their frequency is rarely more than 10–15%. In Amphibia, the hepatocytes are usually diploid, but there are species in which seasonal polyploidization of the liver occurs (Bachmann, Goin & Goin, 1966). In some species of frogs, polyploid cells are consistently found in the liver (Bachmann & Cowden, 1967).

The pigment epithelium of rat retina consists of polyploid cells (see Chapter 2C) but in sheep of diploid cells (Berman, Schwell & Feeney, 1974). Examples are known of diploid and polyploid tissues in organisms of a single species; the types of tissues cannot be distinguished according to their function. Thus, dwarf mice and mice which have been semi-starved in early post-natal life develop livers of low ploidy (Naora, 1957; Geschwind, Alfert & Schooley, 1958, 1960; Swartz, 1967). Thyroidectomy causes a delay in growth and, correspondingly, in polyploidization of hepatocytes, while injection of thyroxine normalizes both of these processes (Mendecki *et al.*, 1978). A study of the properties of the diploid analog of a normally polyploid organ appears to be one of the most promising avenues by which to study the significance of polyploidy.

Examples of excessive growth and correspondingly heightened polyploidization are also known. For example, isoproterenol, injected into growing rats, stimulates additional mitoses in the liver and increases the number of polyploid hepatocytes (Gerzelli & Barni, 1976). Isoproterenol also causes a marked upward shift in the ploidy class distribution in the salivary gland. Likewise, repeated partial hepatectomy induces polyploidization of the hepatocytes to a level not usually observed during development (see Chapter 2A).

During the arrest or stimulation of growth, no changes in the state of hepatocyte differentiation have been noticed. However, the possibility of slight changes in the specialization of the cells as ploidy changes cannot be ruled out. Hepatocytes have many functions and therefore comparative analysis of their differentiation is difficult. Differences in the fractions of non-histone proteins were recently discovered in diploid and tetraploid liver cells (Garber & Brasch, 1979); these may cause functional differences. The suggestion (Barlow 1978b) of possible activation (or blocking) of some alleles in a polyploid genome is another interesting model for the regulation of polyploid cell function.

In the megakaryocyte lineage of bone marrow cells, differentiation and polyploidization seem inseparable; only 8n and higher ploid megakaryocytes are capable of thrombopoiesis in mammals. But, even here, polyploidy does not determine the specific features of the cell type. Diploid megakaryocytes which produce blood platelets, appear in leukemia in humans. These cells are characteristic of normal thrombopoiesis in some lower vertebrates.

Thus, it is not polyploidy that determines the main feature of specialization in such cells as hepatocytes, melanocytes, and megakaryocytes. Nevertheless, we believe that regular polyploidization during the development of certain cell types has a biological significance even though there is hardly likely to be a common role for polyploidy in all these differentiated cell types. An increased genome may have some advantage for, or even be essential to, the viability or functioning of one type of cell; for another cell type some other property of polyploidy may be of greater importance.

B The properties of polyploid cells

The size and surface

One obvious consequence of polyploidy is the increased size of the nucleus and the cell. In hepatocytes, this phenomenon was studied in detail more than 50 years ago (Jacobj, 1925) and is constantly being re-affirmed (Watanabe & Tanaka, 1982). Measurements of hepatocyte volume in rats as a function of ploidy have shown that each picogram of nuclear DNA corresponds to 256 μm^3 of cell volume (Sweeney *et al.*, 1979). But, after treatment with phenobarbital, the latter value increases to 375 μm^3 pg^{-1} of DNA without any overall increase in the nuclear DNA content. The doubling of cell volume leads to the formation of the rule of rhythmic cell growth. It may be noted that nuclear growth can lead to a decrease in its surface-area-to-volume ratio. Every time the volume of geometrically similar bodies doubles, their surface area increases only 1.59 times. Consequently, if the surface-area-to-volume ratio of a 2n nucleus is equal to 1.0, this value for a 4n nucleus will be 0.8, and for a 32n nucleus only 0.4. On the basis of these calculations, it was considered that polyploid cells could not be as active as diploid ones, at least in respect of those functions that depend on the interaction between nucleus and cytoplasm (Alfert & Siegel, 1963; Alfert & Das, 1969; De Leeuw-Israel, van Bezooijen

Table 7.1. *Examples of large high-ploid nuclei with characteristic shapes*

Cell type	Approximate DNA content (C)	Description of the nucleus	Reference
Megakaryocytes	up to 64	Deeply lobed	Odell et al., 1970
Trophoblast	10^3	With folded nuclear envelope	Zybina, 1977
Salivary gland of *Gerris lateralis*	10^3	With irregular contours	Geitler, 1938
Endosperm and suspensor in plants	10^4	Spoon-shaped	Deumling & Nagl, 1978
Giant neurons of the mollusc *Tritonia diomedia*	10^5	With long outgrowths into the cytoplasm	Sakharov, 1965
Esophageal gland of ascarids	10^5	Complex spatula-shape	Anisimov, 1976
Silk gland of *Ephestia kühniella*	10^5	With branches	Traut & Scholz, 1978
Silk gland of *Bombyx mori*	10^6	With twisted and arborescent outgrowths of the nuclear envelope	Nakanishi et al., 1969

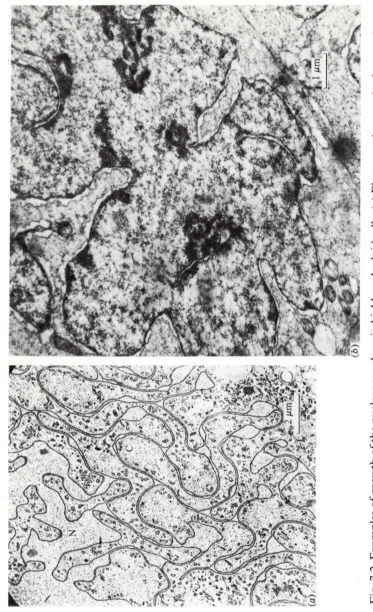

Fig. 7.2. Examples of growth of the nuclear envelope in highly polyploid cells. (*a*) Electron micrograph of protrusions and invaginations of the nuclear membrane in the gregarine *Didymophyes chaudefouri*; N, nucleus; C, cytoplasm (from Baudoin & Ormières, 1973). (*b*) Rat trophoblast giant cell (electron micrograph kindly provided by Dr E. V. Zybina).

& Hollander, 1972). But examples are known of a considerable increase in the surface area of high-ploid nuclei, owing to changes in the shape of the nuclear envelope. Convolutions of the nuclear membrane considerably increase the surface area available for nuclear–cytoplasmic interactions. Sometimes, this can be seen using the light microscope. In other cases, the irregular contours of the nuclear envelope were only discovered by electron microscopy. Some examples have been examined in detail earlier in this volume (pp. 106, 133); many are summarized in Table 7.1, and two are illustrated in Fig. 7.2.

In high-ploid cells, the surface area of the plasma membrane frequently increases, too. Thus, the deep invaginations of the glial cells into the cytoplasm of mollusc neurons (Chapter 3F) undoubtedly facilitate the metabolism of these giant cells. Likewise, the nucleus of these giant cells, which usually occupies a considerable part of the cell volume, ramifies into the cytoplasm (Sakharov, 1965), increasing the efficacy of nucleocytoplasmic exchange.

Thus, the reduced surface-area-to-volume ratio, and, consequently, the diminished contact between the nucleus and the cytoplasm (and also between the cell and the extracellular medium) are overcome in various ways in giant cells. It would be interesting to assess the degree of success of these strategies, for instance by studying such membrane-dependent processes as transport, secretion, and intracellular synthesis (e.g. of lipids and proteins).

The activities of highly polyploid cells may be regulated by changes in the membrane surface. Such changes have been observed during different functional states of the giant nerve cells. Some peculiarities of polyploid cell behavior in culture may result from the relative decrease of the cell surface area. For example, Li *et al.* (1978) could select Novikoff hepatoma cells which had an increased resistance to lysis in an isotonic solution of glycerine; the selected cell line was tetraploid while the original line was diploid. The selection may have been due to the lower surface-area-to-volume ratio of the tetraploid cells. Polyploid cells are not, however, always more viable than diploid ones. Thus, in a monolayer culture of liver cells, selection of the diploid hepatocytes may occur (Gómez-Lechón *et al.*, 1981). The total number of 2n cells does not increase, the polyploid cells die and so the percentage of 2n cells increases. In other cultures, and in other conditions, the polyploid cells may be favoured. Thus, in a suspension culture of Chinese hamster cells under conditions of high cell density and inadequate nutrient supply, tetraploid cells are selected (Olive, Leonard & Durand, 1982).

Probably, the changed surface-area-to-volume ratio is responsible for the increase in membrane potential in multinucleate endothelial cells cultured *in vitro* (in comparison with mononucleate cells; Richter & Halle, 1983).

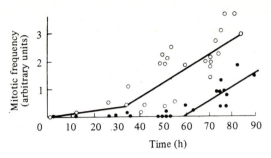

Fig. 7.3. The proliferative response of diploid cells (open circles) and tetraploid cells (solid circles) to the stimulation of mitosis in pea-root segments after placing them on a culture medium containing auxins and kinetin (after Matthysse & Torrey, 1967).

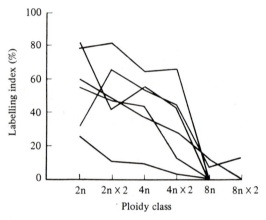

Fig. 7.4. The decrease in the labelling index with increased polyploidization in mouse hepatocytes. 23 h after partial hepatectomy, the mice were injected with [³H]thymidine 1 h before killing. The labelling index was determined separately for each ploidy class. Each line represents one mouse (after Brodsky & Uryvaeva, 1977).

Proliferative response

As the level of ploidy rises, the ability of a cell to proliferate as a rule, decreases. Diploid cells in roots were shown by autoradiography to respond to a proliferative stimulus faster than tetraploid cells (Fig. 7.3). The decrease in the proliferative ability of polyploid hepatocytes is shown by their lower rate of entry into the mitotic cycle. After partial hepatectomy, cells of different ploidy classes synthesize DNA but the rate of increase of the labelling index decreases as the level of ploidy in the liver rises (Uryvaeva & Marshak, 1969; Watanabe, 1970). The low [³H]thymidine labelling index of high-ploid cells during the first wave of DNA synthesis

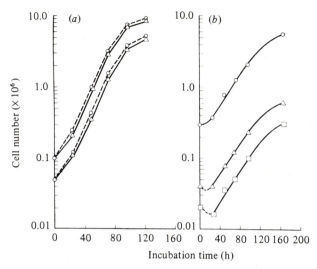

Fig. 7.5. The growth curves of (*a*) hamster cells and (*b*) pig kidney cells of different ploidy *in vitro*. (*a*) Hamster cells; upper curves represent diploid clones where $T_c = 12$ h, lower curves represent tetraploid clones where $T_c = 12$ h. (*b*) Pig kidney cells; upper curve represents diploid cells where $T_c = 29$ h, the middle curve represents octaploid cells where $T_c = 29$ h, and the lower curve represents hexadecaploid cells where $T_c = 28$ h (after Harris, 1971).

(Fig. 7.4), does not mean that these cells do not take part in regeneration; it is caused solely by the prolonged G_1 period of the first induced mitotic cycle. Activation of chromatin (estimated from the binding of acridine orange and actinomycin D) in regenerating mouse liver was delayed in tetraploid compared to diploid nuclei (Kushch *et al.*, 1976).

The decreased rate of response to the proliferative stimulus has been thought to be connected with the reduced surface-area-to-volume ratio in high-ploid cells. However, observations of clones of 2n, 4n, 8n and 16n fibroblasts *in vitro* revealed similar growth kinetics (Fig. 7.5), although (if we proceed from the model of geometrically similar bodies) the relative surface of the 16n nucleus will be half that of the 2n nucleus.

Mitotic cycle

Percent labelled mitosis curves are the same for 2n, 4n and 8n mouse hepatocytes (Fig. 7.6); this indicates similar durations of S-, G_2- and M-phases of the mitotic cycle. Similar curves were obtained for 2n and 4n cells of the mouse bladder (Fig. 7.7).

[³H]thymidine-incorporation kinetics during S-phase do not differ in 2n and 4n nuclei of liver cells (Fig. 7.8). The intensity of DNA synthesis in 4n nuclei is twice as high as in 2n nuclei, if self-absorption of tritium is

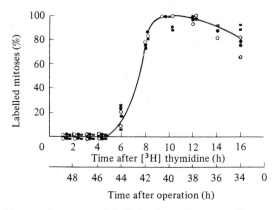

Fig. 7.6. Changes in percent labelled mitoses in mouse hepatocytes of different ploidy (open circles, 2n; filled circles, 4n; squares, 8n). All the experimental data were obtained 50 h after partial hepatectomy and 4 h after colchicine injections; [³H]thymidine was injected at various times after the operation, as indicated on the lower abscissa (after Faktor & Uryvaeva, 1972).

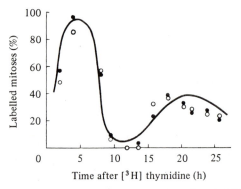

Fig. 7.7. Changes in percent labelled mitoses in cells of different ploidy (filled circles, 2n; open circles, 4n) from bladder epithelium of mice stimulated with 4-ethylsulfonylnaphthalene-1-sulphonamide (ENS). [³H]thymidine was injected 18 h after ENS, and colchicine was given 1 h before killing (after Levi *et al.*, 1969*a*).

taken into account (Uryvaeva & Faktor, 1971*a*). A linear relationship has been found between the amount of DNA in mouse hepatocytes and DNA-polymerase I activity (Fig. 7.9). The number of sites in the nucleus which could react with exogenously added DNA-polymerase I of *Escherichia coli* were doubled when the ploidy class of the hepatocyte increased from 2C to 4C and doubled again with each genome doubling up to 16C. In bean root the growth rate of tetraploid cells was the same as that of diploid cells (Bansal & Davidson, 1978). A comparison of 2n, 4n and 8n nuclei in cells of the onion root (Giménez-Martín *et al.*, 1968), and in diploid and

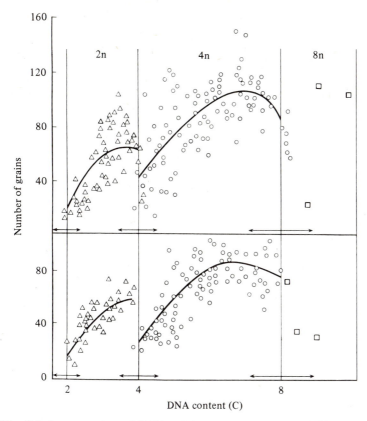

Fig. 7.8. Incorporation of [³H]thymidine into diploid, tetraploid and octaploid hepatocyte nuclei during the S period in regenerating liver of two adult mice. Each point represents the result of a combination of autoradiography and Feulgen–DNA cytophotometry on the same nuclei. Range of 2C, 4C and 8C DNA contents are indicated by the arrows (after Uryvaeva & Faktor, 1971*a*).

polyploid HeLa cells *in vitro* (Firket & Hopper, 1970), also showed the mitotic cycle to be of the same length in each. But, in the polyploid cells of these populations, mitosis took longer than in the diploid cells.

A series of studies has been carried out on diploid and autotetraploid or colchicine-induced tetraploid plants (Van't Hof, 1966; Troy & Wimber, 1968; Ivanov, 1974; Murin, 1976). The length of S-phase and of the entire cell cycle was frequently the same in cells of similar tissues in diploid and polyploid forms of the same species*. There are, however, exceptions. Thus, in *Secale cereale*, DNA synthesis in the tetraploid form lasts twice as long as in the diploid form, and the entire cell cycle is slowed down 1.5 times (Karpovskaya & Belyaeva, 1973). Another example of the

* The length of the mitotic cycle correlates with the mass of the genome (Fig. 4.2).

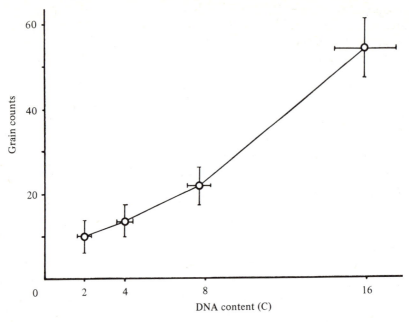

Fig. 7.9. Template–primer activity of 12-month-old mouse hepatocytes of various ploidy classes detected by *E. coli* DNA polymerase I. Template activity was assessed by incubating isolated nuclei with DNA polymerase I in the presence of [³H]TTP and counting the grains over nuclei in autoradiographs. DNA content and grain counts were made on the same nucleus (after Shima, 1980).

considerable lengthening of S-phase and of the entire cell cycle was observed in cells of a near-tetraploid Ehrlich carcinoma as compared with its diploid form (Defendi & Manson, 1963).

A comparison of the cycle in diploid and in high-ploid cells of the trophoblast shows the same length of S-phase, although the nuclear DNA content differed more than 200-fold (Andreeva, 1964). Consequently, the rate of DNA synthesis in the polytene nucleus of the trophoblast is hundreds of times faster than in the small diploid nucleus, although the individual chromonemata take the same length of time to be replicated. However, in polytene cells of the salivary gland of *Drosophila*, DNA replication is slower than in the small diploid cells in larvae of the same age. This conclusion was first drawn on the basis of cytophotometric data (Rudkin, 1972). Recently, using DNA-fibre autoradiography, it was shown that the rate of fork movement in *Drosophila virilis* salivary glands was 3.5 times slower than in diploid brain cells (Steinemann, 1981). As was assumed, on the basis of [³H]thymidine incorporation into various chromosome bands of *Chironomus*, replication time increases in parallel with degree of polyteny (Hägele, 1976). Thus, in many cases, the rate of DNA replication per genome did not change with increases in the number of genomes in the nucleus. In these cases, chromosome structure and organization of the chromatin undergo no changes, and homologous parts of homologous chromosomes replicate in synchrony.

It has been established with computer-aided cytophotometry that, in rat hepatocytes, the ratio between condensed and decondensed chromatin is independent of the degree of ploidy (Romen *et al.*, 1980). According to image-analysis studies, the distribution of chromatin densities is similar in 2n, 4n and 8n nuclei of liver cells of *Microtus agrestis* (Pera & Detzer, 1975). The organization of the chromatin was shown to be similar in all the ploidy classes of mature megakaryocytes (Pelliciari & Prosperi, 1980).

The centriolar apparatus

Numerous centrioles have often been reported to be present in multinucleate osteoblasts and foreign body giant cells (Hertwig, 1928; Mathews, Martin & Race, 1967; Sapp, 1976). Likewise, more centrioles are found in polyploid mitotic fibroblasts than in diploid mitoses (Pera, 1975). Onishchenko (1978) determined a strict correlation between the number of centrioles and the number of chromosome sets by electron microscopy of ultrathin sections of mouse hepatocytes. Two centrioles are found in the diploid interphase hepatocyte, four in tetraploid cells and eight in octaploid cells (Fig. 7.10). In mononucleate and binucleate cells of the same ploidy (4C and 2C × 2, or 8C and 4C × 2) there is an equal number of centrioles. If there is more than one diplosome (a pair of centrioles), then they are all located together in the interphase cell, forming a single centriolar complex. In polyploidizing megakaryocytes, a complex of 8, 16 or 32 centrioles is observed (Fig. 7.11). In the thrombopoietic, mature megakaryocytes there is no centriolar complex (Moskvin-Tarkhanov & Onischenko, 1978). In these cells, which are incapable of mitosis, just a few separate centrioles were found. In the polytene cells of the salivary gland of *Chironomus*, Onishchenko & Chentsov (1977) could not find any centrioles, nor are they found in the polytene follicular cells of *Drosophila* gonads (Mahowald, Caulton & Edwards, 1979).

Centrioles (basal bodies) have a special motor function when associated with flagella and cilia. The movement of chromosomes during mitosis is another function of the centrioles. At each pole of a bipolar mitosis, two centrioles are known to be found: the mature parental centriole and a daughter one. Only the former is a center for the organization of microtubules. In a multipolar mitosis (especially in cultures of heteroploid cells and sometimes, polyploid cells *in vivo*) one or several centrioles may be found at each pole (Onishchenko, Bystrevskaya & Chentsov, 1979). It is suggested that the reason for multipolarity is the prolongation of mitosis which allows the maturation of the daughter centrioles and their divergence from the parental centriole, resulting in the formation of additional centers

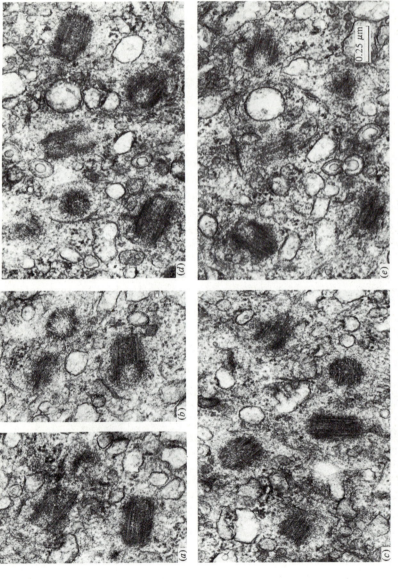

Fig. 7.10. Eight centrioles revealed by serial sections (*a–e*) of an octaploid hepatocyte of mouse (electron micrographs kindly provided by Dr G. E. Onishchenko).

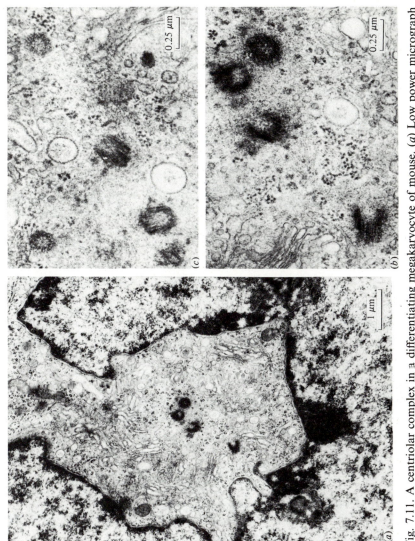

Fig. 7.11. A centriolar complex in a differentiating megakaryocyte of mouse. (*a*) Low power micrograph showing the complex close to the nucleus. (*b,c*) Higher magnifications of the same complex with a mature centriole and a pro-centriole (electron micrographs kindly provided by Dr G. E. Onishchenko).

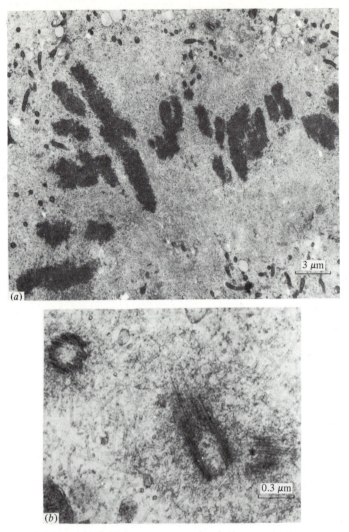

Fig. 7.12. Electron micrograph of a multipolar mitosis (*a*) in a hybrid cell (mouse × Chinese hamster), (*b*) one of the four poles, containing three centrioles (micrographs kindly provided by Dr G. E. Onishchenko).

for microtubule organization. Multipolar mitoses are common in hybrid cells arising by experimentally induced cell fusion (Fig. 7.12) and, here, at some poles only one active centriole can be seen. The modal number of the chromosome sets in hybrid cells (Chinese hamster cells and murine 3T3 cells) is similar to the modal number of centrioles (Onishchenko & Volkova, 1983). Hence, for the distribution of a haploid chromosome set, one centriole is needed. In cells which have irreversibly left the cycle, the centrioles degenerate.

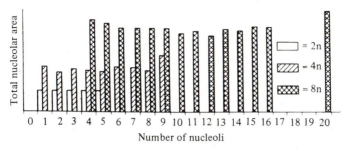

Fig. 7.13. Mean total nucleolar area (arbitrary units) in rat hepatocyte nuclei with different number of nucleoli. The mean total nucleolar area of the diploid nucleus is 8.2 ± 0.1 arbitrary units, 16.7 ± 0.2 for the tetraploid, and 32.9 ± 0.3 for the octaploid (after Meinders-Groeneveld & James, 1971).

RNA synthesis

This section describes the genetic activity of polyploid nuclei. A number of studies has been made on the cells of rat liver. In some works, it has been determined that the amount of RNA transcribed doubles as the quantity of DNA in the nucleus doubles. Thus, a biochemical analysis of the fraction of 2n, 4n, and 8n nuclei of hepatocytes after the incorporation of [14C]orotic acid into the liver *in vivo* reveals differences in the radioactivity of RNA (in the ratio 1:2.1:4.2), in the activity of RNA polymerase (1:2.2:5.9), and in the synthesis of the ribosomal RNA in 2n and 4n nuclei (1:1.7; Johnston & Mathias, 1972). This means that the total expression of the genome corresponds to the gene dosage. A study of orotic acid incorporation into isolated liver nuclei led to the same conclusion (Roszell, Frodi & Irving, 1978). Cytophotometry revealed the doubling of RNA content in the tetraploid hepatocytes compared with diploid cells (Morselt & Wijgerden, 1975). In the same way, the total area of the nucleoli (but not their number) in hepatocytes of different ploidy changes in the proportion 1:2:4 (Fig. 7.13).

In autoradiographic research into the incorporation of [3H]uridine into mouse liver (Jehle & Kiefer, 1979), a strict correlation between RNA synthesis and the ploidy of the hepatocytes was shown (Fig. 7.14). Mononucleate and binucleate polyploid cells with the same total genome (4n and 2n × 2, or 8n and 4n × 2) had the same degree of labelling. The similarity of the DNA-reassociation kinetics in these cells points to the fact that the tetraploid hepatocyte is a genetic analog of two diploid cells (Ordahl, 1977; Lebedeva & Diment, 1982).

Results of two investigations contradict the data on the proportionality of total transcription to gene dosage. After selection in a flow cytofluorometer of hepatocytes of different ploidy labelled with [3H]uridine, it appeared that they all had approximately equal radioactivity (Collins, 1978). It may be imagined that during the selection of the hepatocytes in

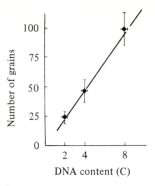

Fig. 7.14. Incorporation of [¹⁴C]uridine into mononucleate diploid, tetraploid and octaploid hepatocytes of mice. Estimates of radioactivity made from grain counts (means ± standard deviation) in autoradiographs (after Jehle & Kiefer, 1979).

the flow cytometer by DNA fluorescence, mononucleate and binucleate cells were not sorted. Moreover, after brief incorporation periods, differences in the rate of precursor transport into cells of different sizes might be revealed (rather than differences in the rate of RNA synthesis). The RNA content in the 2n, 4n and 8n hepatocytes, isolated by a micromanipulator and treated with ribonuclease, was not found to double. The RNA contents of these cells were 1 : 1.37 : 2.32 respectively (Roozemond, 1976). The method of mechanically isolating the cells is not perfect. The boundaries were probably defined inaccurately, and the smaller the cell the greater the error. A study of other systems definitely showed a doubling of the total transcription in the polyploid cells as compared with diploid cells. In the wall of locust testis (see Chapter 3D) 4n cells incorporated twice as much [³H]uridine as diploid ones (Fig. 7.15), and 8n cells incorporated four times more. A biochemical study of RNA in the *Drosophila* gonad revealed a 4.7-fold increase in the rate of synthesis during oogenesis. The number of ribosomal genes increased in that time in the polytenizing trophocytes by 4.9-fold (Mermod *et al.*, 1977).

The incorporation of [³H]uridine in the 2n, 4n, and 8n cells of the root of *Allium carinatum* during S-phase can be expressed by the ratio 1 : 2.08 : 4.35, and (in pre-prophase 2n and 4n cells) by 1.91 : 3.85 respectively (Fig. 7.16). In this case, just as in those noted earlier, the doubling of the DNA content did not lead to an increase in RNA synthesis per genome unit. This conclusion could have been drawn during the study of RNA synthesis in the diploid and polyploid cells of cotyledons of various species of plant (Scharpe & van Parijs, 1973; Davis, 1976; Manteuffel *et al.*, 1976). RNA-polymerase was studied in one species, and the increase in its activity was noticed to be proportional to gene dosage. The ratio of RNA to DNA also remained stable.

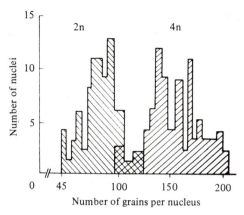

Fig. 7.15. Incorporation of [³H]uridine (from grain counts in autoradiographs) into diploid and tetraploid nuclei of testicular wall cells in the locust *Chrysochraon dispar* (after Kiknadze & Tuturova, 1970).

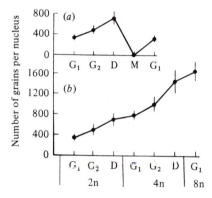

Fig. 7.16. [³H]uridine incorporation into cell nuclei of *Allium carinatum* during (*a*) the mitotic cycle and (*b*) the endocycle. G_1 and G_2, the interphase stages; D, dispersion (or Z) phase; M, metaphase; n, nuclear ploidy. The bars indicate standard error about the mean (after Nagl, 1973).

An increase in the intensity of RNA synthesis was noted in polytene cells of the bean suspensor (Clutter *et al.*, 1974). The number of cells, approximately 200, did not change during the development of the suspensor. The cells accumulate huge quantities of DNA – up to several thousand haploid units (see Chapter 4). In the embryonic tissue proper, complete mitoses continue. At some stages of development, the RNA synthesis per genome (2C quantity of DNA) in the suspensor was twice as high as in the embryo. There was thereby no DNA amplification.

Protein synthesis

The dry weight of the hepatocytes of rats, mice, and humans, on average, doubles in each polyploidization cycle. Twice as many proteins are found in 4n nuclei and cells as in 2n ones (Rigler, 1962; Ranek, 1976; Frederiks, Slob & Schröder, 1980; Engelmann *et al.*, 1981; Shalakhmetova, Kudryavtseva & Kudryavtsev, 1981*a*). The rate of protein synthesis is proportional to the degree of ploidy in cultured hepatocytes (Le Rumeur *et al.*, 1982, 1983). Fractionation of isolated hepatocytes by velocity sedimentation shows that albumin synthesis increases as the cells grow in size (ploidy). A polyploid liver of a 3-month-old rat produces on average twice as much albumin per hepatocyte as in a 3-week-old animal with mainly diploid hepatocytes (Deschênes *et al.*, 1981; Marceau, Deschênes & Valet, 1982). In hepatocytes separated by counterflow centrifugation the 4n cells secrete, on average, twice as much albumin as 2n cells (Le Rumeur *et al.*, 1981). Immunochemical technology has revealed albumin in all hepatocytes. In the first 3 days after partial hepatectomy, nuclear proteins increase considerably and then decrease; the twofold difference between 2n and 4n nuclei is maintained no matter what changes take place (Morselt & Frederiks, 1974). The same ratio is also found in cellular proteins during the mitotic cycle (Morselt & Wijgerden, 1975). Collagen and other proteins are synthesized twice as intensively in tetraploid fibroblasts *in vitro* as they are in diploid ones (Terzi, 1974). The study of histones in isolated diploid and highly polytene nuclei from *Drosophila* salivary gland shows identical fractions in both nuclear types (Cohen & Gotchel, 1971).

Thus, some investigations have shown that there is a correlation between the intensity of protein synthesis and gene dosage. It is, however, essential to note that in these works the average amounts have been evaluated. Important data have been obtained in the study of protein fractions (Garber & Brasch, 1979). It was established that, in isolated liver nuclei, the average amounts of RNA, histones, and non-histone proteins double in accordance with the gene dosage. But, depending on the ploidy of the nuclei and on the age of the rats, the ratio of two fractions of non-histone nuclear proteins differed. Later, Brasch (1982) confirmed these protein patterns in 2C, 4C, and 8C nuclei of rat hepatocytes. Relatively more of the non-histone proteins were saline-soluble in 2C nuclei from young rats than in 2C and other nuclei in mature and old rats. It is essential that these studies be continued; they will possibly determine the actual degree of the functional heterogeneity between cells of different ploidy.

Recently information has also appeared pointing to the fact that, when the rat liver cells polyploidize during the first few weeks after birth, the overall protein content in tetraploid hepatocytes exceeds the diploid level by less than twofold. At the third month of postnatal life the 4C and 2C × 2 cells have achieved double the diploid mass (James *et al.*, 1979; Shalakhmetova *et al.*, 1981*a*).

Other indices of functional activity

The result of polyploidy is the doubling of the number of genes, including those for tissue-specific functions. Many studies have noted the correspondence between transcription and gene dosage. There are also other observations on the cell-functional capacity during polyploidization. Thus, the secretory activity of polyploid cells is greater than that of diploid ones (Merriam & Ris, 1954; Anisimov, 1974; Bennett, 1974). The suggestion has been made that the increase in the level of DNA in the cells of the fat body of the adult female migratory locust allows the production of mRNA for vitellogenin, and possibly for other proteins, to be accelerated (Nair, Chen & Wyatt, 1981).

Polyteny possesses potentially greater possibilities than polyploidy for the intensification of cell functions. Thus, on the basis of the number of haploid genomes, the silk gland cell of *Bombyx mori* shows approximately the same rate of transcription and translation as the other cells studied. For the haploid genome 4.5×10^5 molecules of fibroin are synthesized per minute in the silk gland of *Bombyx mori*, 3.0×10^5 molecules of chymotrypsinogen in the rat pancreas, and 3.1×10^5 of globin in the erythroid cells (Kafatos, 1972). But the number of haploid genomes in each cell of the silk gland of *B. mori* reaches hundreds of thousands, while in the pancreas of the rat there are a maximum of eight, and in the erythroblast only two.

In rat hepatocytes the glycogen content grows in proportion to ploidy (Kudryavtsev, Kudryavtseva & Shalakhmetova, 1979). The proportion was similar in rats of different ages. During the S-phase of the cell cycle (both mitotic and polyploidizing) the glycogen content falls. The decrease in the glycogen content is proportional to the ploidy of the cells (Shalakhmetova *et al.*, 1981*b*).

In the hepatocytes of 1-year-old rats, in each polyploidization cycle the activity of the acid phosphatase doubles (Middleton & Gahan, 1979, 1982). The same occurs in old animals with the transition from a diploid to a tetraploid level. However, in the shift from tetraploidy to octaploidy, the enzyme activity increases by only about 50% in old rats. The proportional relationship between ploidy and glucose-6-phosphatase activity was shown for $2C \times 2$, $4C$ and $4C \times 2$ hepatocytes (Van Noorden *et al.*, 1984). But $2C$ hepatocytes exhibit extremely high activity, about 1.5-fold higher than expected. It has also been shown that bile formation correlates with hepatocyte ploidy (Deschênes *et al.*, 1981).

An increase in the number of chloroplasts was discovered in cells of several species of plants as a result of a doubling of their chromosome

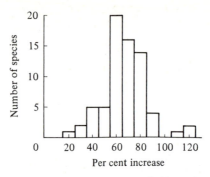

Fig. 7.17. Factor of increase in the number of chloroplasts in guard cells, caused by doubling the chromosome number (from diploid to tetraploid). 70 different plant species were examined (after Butterfass, 1973).

number (Butterfass, 1973). But the number of chloroplasts is not usually doubled, the average increase being 60–70% (Fig. 7.17).

As mentioned earlier in diploid and tetraploid forms of the same plant species, both S-phase and cell cycle are usually of similar duration. But, in some species, this is not the case; there may be a change in the activity or structure of the chromosomes during polyploidization. This is suggested by a study of flavin synthesis in diploid and tetraploid races of the grass *Briza media*. Different flavins accumulate in tetraploid plants compared with diploid ones. Polyploidization leads to a four-fold increase in esterase activity, while peroxidase activity remains constant (Murray & Williams, 1973). In tetraploid murine embryos the activity of malate dehydrogenase and [³H]uridine incorporation does not double compared to the levels in diploid embryos (Eglitis & Wiley, 1981). Cell number in the 2n and 4n embryos is the same, although perhaps the viability of 4n embryos is reduced, or there may be other reasons for the absence of a balanced gene dosage. It has already been noted that there are small differences in the protein pattern in diploid and tetraploid hepatocytes. Cytomegalovirus detects another difference between cells: virus infection is, for some reason, connected mainly with diploid and not polyploid hepatocytes in mice (Papadimitriou & Schellam, 1981).

The properties of mononucleate and binucleate cells of similar ploidy

Mononucleate and binucleate hepatocytes with the same overall DNA content do not differ in DNA-reassociation kinetics, genome replication rate, level of RNA synthesis, enzyme activity, overall protein content or in cell size. 4C and 2C × 2 (i.e. tetraploid) cardiomyocytes have identical mass and similar growth rates (Fig. 7.18). Mononucleate and binucleate

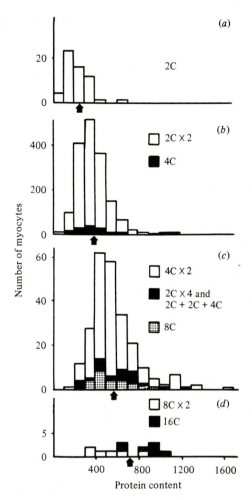

Fig. 7.18. Histograms of protein content (in arbitrary units) in the mouse heart myocytes of different ploidy classes. (*a*) Diploid, (*b*) tetraploid, (*c*) octaploid, (*d*) hexadecaploid. Arrows indicate the mean protein content (after Brodsky, Tsirekidze & Arefyeva, 1984).

cardiomyocytes of higher ploidy – 8C and 4C × 2 or 16C and 8C × 2 – do not differ in these properties either.

2C × 2 cells are produced by acytokinetic mitosis, after complete doubling of the diploid chromosome number (see Chapter 2). All four gene sets are active. Binuclearity, in this case, signifies real polyploidy.

Mononucleate and binucleate polyploid cells may, however, possess some functional differences. Two nuclear sites of transcription may be advantageous. In any case, where there are two nuclei (instead of one) with the same chromosome set, a larger nuclear surface-area-to-volume ratio may be expected. Therefore, binucleate cells may have certain growth and functional advantages over their mononucleate analogs. This is possibly why $2C \times 2$ Syrian hamster cells traverse G_1 more rapidly than do 4C cells (Fournier & Pardee, 1975).

C Cell and tissue growth

Cell-cycle-dependent and mitosis-independent cellular growth

All tissues at some stage in their development grow by means of cell division, i.e. by an increase in the number of cells. In some tissues, cell multiplication is the only mode of growth, in others it is only an early stage in growth. Another mode of tissue growth is cell polyploidy and hypertrophy. Tissue enlargement may arise by an increase in the mass of individual cells. Two kinds of cellular enlargement are discussed below. The first resembles the mitotic mode of growth but is accompanied by polyploidization and polytenization. The second is independent of mitosis. There are some differences in control mechanisms governing these two modes of cellular growth.

A convenient model of the life cycle of a proliferating cell is to consider it as consisting of two complementary cycles: the mitotic (chromosome) cycle and the growth cycle (Mitchison, 1971; Prescott, 1982; Prescott, Liskay & Stancel, 1982). These cycles are not strictly co-ordinated and examples are known where the dissociation of DNA synthesis and mitosis from growth in cell size can be artificially induced (Baserga, 1984).

In the course of the mitotic cycle, the cell doubles all its constituents and then halves its size by division. If the mitosis is polyploidizing, cell growth proceeds unchanged and the cell doubles its mass in synchrony with each round of nuclear reproduction. This is why tetraploid cells are, as a rule, twice as large as diploid cells, octaploids are twice as large as tetraploids, and so on. The reason for the proportionality of the polyploid cell mass to their ploidy level is that the cells repeatedly pass through all the events of mitotic growth cycles (except for the division phase) parallel with the chromosome reproduction cycles.

Another component of cellular growth is the increase of cell constituents in the absence of mitotic cycling, polyploidization or polytenization. Continuing growth, leading to hypertrophy of the cytoplasm is found in hepatocytes and heart muscle cells after the genome has stopped reproducing itself.

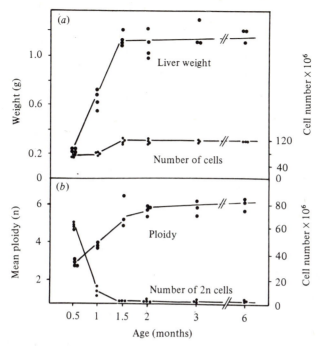

Fig. 7.19. Changes (*a*) in liver weight, total number of hepatocytes and (*b*) number of diploid hepatocytes, and mean hepatocyte ploidy during postnatal growth of the mouse. Each datum point represents a result from one mouse.

Polyploidy and polyteny as factors in tissue growth

During polyploidization, each doubling of the chromosome set is accompanied by a doubling of all the cell constituents. Since one polyploid cell is functionally and structurally equivalent to several diploid cells, polyploidization may be regarded as a process similar to cell multiplication. Moreover, polyploidization does not appear to change the differentiated properties of the tissue. We do not yet know, however, whether the strict coordination of cell growth and chromosome cycles is also retained in the polytene cycle, or whether the doubling of cell mass is the invariable result of each cycle of genome multiplication; we are particularly ignorant of the situation in high-ploid and polytene cells. But, it is known that polytene cells increase considerably in size with each nuclear reproduction cycle. When polyploidy and polyteny occur in many cells, the overall mass of the tissue increases.

In mouse liver (Fig. 7.19) rapid division of diploid hepatocytes occurs only in the early postnatal life of animals. Later, cell multiplication continues at a much slower rate. The number of diploid hepatocytes decreases sharply, owing to their polyploidization. The weight of the liver increases roughly tenfold from early postnatal life to old age in the mouse, while hepatocyte number changes to a lesser extent. According to our measurements, the liver of 2-week-old CBA/C57BL mice contains approximately 8×10^7 hepatocytes. This value increases to 1.2×10^8 in 3-month-old mice and then hardly changes at all. Thus, the total number of hepatocytes increases only 1.5-fold. The postnatal growth of the mouse liver correlates with hepatocyte ploidy, rather than with number.

Polyploidy and hypertrophy are the main factors in the postnatal growth of the pigment epithelium of retina (see Chapter 2C) and myocardium. In mice, approximately 75% of the final number of ventricular cardiomyocytes appear before birth. Then myocardium mass doubles by means of cell polyploidization. This is followed by extensive mitosis-independent growth, i.e. hypertrophy of the cytoplasm.

A series of incomplete mitoses results in growth of the decidual tissue in the uterine endometrium (Chapter 2G). And, by means of polyploidizing mitoses, the epithelial surface of starfish stomach (Chapter 3E) is formed, as is the ascarid intestine and uterus (Chapter 3G). The overall volume of the ascarid uterine epithelium increases in proportion to cell growth (while the cell number actually decreases because of a certain amount of cell death).

Polyploidization, as a mode of active protoplasmic growth, is the equivalent of cell multiplication in regenerative as well as in normal tissue growth. Various definitions of regeneration exist; the essence of regeneration, however, is the restoration of what has been lost. When cells are lost, the functions of the affected organ are undermined. The essential thing is the restoration of these functions; restoration of the original organ mass or cell number is of less importance. Regeneration which restores the appropriate level of function to the damaged organ, may be achieved in two ways: by reproduction of the remaining cells, or by reproduction of the remaining genomes in non-dividing cells. In both cases, the number of genes needed for specific syntheses is restored and the activity of the tissue is thereby normalized.

After injury to the mouse liver, the ploidy of the hepatocytes increases. As a result, both the mass and also the activity of the organ is restored. After three successive partial hepatectomies, the average ploidy of the hepatocytes in 5–6-month-old mice reaches the level typical of 2-year-old animals (Fig. 7.20). In the latter, the mean ploidy often corresponds to 5–6C per nucleus and 9–10C per cell according to our measurements. The same effect is brought about by eight injections (at monthly intervals) of carbon tetrachloride.

Tissue growth is particularly pronounced where the component cells are polytene. In polytenization, organ weight may increase dozens or hundreds of times while cell number remains constant. A classical example of this growth is the dipteran salivary glands. It is hard to imagine the development

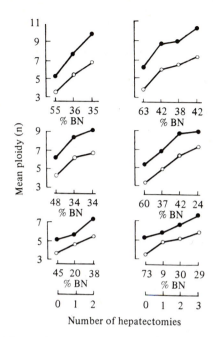

Fig. 7.20. Changes in mean ploidy (n) per nucleus (open circles) and per hepatocyte (filled circles) 1 month after the first (1), second (2) and third (3) partial hepatectomies. 0, control (data obtained from the lobe ablated by the first operation). Each pair of the curves represents hepatocytes from the same mouse. The numbers (% BN) below each graph record the per cent of binucleate cells (after data of Faktor & Uryvaeva, 1975).

of this organ by means of cell multiplication. The salivary gland of *Chironomus thummi*, for instance, is formed by approximately 32 cells each with a mean ploidy of 4000C, i.e. the gland's total ploidy is 128 000C. The same total genome could produce a tissue consisting of 64 000 normal diploid cells. This diploid organ would be considerably larger than the polytene one (owing partly to additional intercellular spaces) but, more significantly, it would need a larger support system of trophic cells (stromal elements, blood vessels, etc.). One may add that this hypothetical organ would form much more slowly than the actual polytene organ. Besides polytene tissues of Diptera, many other examples of similar growth have been found (see Chapter 3).

The mechanisms controlling the growth that accompanies somatic and generative polyploidy are not always comparable. In the former case, the number of genomes in the tissue is apparently regulated. In the latter case, where all the cells are polyploid from the outset, the regulations may be more complicated, but also include the stabilization, in the soma, of the number of genomes. Thus, differences in the weight of diploid, triploid and tetraploid caterpillars of *Bombyx mori* become less marked with age. A larger number of cells are polytenized in the diploid form than in polyploid ones (organs of the diploids are formed by a larger number of cells). As a result, the total quantity of DNA in caterpillars of different ploidy (and, similarly, in their silk glands) evens out (Klimenko, 1971). The same is true of triploid chickens, which have 1.5 times more DNA per single erythrocyte but a decreased number of erythrocytes in the blood, compared to a diploid. All in all, the concentration of hemoglobin is the same in the diploids and triploids (Abdel-Hameed, 1972).

Cell growth in the absence of nuclear reproduction

One component of postnatal growth of hepatocytes and cardiomyocytes is polyploidization; another component is the growth of cells that have ceased their polyploidization.

Hepatocytes continue to grow without any change in the number of the chromosomes (James *et al.*, 1979; Shalakhmetova *et al.*, 1981*a*). From the second week after birth, up to 2–3 months of age, the protein content in the diploid and tetraploid hepatocytes increases by 30–50%. This rate of growth is not on the same scale as that found during subsequent polyploidization. Here, after achieving a 16n level, the hepatocyte increases its mass eight-fold (equivalent to three doublings). Likewise, polyploidy and cytoplasmic hypertrophy make different contributions to the growth processes of cardiomyocytes; but, in this case, the relations are the converse of those in hepatocytes.

Most myocytes in the heart ventricles of new-born mice are small, diploid cells. Then, in just 1 week, approximately 80% of these cells become polyploid. The predominant group of ventricular cardiomyocytes in mice of any age (beyond 7 days after birth) is always $2C \times 2$ cells; $4C$, $4C \times 2$, $8C$, $8C \times 2$, and $16C$ myocytes are also found in small numbers in the ventricles (see Table 2.8). Mitoses occur at a sharply reduced rate by 2–3 weeks after birth in mice. The overall number of cardiomyocytes increases by 15–20% from birth to day 3; from day 3 to 21 it increases by another 5–7% and then does not change until the animal reaches 1 year of age. Cell size and mass increase considerably in a population that is stable in cell number and in genome weight (Figs. 7.21 and 7.22).

All the cells in the ventricles grow, both the polyploid and the few remaining diploid ones (approximately 2–4% of myocytes are $2C$, even in 1-year-old mice). It is noteworthy that, in 1-month-old animals, the ratio of the masses of cardiomyocytes of different classes – $2C$, $4C$, $8C$, and $16C$ – is not a multiple of two (Fig. 7.18). The cells with the same total genome (for example $8C$, $4C \times 2$, and $2C \times 4$) grow at a similar rate. But the masses of the cells with a different DNA content (for example $4C$ and $8C$) are not doubled after each DNA replication. Instead of the expected

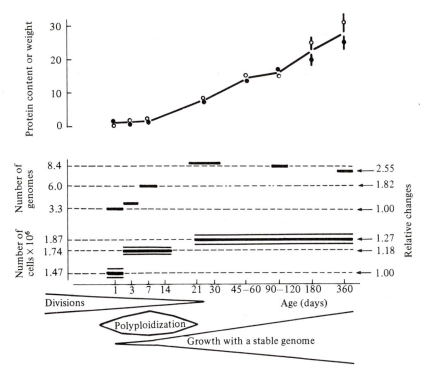

Fig. 7.21. Changes in the number of the heart myocytes, their ploidy, mean protein content per myocyte (arbitrary units, open circles) and weight of heart ventricles (filled circles) during postnatal growth of the mouse (after Brodsky *et al.*, 1984).

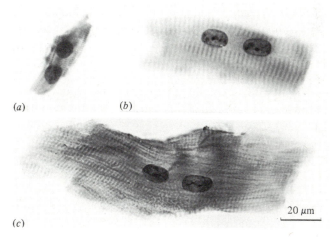

Fig. 7.22. Binucleate cardiomyocytes from the heart ventricles of (*a*) a 1-day-old, (*b*) a 1-month-old, (*c*) a 1-year-old mouse, each photograph at the same magnification.

proportions of the protein content in polyploidizing cardiomyocytes (2:4:8:16), ratios of 2:3.4:5.0:6.3 are found. These average ratios hardly changed from the ages of 1 month to 1 year, although the absolute cell masses increase several-fold during that time. Thus, the degree of growth of polyploid cardiomyocytes is less than that of diploid cardiomyocytes. Non-cycling tetraploid cells fall short of the expected mass by 15%, the octaploid cells by approximately 40%, and the hexadecaploid cells by 60% (on average).

This feature of the growth of cardiomyocytes with various ploidy levels may explain part of the advantage of $2C \times 2$ cells over other classes of cardiomyocytes. For an elongated and isomorphous cell such as the cardiomyocyte, binuclearity would be a most useful property. Perhaps it is not accidental that the nuclei of cardiomyocytes are found far away from one another, located near each end of the cell where they can 'control', or maintain the surrounding cytoplasm. Theoretically, multinuclearity would be even more advantageous. However, there are few ($< 0.5\%$) tetranucleate cardiomyocytes in the murine heart ventricles; binucleate ($2C \times 2$) cells comprise the majority (80–85%) of all cells. This may be explained by the fact that the octaploid cells (including those with $2C \times 4$ nuclei) grow more slowly than the tetraploid ones (such as $2C \times 2$). The latter cells grow at almost the same rate as the diploid cells. But, at the same time, the tetraploid cell is larger than the diploid at all stages of growth. Possibly, accumulation of low-ploid, binucleate cells occurs in the mouse myocardium to optimize the amount of cell growth possible and to couple it with the functional advantages of spatially separated genomes.

The accumulation of binucleate ($2C \times 2$) cells is also a characteristic of the rat, dog and guinea pig myocardia. Many tetranucleate cells ($2C \times 4$) have been found in the pig and cow myocardia; human heart ventricles contain many mononucleate, high-ploid cells. It would be interesting to learn whether human or pig myocytes differ from those of mouse in their growth rate.

The growth of cells with a constant genome is typical not only of normal ontogenesis of the myocardium. During hyperfunction of the cardiac muscle, myocyte size and mass increases more than during normal development. This has been observed, for example, in animals in high-altitude, hypoxic conditions (Arefyeva *et al.*, 1982), after coronary-artery ligation, and after infarction of the left ventricle in rats (Oberpriller *et al.*, 1983).

It may be inferred from the examples examined above that the growth of hepatocytes and cardiac myocytes occurs in parallel with the growth of the liver and heart, and continues while the whole organism is growing. Mammalian neurons are the other example of fully differentiated cells which have a stable genome and possess a great capacity for cytoplasmic enlargement. Mammalian neurons are, however, diploid cells. In the hepatocyte and cardiomyocyte populations, the processes of hypertrophy of the cytoplasm are concomitant with polyploidization.

Apparently, the growth of a cell with a constant genome is characteristic of stable and expanding (i.e. non-renewing) cell populations, not only in mammalian, but also in other animal and plant tissues.

D The resistance of the polyploid genome to injury

In this section, attention is paid to an important feature of the polyploid genome – the redundancy of the genetic material. Duplication of an information system is extensively used in communication technology; it is the simplest means of improving reliability. Theoretically, it may be expected that the genetic defects of different types (such as gene mutations or losses of genetic material) should be compensated by the genetic redundancy of the polyploid genome.

The general idea of polyploid-genome resistance was confirmed by comparing the mutation rate and the damage caused by effective mutagens and ionizing radiation on diploid and polyploid species of plants and animals. As the number of chromosome sets in a polyploid series of *Avena* and *Triticum* species increases, there is a definite decrease in the yield of induced mutations (Stadler, 1929). The resistance of the embryo of the silk worm, *Bombyx mori*, to radiation damage becomes greater with increase in ploidy level (Tultzeva & Astaurov, 1958). Autotetraploid seeds of buckwheat show a greater stability to gamma radiation than diploids (Sakharov *et al.*, 1960). In the two latter works cited, interesting information is also provided on the increased resistance to X-irradiation of tetraploid forms of yeast compared with diploid forms, and of diploid forms compared with haploid ones. Further experiments carried out on plants showed that polyploids are, in general, more radioresistant than diploids (Wangenheim, 1970; Raghuvanshi, 1978). However, radioresistance in plants does not always increase with ploidy level; this is associated with the influence of other important parameters such as the interphase-chromosome volume, nucleolar volume etc. (Boyle, 1968; Bhadra, Shaikh & Mia, 1979). Polyploid species of plants are also more resistant than diploid ones to the lethal and mutagenic effect of alkylating agents (Simon, 1965; Zutschi & Kaul, 1975).

In the following works, differences were noted in the radiosensitivity of cell populations of common origin, but differing in their degree of ploidy. Revesz & Norman (1960) compared two sublines of Ehrlich ascites tumors in this respect, a hyperdiploid and a subline with a doubled hypertetraploid number of chromosomes growing in diffusion chambers. X-irradiation completely arrested the growth of hyperdiploid cells. Hypertetraploid cells, treated with the same dose, continue to grow in number. Bauer (1974) obtained a stable hypotetraploid subline of Chinese hamster cells with the help of prolonged treatment with colcemid. The dose of X-rays to this subline had to be increased 1.3-fold in order to obtain the same degree of cell killing (measured as the percentage of surviving cells) as in the parent diploid line.

Data on the influence of ploidy on radioresistance of tumors *in vivo* are of special interest. Hyperploidy, which frequently accompanies neoplastic transformation (see section F, of this chapter) may form the basis for the resistance of tumors to radiotherapy and the lack of success in treating them with cytostatics (De, 1961; Berry, 1963; Benedict & Jones, 1979).

The basis of the resistance of polyploid cells to ionizing radiation may be due to either a greater tolerance of damage as well as an increased survival capacity. There can be no doubt that the number of sites for primary damage (in the form of breaks, cross-links, and other disturbances in the DNA) in the genome of polyploid cells is no less than in diploid cells; it is most likely to be proportional to the number of chromosomes. No differences were found in the radiation-induced division delay between a pseudotetraploid hamster cell line and the pseudodiploid parent strain (Kimler, Leeper & Schneiderman, 1981). Repair processes are believed to take place during this period of the division delay.

The most frequent result of radiation damage to the cell is reproductive death – i.e. death in the course of mitosis or after mitosis. Therefore, the effect of ploidy on radiosensitivity is only manifest in proliferating cell systems. The repair (or lack of repair) of damaged DNA may not substantially influence the viability of a cell while it is not proliferating. But the recruitment of cells into the mitotic cycle, and subsequent mitosis, reveal latent alterations in the genetic material which manifest as chromosomal aberrations. Chromosomal aberrations are the main cause of reproductive death of cells (Tremp, 1981) since they lead to losses, during mitosis, of parts of chromosomes and the formation of unbalanced, incomplete genomes in the daughter cells. Polyploid cells, which have multiple chromosome homologs, are protected from the consequences of aberrant mitosis. In diploid cells, one of the structural aberrations – the dicentric or fragment type – has a high probability of being lethal (Carrano, 1973). Polyploid cells are more capable of surviving with an unbalanced, incomplete genome; genetic lesions become manifest later on in the form of functional deterioration or malignant transformation.

The mechanisms of the protective effect of polyploidy with respect to aberrant mitosis has been examined in liver cells (Uryvaeva, 1981). In mammals, the liver is a highly radioresistant organ. This resistance, which extends to any mutagenic agent, is not only manifest in resting cells but also in growing or regenerating liver. In mice and rats, treatment with ionizing radiation or alkylating agents was followed by a partial hepatectomy (Albert & Bucher, 1960; Faktor *et al.*, 1980). After delayed progression

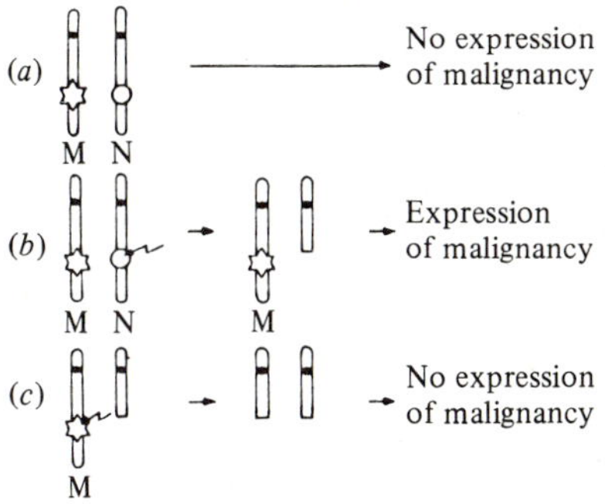

Diploid cell carrier of recessive
malignant genotype

Fig. 7.23. Schematic representation for induction by mutation of malignancy in a diploid cell with a recessive malignant genotype. In (a), one homolog of a chromosome pair carries a malignant factor or allele (M and star), the other carries a normal controlling allele (N and circle) that suppresses malignancy. In (b), mutation (jagged arrow) causes loss of the normal controlling allele and malignancy is expressed. In (c), the controlling allele (N) has been lost and mutation (jagged arrow) causes loss of the malignant allele (M); this results in no malignancy being expressed (after Chadwick & Leenhouts, 1981).

of the cells through the cell cycle and prolonged arrest in G_2, the hepatocytes entered mitosis. Practically all the hepatocytes underwent abberant mitosis, with chromosome fragments and bridges. A month after the combined treatment with a mutagen and the stimulation of the mitosis, the liver consisted exclusively of polyploid cells. Many of them (sometimes as many as 35%) bore traces of chromosomal aberration in the form of nuclear bridges and micronuclei, but preserved their viability and evidently their functional activity. The aberrant diploid cells were eliminated, although individual, apparently non-cycling, diploid hepatocytes remained (Uryvaeva & Faktor, 1982). These experiments demonstrate fairly well the part played by polyploidy in the functional compensation for the loss of chromosome parts and in the maintenance of cellular viability despite an unbalanced genome.

An understanding of the nature of the protective effect contributed by the 'redundant' genes of the polyploid cell is obtained by examining experiments on mutagenesis in polyploid cell lines cultivated *in vitro*. The mutations may be dominant or recessive. Dominant mutations should be expected to appear as frequently in the diploid as in the polyploid lines (or even more often in the latter). If recessive mutation has occurred, phenotypic expression requires the mutation to take place in all the chromosome sets. If the conventional mutation rate in eucaryotic cells is 10^{-5}–10^{-7} for a single gene per cell generation, then for two alleles of the

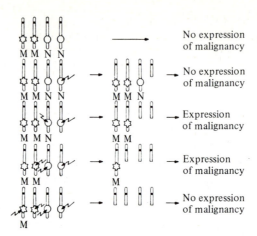

Fig. 7.24. Schematic representation for induction of malignancy in a tetraploid cell with a recessive malignant genotype. Same general scheme as in Fig. 7.23 except that, because the cell is tetraploid, each malignant allele (M and star) and each normal controlling allele (N and circle) is expressed twice (after Chadwick & Leenhouts, 1981).

diploid cell it is 10^{-10}–10^{-14}. In tetraploid cells, the probability of four independent mutation events at the same locus is 10^{-20}–10^{-28} making the expression of such a mutation most improbable. In Figs 7.23 and 7.24 malignancy is presented as an example of a genetically recessive trait, to illustrate the part played by the additional number of controlling-gene alleles in the tetraploid cell in the expression of the malignant phenotype.

Some papers may be cited to confirm the calculations for the mutation rates given above. McBurney & Whitmore (1974) studied the temperature-sensitive (*ts*) variants in the near-diploid Chinese hamster ovary cell line and in the tetraploid line derived from it (by treatment with colcemid). The *ts*-mutation is recessive to the wild-type allele since the *ts*-mutants form non-ts hybrids with other cells. Although *ts*-mutants were found in the diploid line, attempts to isolate such a mutant in the tetraploid line were unsuccessful. Furthermore, a recessive mutation for resistance to 6-thioguanine was 25 times less frequent in tetraploid than in diploid Chinese hamster ovary cell lines. However, the ploidy effect is absent with respect to the frequency of a mutation from glycine auxotrophy to glycine independence, as would be expected for a dominant mutation (Chasin, 1973).

In the above-mentioned example, the frequency of mutants resistant to 6-thioguanine, did not correspond to the calculated probability of two independent mutation events. It is suggested that gene mutations and other non-mutational events (such as mitotic recombination and chromosomal changes) affect the frequency of recessive traits for drug resistance.

Moreover, during lengthy cultivation, and in repeated passages, the initially induced tetraploid lines may be transformed into hypotetraploid ones, possibly losing thereby a functionally important chromosome.

In a number of studies the effect of ploidy on mutation rate was not shown. Harris (1971b) found a spontaneous mutation rate of 10^{-6} per cell generation for resistance to heat shock and 10^{-5} for resistance to 8-azaguanine in diploid, tetraploid and octaploid substrains of Chinese hamster cells.

The frequency of the mutant variants does not differ substantially between haploid and diploid cell lines of the frog (Mezger-Freed, 1977). It is suggested that, in these cases, it is not gene mutations but rather epigenetic events which are the origin of spontaneous phenotypic variants.

If we proceed from conventional genetic concepts, then polyploidy may be a genetic-defence measure against malignant transformation (Strunnikov, Uryvaeva & Brodsky, 1982). This idea comes from a theory of carcinogenesis, according to which cancer appears as a consequence of a change in a specific recessive gene controlling the neoplastic transformation. Malignant transformation of a diploid cell is thought to be the result of at least two mutation events. Theoretically, if the tumor phenotype is to be expressed in a tetraploid cell, four independent mutations to the allele of the critical gene would be necessary. The low probability of realizing these four independent events in a single cell, or its descendants, makes the malignant transformation of a tetraploid cell improbable.

Knudson (1971) was one of the first to suggest the two-phase, two-mutational model of cancer. Other workers later upheld the concept of recessiveness in tumor genes on the basis of data obtained in other fields of oncology and biology (review Stich & Acton, 1979; Chadwick & Leenhouts, 1981; Kinsella, 1980; Strunnikov et al., 1982). It should be emphasized that many of the authors of two-mutational models have postulated the existence of special genes in the cell which are responsible both for its normal growth and development and also for the emergence of the tumorous potentiality. A fundamental similarity in the mechanisms of normal differentiation and carcinogenesis was thereby proposed.

These hypothetical speculations were well confirmed recently by the discovery in vertebrate cells (including human cells) of oncogenes – a DNA sequence homologous to genes of oncoviruses. It is suggested that oncogenes are normally tissue-specific regulators of cell growth and differentiation. Malignant transformation of a cell is precipitated by intensified expression of these genes; this may, in turn, be the result of infection by oncogenic viruses, of amplification of the oncogene, or of any process leading to an increase in the quantity of their transcription product (see reviews: Astrin & Rothberg, 1983; Weinberg, 1983).

Experiments where the malignant phenotype of tumors is transmitted to normal murine cells by transfection with tumor DNA show that the oncogenes manifest themselves as dominant genes (Cooper, 1982). On the other hand, in cell-fusion experiments, the malignancy of the tumorous human cells is suppressed by hybridization with normal cells. One can suppose on the basis of these results that, in a tetraploid hybrid cell, there is compensatory influence of the additional normal alleles (Sabin, 1981).

As can be seen, at the present time there are two groups of observations, some testifying to the dominance of tumorous genes and others to its recessiveness. Nevertheless, the concepts resulting therefrom are hardly likely to be mutually exclusive. It is sufficient, for example, to allow that there is a negative feedback control over the oncogenes or their products by means of recessive-gene regulators (as is done in the model described by Stich & Acton (1979). It is possible that the dominant genetic alteration is only one of several events involved in carcinogenesis.

Further, the question should be posed whether there are any clinical observations or experimental data which testify to a protective effect of polyploidy against malignancy. There are more specific questions related to this problem. Do tumors emerge in tissues composed of diploid cells as frequently as they do in other tissues with polyploid cells? Which cells – diploid or polyploid – are the target of experimental carcinogenesis and are the source of tumors in tissues with pronounced polyploidization (for example, in the liver of rodents)?

Clinical oncology points to the fact that, in humans, virtually all types of tumors occur in diploid tissues which do not contain many polyploid cells. Cell populations such as cardiomyocytes, cells of the pigmented retina, hepatocytes, and megakaryocytes of bone marrow are very rarely the site of tumors. However, in these cell populations there are either very few polyploid cells (as in human liver), or they make up the entire population (e.g. mature megakaryocytes). It is hardly likely that the lack of tumorigenesis in these latter tissues can be attributed solely to the protective effect of polyploidy itself. It is more likely that, besides polyploidy, other properties of the cell population are of great importance in carcinogenesis, such as the incidence of cell proliferation and DNA-synthesis relative to the numbers of immature stem cells and those undergoing differentiation, and the level of cell renewal. We shall return to some of these questions in section F of this chapter.

E The advantages of polyploidy and polyteny

One of the main consequences of genome multiplication is the increase in cell size. In some tissues cell growth may be advantageous and play a decisive part in establishing and fixing the level of polyploidy. Polyploidy is evidently the result of the evolution of cells of an especially large size (Levi *et al.*, 1969a). A large cell size may be useful in organ function where great mechanical tension or pressure is experienced. This was suggested for the retina pigment epithelium (see Chapter 2C). Large cells, which continue to grow even after polyploidization, may also favor the functioning of the heart.

Cell giantism is not merely beneficial to the epithelium of the urinary bladder, but also appears to be necessary for its normal functioning. In mammals, the epithelial layer is covered with a thick impermeable membrane which serves as a barrier between the tissue fluid and the urine. The membrane may be formed only by superficial, large, polyploid cells and not by diploid ones (even following their pathological hyperplasia). Fragments of the membrane are synthesized as thickened areas in the membranes of the Golgi apparatus. Evidently, increased cytoplasmic volume is needed both for the development of the powerful Golgi apparatus, as well as for the storage of reserves of the membrane components (these are included in the composition of the superficial lamina) which are required to increase membrane area when the urinary bladder is stretched (Hicks, 1975).

An extraordinary example of cell giantism occurs in some neurons of opisthobranchian molluscs (Chapter 3F). Unlike other highly polytene cells, which live for a short time, the giant neurons live for years and continue endocycling into old age. Some of these cells reach 1 mm in diameter. The multifunctionality of these neurons has been noted. As a matter of fact, one such neuron may function as a whole ganglion. Multiplication of the genome may have a particular meaning in this case.

A giant polytene cell may regulate the activity of surrounding small diploid cells (Barlow, 1978*b*). This model is particularly apt in nerve tissue. Thus, the giant neurons of molluscs may make contact with a huge number of glial cells and influence their activity.

The role of polyploidy in the regular distribution of structural elements of an organ is manifest in the development of the butterfly wing (Chapter 7A). Three types of scales, differing in size and position on the wing surface, are outgrowths of epidermal cells of different ploidy. Cell giantism is necessary for the expression of the organ's function. The specific mode of participation by the cells (scales) in wing construction is determined by the genome according to whether they are 8, 16 or 32n.

The functional advantages of polyploidy can also be seen in the ciliated cells of the starfish stomach (see Chapter 3E). In species which do not have special pump organs, the density of cilia on the superficial epithelial cells of the stomach increases. Cilia number is proportional to cell ploidy.

The small number of cells in a polyploid tissue compared with a diploid one of the same overall activity may, in itself, be an advantage. The smaller number of working elements (i.e. the reduction in surface contacts, the number of nerve endings, receptors, etc.) may facilitate the regulation of function in the polyploid organ. This may be the special advantage of polyploidy for the myocardium.

Polyploidization of the macronucleus of ciliates and the nuclei of other protozoans is regarded as an important factor in their evolution (Polyansky, 1972). Polyploidy ensures the functioning of the giant cell by determining a sufficient volume of cytoplasm in which may be gathered all the specialized organelles necessary for a unicellular organism.

Yet another consequence of polyploidy and, especially, of polyteny is the lengthening of the active life of the cell owing to the reduction in, or even the complete omission of, mitosis. Polytene chromosomes are permanently in interphase and are consequently always active. The growth of the cell by polyteny is economical in other respects, too. In the polytene cycles, the nuclear envelope and the nucleolus are not destroyed and then reconstructed, no spindle is assembled, nor do the chromosomes condense and then decondense. Accordingly, many structural and regulatory proteins associated with mitosis do not need to be synthesized and so less energy is expended than if tissue growth is accomplished by cell proliferation. The great potential of the polytene cell can then be directed towards cellular growth and the production of huge quantities of substances which are needed by the organism.

Interesting examples of the advantages of a curtailed cell cycle for the acceleration of growth have been examined by Barlow (1978*b*). At certain stages of plant development, the suspensor has to grow much faster than the embryo. The most economical use of resources is achieved by the transition of the suspensor cells to a polytene cycle; this ensures its development within a short period. Similarly, the antipodal cells are rapidly polytenized and accumulate large quantities of ribosomes and other structures. During the subsequent degeneration of these cells, these products promote endosperm development. The degeneration products of highly polytene nurse cells of the ovary of insects are used by the oocyte for its own growth; earlier the oocyte depends on the synthetic activity of these nurse cells.

Proposals regarding the advantages of enlarged cell size and shortened cycle are inapplicable to many cases of polyploidy. For example, relative to their size, giant hepatocytes have a decreased surface area and so are obviously less fit to perform liver functions than are smaller diploid cells. But, for the mammalian liver, other advantages of cell polyploidization may outweigh this seeming disadvantage.

The switch from growth by cell multiplication in the liver to growth by polyploidization may be explained by the high level of spontaneous damage to hepatocyte chromosomes (Fig. 7.25). A model for the biological significance of hepatocyte polyploidy (Uryvaeva, 1979, 1981) is based on this and on the theory, already examined, concerning the resistance of the polyploid genome to injury. Structural rearrangements of chromosomes which are incompatible with mitosis are characteristic of hepatocytes (Curtis, 1963). During division of these cells, anaphase bridges and

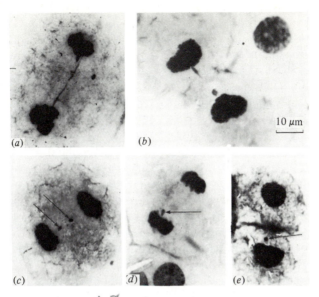

Fig. 7.25. Chromosome abberations in mitotic hepatocytes from regenerating mouse liver. (*a,b*) Chromosomal bridges; (*c,d*) lagging chromosomes or acentric chromosome fragments (arrows); (*e*) micronucleus formed from a chromosomal fragment in a postmitotic hepatocyte (arrow) (from a preparation of isolated cells).

chromosome fragments are formed and unequal division, with loss or gain of genetic material in the daughter cells, occurs. These aberrant diploid cells usually die and are eliminated from the population. The hepatocyte population does not have a special reserve of cambial cells from which to restore the lost cells. Growth and regeneration of the liver in mice or rats is accomplished by repeated reproduction cycles of differentiated cells. Polyploidizing mitoses, which conclude the cycles, do not permit the defective daughter genomes to separate. According to this hypothesis, polyploidization of the hepatocytes offers protection against the functional defects resulting from aberrant-genome formation.

The role played by DNA lesions in hepatocyte polyploidization is taken into account by Nakanishi & Fujita (1977). These authors believe that polyploidization is caused by cross-links between the double strands of DNA. If the number of cross-links is small or if they are weak, they break and do not interfere with the normal division of the daughter nuclei during mitosis. If the number of cross-links is large, they may hinder the separation of the daughter nuclei and, as a result, binucleate cells are formed with linked nuclei. A mononucleate polyploid cell is formed when the cross-links impede separation of the sister strands.

It is hardly likely that the real mechanisms of formation of binucleate and polyploid hepatocytes are similar to the scheme described by Nakanishi & Fujita. It is particularly important to emphasize that, in every hepatocyte mitosis, separation of sister chromatids has to occur and genuinely polyploid cells with an increased number of chromosomes are formed.

The thesis that ontogenetic polyploidization of hepatocytes is connected with genome damage has allowed us, for the first time, to imagine the reasons for species-specific differences in liver polyploidy. As frequently noted, in mice and rats the hepatocytes are of especially high ploidy and they are particularly numerous; in the liver of guinea pig there are far fewer polyploid cells and, in the dog, fewer still. It appears that these species of animals differ in the frequency of spontaneous chromosome aberrations in the liver and that this corresponds directly to the level of polyploidy. Species differences in mutation rate and in the frequency of the chromosomal aberration may be explained by some metabolic peculiarities responsible for the DNA-damaging effect. For example, the level of activity of the repair enzyme uracil-DNA-glycosylase (UDG) in the liver of various species of mammals (see Table 2.2) was examined recently (Kudryavtsev *et al.*, 1984). A direct correlation was observed between the mean level of cell ploidy characteristic for a given species and the activity of UDG per μg of DNA. As replicative DNA synthesis is negligible in livers of adult mammals, these data represent the increased activity of antimutagenic systems of DNA repair in polyploid liver. The meaning of these findings is not yet completely clear.

It is difficult to say whether there are other tissues in which chromosome aberrations are responsible for cell polyploidization. Anaphase and telophase bridges and fragments are typical, for example, of dividing cells of the trophoblast. We can also recall the dumbbell-like nuclei in the cells of the mesothelium and the bladder transitional epithelium. These forms are sometimes interpreted as amitosis, although they are more likely to be the consequence of chromosomal aberrations. The rare dumbbell-like nuclei in the liver are due to chromosome bridges which persist into interphase, as mentioned by Wilson & Leduc (1948). Polyploidization participates in the process of growth and development as a means of protecting other cell populations where these consist (like hepatocytes) of long-lived and slowly proliferating cells. Polyploidy ensures the preservation of cells with aberrant chromosomes in the population. This goal may also be achieved by uniting the genomes by cell fusion and the formation of a symplasm (see Chapter 6E). The tissue functions of such multinucleate elements in tissues such as osteoclasts, giant cells of foreign bodies, and cells of the syncytiotrophoblast are, at present, unclear (Fischman, 1979). Genetic protection should also be considered among the reasons for their formation. Frequent chromosome anomalies have been discovered in fusion macrophages (Mariano & Spector, 1974) and in mesothelial cells, creating giant multinucleate cells (Dreher *et al.*, 1978). In the opinion of these authors, the result of fusion is the integration of genetically defective cells into the symplasm and the neutralization thereby of potential sources of malignancy.

Thus, the biological significance of genome multiplication is determined by the particular properties of polyploid and polytene cells and their functional advantages. However, firm data relating to these phenomena in many plant and animal tissues is still wanting. According to Barlow (1978*b*), polyploidy may be regarded as a stratagem of evolution. It is possible that polyploidy in some tissues is, at present, a neutral feature without advantages or disadvantages. Under conditions of increased environmental pressure, the properties of polyploid cells may become important to the success or even the survival of the species.

F Polyploid cells in tumors

Quantitative changes in the genome of tumorous cells

It has long been established that many tumor cells possess chromosomal abnormalities; aneuploidy, polyploidy and an elevated DNA content are common. Hsu (1961) and Oberling & Bernhard (1961) published reviews on the chromosomes of tumor cells which covered much of the literature on this subject.

Most extensive studies of karyotype abnormalities of tumour cells have examined structural chromosomal changes using cytogenetic analysis, usually for the purpose of discovering the marker chromosomes specific to each type of tumor and so which may be concerned with the malignant transformation of the cell (see reviews by Rowley, 1983; Sandberg, 1983). Until recently, genome mutations resulting in polyploidy or aneuploidy did not attract so much attention. They were considered to be one of the numerous characteristics of cancerous cells also common to normal cells and therefore hardly likely to be connected specifically with malignancy.

Numerous new data regarding the heteroploidy of cancer cells were obtained by means of DNA cytophotometry which, unlike cytological karyotyping, allows a wide range of cells to be studied (including those which are not dividing). Studies of the DNA content in individual tumor cells during the last few years have been especially stimulated by the introduction of a high-speed method of flow fluorometry and the promise of its clinical application for diagnosing human tumors.

Malignant tumors are highly individual with respect to their genome composition. There may be great differences, not only between neoplasms from different sites and tissue origin but also in the range of their histological forms. The cells within a single tumor also form a heterogeneous population.

Results of quantitative DNA determination and analysis of chromosomes of cancer cells show pronounced aneuploidy of cells and a heightened DNA content in the overwhelming majority of human neoplasms and in experimental animals (Atkin, Mattinson & Baker, 1966; Böhm & Sandritter 1975; Avtandilov, 1982). According to Olszewski *et al.* (1981) and the literature cited therein, 52–92% of solid human malignant tumors are aneuploid. In the same work, 85 of the 92 cases of breast cancer studied showed evidence of hyperdiploidy as determined by flow cytometry. An abnormal nuclear DNA content is found in solid tumors from different localities in the body and in hematological cancers, with a frequency of 67% among 4941 patients (Barlogie *et al.*, 1983). Many highly malignant tumors do not deviate from the diploid DNA content; this is true, for example, of leukemia and malignant brain tumors. In the latter, 66% of cases showed euploid DNA quantities (Linden, 1982). In these cases also, an abnormal chromosomal constitution is the rule rather than the exception. DNA determination can reveal only obvious changes of chromosomal material. Low-level aneuploidy and some types of chromosomal rearrangements may not affect the DNA content.

Although the general tendency towards heteroploidy in tumor cells appears sufficiently clear, various tumors differ in their cellular DNA-content distribution profiles. Unimodal DNA distributions are cited with the peak around the hyperdiploid value (Barlogie *et al.*, 1978), near the triploid (Patek *et al.*, 1980) or, rarely, at the pentaploid value (Olszewski *et al.*, 1981). However, bimodal DNA distribution patterns are also found, with one peak around the near-diploid value and another in the triploid or hypotetraploid region. These DNA histograms have been obtained by densitometry of Feulgen-stained nuclei in smears of primary human tumors from many localities and of many different histological types (Atkin & Kay, 1979). A bimodal, and even a trimodal distribution, which spreads to the 8C DNA content level, was described for malignant breast tumors (Auer, Caspersson & Wallgren, 1980), cervical carcinoma (Jakobson, Bichel & Sell, 1979) and experimentally induced lung adenoma in animals (Digernes, 1981). For adenomas induced by urethane in mouse liver, a multimodal distribution of DNA is typical (Faktor & Uryvaeva, 1983).

Thanks to studies of biopsies and surgically removed tumorous material, it has been established that the diploid quantity of DNA is usually characteristic of both benign and highly differentiated tumors. Elevated DNA contents are characteristic of malignant tumors. The hyperdiploid DNA contents measured by cytophotometry evidently often reflect an increased fraction of cycling and DNA-synthesizing cells and point to enhanced proliferative activity, rather than to the existence of aneuploidy or polyploidy. Nevertheless, an atypical DNA distribution is an important criterion in the diagnosis and prognosis of tumorous disease in mass screening for cancer (Linden, 1982). In clinical applications aneuploidy and polyploidy of neoplastic cells is always taken as a sign of malignancy. Even during early preneoplastic changes, pronounced aneuploidy suggests the likelihood of subsequent development of malignancy (Sugar, Szentirmay & Decker, 1982).

While abnormal profiles of DNA distribution are of great practical significance, in this volume we shall confine our attention only to those cases of polyploidization which are observed during tumor progression. Recently, polyploidization was shown to occur at the earliest stages of carcinogenesis, in the establishment of malignancy. Data on the evolution of tumor-cell karyotypes and on the interrelationships between polyploidy and malignancy deserve special attention since they touch on the question of the properties of the polyploid genome and facilitate an understanding of polyploidy as a general biological phenomenon.

Polyploidy as a sign of tumor progression and malignancy

Thanks to the fact that clinical material has been so thoroughly studied, it has become possible to compare DNA content with histological structure, the extent of tumor differentiation and the course of the disease. It has become clear that polyploidy in a tumor is a sign of particular malignancy.

The existence of a good correlation between the extent of tumor differentiation and the ploidy level of its cells has been recorded. Highly differentiated tumors have a diploid level of DNA, while tumors with a low level of differentiation frequently show a high level of ploidy. Among breast tumors, the well-differentiated colloid and tubular carcinomas are diploid or near-diploid. The non-estrogen-binding tumors are histologically more anaplastic and they may also be triploid, tetraploid and sometimes pentaploid (Olszewski *et al.*, 1981). The shift of the modal DNA values towards triploidy in the case of anaplastic breast carcinomas, has also been described by other authors (Patek *et al.*, 1980). The modal polyploid class represents the stem cell line for these types of tumors. The DNA distribution of cells in cervical cancer also correlates very well with the histological classification of the tumor according to the extent of their differentiation. As a rule, poorly differentiated cancers contain tetraploid and hypertetraploid cell populations (Jakobson *et al.*, 1979).

Among the cartilaginous tumors, all the benign chondromas are diploid, but the chondrosarcomas are both diploid and hyperploid; the latter are especially difficult to treat (Cuvelier & Roels, 1979; Kreicbergs, Zetterberg & Söderberg, 1980).

It has been shown by comparing laryngeal tumors that the benign papillomas differ very little from the normal diploid larynx mucosa with respect to DNA content in the individual cells. Intraepithelial cancers of

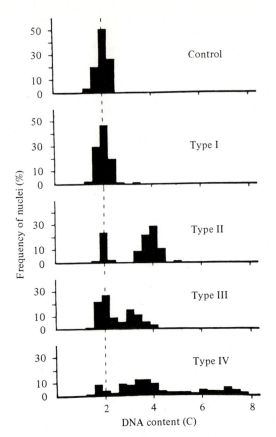

Fig. 7.26. DNA distribution patterns, determined by Feulgen cytophotometry, from non-proliferating normal mammary epithelial cells (control) and from a series of mammary carcinomas – Types I–IV (after Auer *et al.*, 1980).

the larynx, especially with metastases to regional lymph nodes, exhibit a significantly elevated DNA content, indicating aneuploidy (Avtandilov *et al.*, 1975).

Experimental tumors in animals show a tendency to polyploidization in parallel with the development of malignancy. Chemically induced lung tumors in mice differ in DNA distribution and in ploidy of the stem cell line. Benign adenomas have a diploid unimodal distribution, papillary cancers have a near-tetraploid stem cell line in addition to a diploid line (Digernes, 1981).

There is an interesting description of chromosomal evolution during development of sublines of spontaneous adenocarcinoma of the rat prostate. The original parent tumor, with a near-diploid karyotype, was well-differentiated, slow-growing and androgen-sensitive. After a few

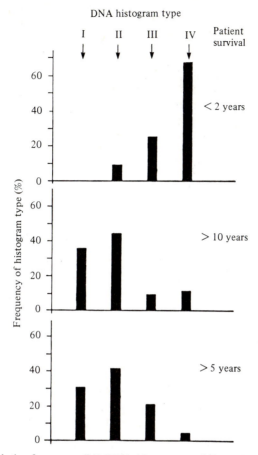

Fig. 7.27. Relative frequency (%) DNA histograms of Types I–IV (see Fig. 7.26) in three groups of patients with breast cancer, each group showing a different survival time following diagnosis (after Auer *et al.*, 1980).

passages *in vivo*, androgen-insensitive tumors with modal tetraploid chromosome numbers were formed. The tumors later acquired the ability to metastasize in parallel with the development of hypotetraploidy by means of loss of particular chromosomes (Wake, Isaacs & Sandberg, 1982).

Polyploid tumors present the least-promising prognosis. A retrospective analysis of material from patients with mammary carcinoma showed that the survival of the affected patients (following diagnosis) depends on the DNA distribution pattern (Figs 7.26, 7.27). The most favorable forecasts correlate with distribution patterns consisting of a single diploid class, and also with bimodal patterns having well-defined diploid and tetraploid peaks. Distributions revealing pronounced aneuploidy and high DNA values (up to 8C) carry unfavorable prognoses (Auer, Caspersson &

Wallgren, 1980). The formation of a polyploid stem line in the tumor determines the least promising prognosis for carcinomas of other organs, too (Tavares *et al.*, 1966; Atkin, 1976).

It is thus concluded that malignant tumors may be diploid and polyploid, with aneuploidy of one of the modal classes. A clear-cut impression is created that polyploid tumors are more malignant than diploid ones. This means that they are more invasive, grow more rapidly, metastasize more readily, and that their cells possess greater viability and adaptability to the unusual conditions which they create in their micro-environment.

The connection between acquisition of highly malignant properties and polyploidization is revealed in the discussion of clonal evolution of tumorous cells (Nowell, 1976). According to Nowell, tumors (which originate from a single cell and consequently consist of genetically identical cells right from the start) undergo a permanent developmental change in their genetic properties. Genetically unstable cells provide material for the selection of subpopulations with properties of heightened malignancy. Tumors with a diploid DNA content, or with minimal aneuploidy, may be regarded as having not yet undergone complete clonal evolution and having not yet acquired high malignancy. Heteroploid cells are a product of a multistage selection process. The cells which exhibit the highest growth rate and the greatest adaptability to any unfavorable conditions of the medium become the dominant and most malignant population.

Elevated ploidy levels may play a positive role in tumor progression and in the acquisition of great aggressiveness and the ability to metastasize. With the help of chromosome analysis, changes were traced in the karyotype of ovarian carcinoma cells, taken not only from their site of origin but also from secondary sites of metastases in the same patient. In the primary site, near-diploid cells predominated. In the metastases, near-tetraploid cells comprised 34% of all metaphases; in the ascitic tumour, the figure rose to 66% (Kusyk *et al.*, 1981). These data indicate that the near-tetraploid cells are better able than the near-diploid cells to grow in places which are unusual for them. Polyploidy in the metastases has also been discovered in other tumors, although it does not always occur (see Kusyk *et al.*, 1981).

The progression of a bladder cancer, also studied with the help of the chromosome analysis of serial biopsies taken in the same patient, was accompanied by an increase in ploidy level and by the appearance of new chromosomal markers (Sandberg, 1977).

The development of malignant features occurred parallel with the increase in DNA content in clones isolated from chemically induced murine fibrosarcomas. The clones were grown in a soft agar medium and then transplanted to syngeneic mice. The clones varied greatly in their properties but those in which a doubled DNA content had been measured (evidently tetraploid ones) possessed a more pronounced ability to form lung colonies, to metastasize and to display other manifestations of malignancy (Suzuki *et al.*, 1980). Moreover, during cloning of rat rhabdo-myosarcomas *in vivo*, tumors grown in lungs were preferentially hyper-diploid, while those propagated subcutaneously developed tetra- and octaploid nuclear classes and became more malignant (Kaminskaya, Stepanyan & Vakhtin, 1981).

Hence, polyploidization is involved in the process of tumor progression towards the acquisition of greater aggressiveness. The cause of increased viability of malignant cells with an increased DNA content should be sought in the properties of polyploid cells. As discussed in section B of this chapter, polyploid cells of normal tissues do not exhibit an increased rate of cycling and proliferation. Consequently, an increased growth rate is not an automatic property of polyploid cells (as is the change in the surface-area-to-volume ratio, the intensification of synthetic processes etc.). Polyploidization may play a role in the maintenance and development of malignancy, most probably thanks to the resistance of the polyploid genome to injury (see section 7D).

Features of carcinogenesis such as aneuploidy and genetic instability mean that chromosomal reorganization is continuously taking place in the tumors. The nature of this reorganization is such that, as a result, the tumor cells tend to lose or gain part or the whole of a chromosome. Cells with an unbalanced, often incomplete, genome (belonging to the diploid class) evidently are usually either non-viable, or less well adapted to the conditions of the environment. Tetraploids are resistant to genetic imbalance because of their gene and chromosomal redundancy. As a result, the evolution of malignancy in many tumors is a transition from diploidy, with or without slight aneuploidy, to the tetraploid state. This represents the basis for further, more extensive genetic changes leading to the progression of the cancer. This process was understood by Levan (1956) on the basis of cytogenetic studies of two strains of mammary adenocar-cinoma with different ploidy levels.

The individual stages in karyotype evolution have been traced in recent studies of transformation *in vitro* with chemical carcinogens, using cell lines derived from normal mouse salivary gland, bladder epithelium and fibroblasts.

The development of the cell from diploidy and non-malignancy to hypotetraploidy and the acquisition of a malignant phenotype is illustrated in the work of Cowell & Wigley (1980, 1982). Mouse submandibular gland was explanted into culture and treated with carcinogens. After a number of passages, the culture underwent malignant transformation (Figs 7.28–7.31). In the preneoplastic stage, foci of morphologically altered and rapidly growing cells appeared. 30–35% of the cells in these foci were tetraploid in DNA content. They were not the post-DNA-synthetic (G_2)

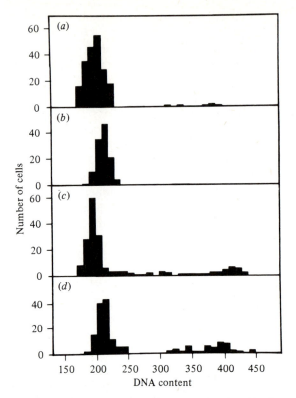

Fig. 7.28. Nuclear DNA content (arbitrary units, in this figure and in Figs 7.29–7.31) derived by fluorometry of explanted mouse salivary gland cells. (*a*), Bone marrow control; (*b*), salivary gland cells, diploid population; (*c*, *d*), explant and primary outgrowth, dividing diploid populations (after Cowell & Wigley, 1980).

compartment of the diploid population; they formed a genuinely tetraploid subline with signs of proliferation. The tetraploid cells then initiated hypotetraploid cell lines, which were not yet tumorigenic in syngeneic hosts. The malignant cell lines which developed later were very often hypotetraploid and sometimes hexaploid along with many other high-ploid and aneuploid cells (Figs 7.29, 7.31). Cell lines derived from carcinomas of the mandibular gland of mice, induced by chemical carcinogens *in vivo*, also had similar DNA distribution patterns (Fig. 7.31*a*).

During neoplastic transformation of cell lines originating from murine bladder epithelium, selection of tetraploid cells was also traced. This was followed by a period of chromosomal instability and the formation of a hypotetraploid stem cell line (Cowell, 1980). Among the transformed lines there were near-diploid and near-tetraploid ones. Judging from the duplication in the latter of a chromosomal marker which was found in the diploid lines, the tetraploid lines were formed either by c-mitosis

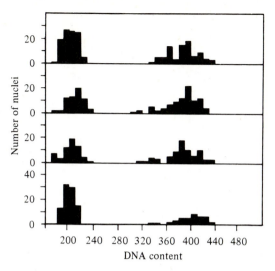

Fig. 7.29. Preneoplastic growth of the explanted mouse salivary gland: formation of a tetraploid population. DNA content distributions are illustrated for four foci of fast-growing cells (after Cowell & Wigley, 1980).

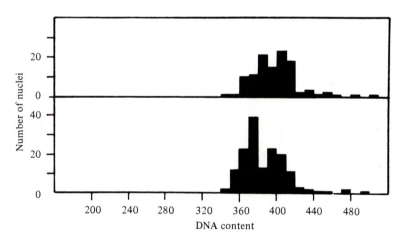

Fig. 7.30. Development of preneoplastic growth (see Fig. 7.29) with formation of a hypotetraploid population. Nuclear DNA content distribution of two cell lines are illustrated (after Cowell & Wigley, 1980).

(polyploidizing mitosis) or by endoreduplication of a diploid cell. Consequently, the target of the primary genetic changes was the diploid and not the tetraploid cell.

The work of Saxholm & Digernes (1980) draws attention to the progressive decrease in chromosome number and DNA content, compared

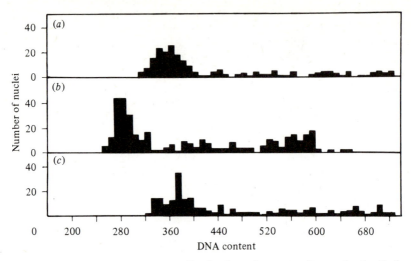

Fig. 7.31. Nuclear DNA content distribution of mouse salivary gland cells from tumorigenic lines. (*a*), Carcinogen-induced, *in vivo*. (*b*) A hypotetraploid stem line and (*c*) its further polyploidization are shown (after Cowell & Wigley, 1980).

with the initial near-tetraploid level, which occurred parallel to the development of the oncogenic potential of transformed mouse fibroblasts.

Cowell (1980) considers that the structural changes and losses of chromosomes may be the reason for the expression of the recessive gene for malignancy (see section D), once the wild-type alleles are lost and the cell becomes haploid with regard to this gene. Chromosomal imbalance may also be achieved on a diploid level. In actual fact, there are many cases of highly malignant tumors with minimal karyotypic deviations. The logic of further discussion will lead us to a question already examined concerning the advantages of polyploid cells in situations where the compensatory possibilities of the unbalanced genome are needed. The production of chromosomal imbalance by tetraploidization and subsequent loss of chromosomal material contributes (from a theoretical point of view) to the generation of viable karyotypic variants.

The above-mentioned pattern of behavior in neoplastic cell populations, also presented in papers by Levan (1956) and Cowell & Wigley (1980, 1982), is similar to that observed by Levan & Biesele (1958) in mouse skin fibroblasts *in vitro* (see section G of this chapter). The shift from diploidy to heteroploidy, via tetraploidy, is considered as a way of evolution for any autonomous, growing cell poulation (Hsu, 1961, Vakhtin, 1980).

Here in this section it would appear expedient also to cite some observations which have not been interpreted in terms of the compensatory influence of the increased number of gene alleles. Thus, Theile (1982) discusses the possible role played by polyploidy in viral transformation.

Cultured cells infected by oncogenic viruses form colonies of near-tetraploid cells. The formation of tetraploids seems to be a necessary stage in the process of tumor transformation. The selection of tetraploid forms possessing increased viability should be considered, although this factor may not be connected with the multiple genome, but rather with other properties of polyploid cells (see section D of this chapter).

Of interest is the ploidy composition of the tumors of mammalian tissues in which polyploidization is a regular phenomenon of normal ontogenesis, or associated with terminal differentiation (e.g. in the rodent liver, the mammalian bladder, and the megakaryocytic series of the bone marrow). Are the tumors which develop in these cell populations diploid or polyploid? Does the polyploidy already present correlate here with the subsequent degree of aggressiveness of the tumor?

Bladder tumors are no exception to the general rule, and are often polyploid (Levi *et al.*, 1969*b*; Spooner & Cooper, 1972). Clinical material (e.g. biopsies or exfoliated cells produced by bladder irrigation) testifies to the fact that well-differentiated papillomas are usually diploid or near-diploid; invasive, poorly differentiated tumors exhibit a whole series of ploidy values (from diploid to octaploid) with pronounced aneuploidy and an increased number of chromosomal markers (Sandberg, 1977; Klein *et al.*, 1982). Moreover, a correspondence is noted between the level of ploidy and both the degree of differentiation (diagnosed histologically) and the clinical stage of the cancer. In the initial states, the tumors are usually diploid; then, tumor expansion is frequently accompanied by the development of triploidy, tetraploidy, and even higher degrees of ploidy (Andersson *et al.*, 1980).

As regards human liver tumors, we can find in the literature no studies of DNA content in connection with the extent of anaplasia. The reasons for this are obviously the relative infrequency, until recently, of diagnosing primary tumors of the liver. In biopsy samples, taken from tumorous hepatic nodes, more than 80% of nuclei are polyploid and aneuploid; in the liver of healthy individuals of the same age, polyploid nuclei are uncommon (Koike *et al.*, 1982). Hepatocellular tumors, induced by chemical carcinogens in animals, very often reveal polyploid classes. These results give the impression that the general tendency for polyploidy to be correlated with great malignancy also applies to the case of the liver. Thus, in the most benign murine adenoma the modal class of DNA content is diploid, but in malignant carcinomas the quantities of DNA per cell varies from diploid to hyperhexaploid (Inui, Takayama & Kuwabara, 1971). Transplants of induced liver tumors in rats show a fairly good correspondence between the level of ploidy and the degree of differentiation, as judged by histological structure, growth rate, and the production of the embryonic alphafetoprotein (Becker *et al.*, 1973). Evidently, in liver as in other tumors, the parallelism between polyploidy and malignancy reflects

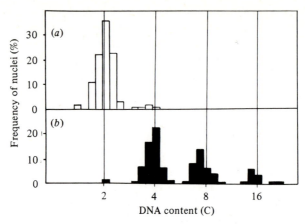

Fig. 7.32. Near-diploid stem cell line in a urethane-induced liver adenoma in mouse. (*a*) Distribution of nuclear DNA content (C), determined by Feulgen cytophotometry in cells isolated from an adenomatous node; (*b*), DNA content of hepatocytes from surrounding liver tissue (after Faktor & Uryvaeva, 1983).

the degree of evolution of the originally diploid tumor. This evolution from normal hepatocyte to a tumor cell is, in principle, similar to the development of transformed cells from explants of the salivary gland, mentioned earlier. The hypotetraploid stem cell line is also formed in the liver as a result of lengthy selection from a genetically unbalanced cell population. The cells of the latter are, in turn, progeny of the normal actively proliferating diploid cells (Stich, 1960).

Interpretation of cell polyploidy of liver tumors is complicated by the fact that, during hepatocarcinogenesis, polyploidization may have also occurred as part of normal differentiation of hepatocytes. Many features of normal differentiation – the synthesis of special proteins, the formation of specific cell-to-cell contacts, etc. – are retained in the tumors. This applies particularly to the early stages of carcinogenesis, but is perhaps also characteristic of tumorous nodes. Fig. 7.32 shows the types of DNA class distribution in hepatocytes isolated from adenomatous nodes; these were induced in mice by urethane and carbon tetrachloride. These animals did not have metastases in the lungs, but this is not, however, absolute proof of the benign nature of the tumor. Some of the adenomas had a purely diploid stem cell line with an admixture of a small number of binucleate cells. Besides these types, there were nodes of cells in which the distribution of ploidy classes differed little from that of the surrounding liver tissue, and from which aneuploidy was absent (Fig. 7.33*a*). Evidently high-ploid, mature adenomas present a variant of a more favourable outcome of proliferation induced by carcinogens. Fig. 7.33*b* presents results from a variant node which has a hypotetraploid stem cell line. The DNA

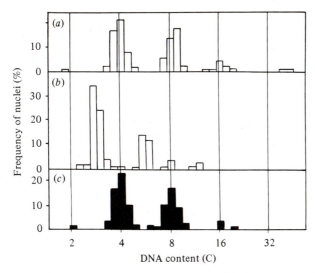

Fig. 7.33. Nuclear DNA content (C) of cells (*a* and *b*) from two urethane-induced adenomatous nodes in the same mouse liver; and (*c*), hepatocytes from surrounding liver tissue (after Faktor & Uryvaeva, 1983).

distribution in this tumor is reminiscent of that in advanced malignant tumors from other organs.

Thus, polyploidization of the cells facilitates the progression of the tumor and also, sometimes the coming into being of malignancy as a genetically determined feature. Frequently, the tumor stem line is polyploid. Does it always emerge from an initially diploid tumorous population, or are precursor tumor cells sometimes tetraploid? If this is so, then can the polyploid cells of normal tissues be a ready target for the impact of carcinogenic factors and the point of emergence of a tumor?

This question was examined earlier from the point of view of its correspondence with existing theories of the genetic basis of carcinogenesis (see section D of this chapter). The data indicate that although cell populations in the liver are polyploid, tumors develop from diploid clones. The multiplication and accumulation of diploid murine hepatocytes has been observed during the initial stages of chemical carcinogenesis (Faktor, Poltoranina & Uryvaeva, 1982). Even if the tetra- and octaploid hepatocytes proliferate rapidly in a carcinogen-treated liver, they are evidently not the target of malignant transformation. Of crucial importance, besides polyploidy itself, is the fact that polyploid hepatocytes are highly differentiated cells, which are not considered to be cancer-prone. The tetraploid and octaploid hepatocytes are evidently not involved in the development of malignancy. Whether the resilience of the polyploid cells to genetic lesions (which might lead to the development of malignancy) is of any importance, or whether their properties (for example, their high level of differentiation) play a role, is still unclear.

G The relationship of polyploidy to cell aging

Ontogenetic polyploidization in mammals is sometimes considered to be a sign of aging. This opinion is often supported by the correlation of polyploidy with age. This correlation is characteristic of liver, myocardium and salivary glands. But the association of polyploidy with age has no causal foundations. It is rather a consequence of the additional correlations between the cell ploidy level and organ weight on the one hand, and between the organ weight and the age of the animal on the other. Therefore, the number of polyploid cells in liver or myocardium is a function of the weight, rather than the age of the animal. The examples of artificial arrest and stimulation of growth (Chapters 2A, 2F, 7C) are most illuminating here. Thus, in thyroidectomized animals, there is inhibition of growth, mitotic activity and the emergence of high-ploid hepatocytes. Any disturbances of the environmental conditions which influence body weight also inhibit development of polyploid hepatocyte classes. Stress, hypokinesia and malnutrition are examples of such influences. Isoproterenol treatment of rats results in accelerated growth and polyploidization of the liver and (especially) of the salivary glands.

Pronounced and stable changes in hepatocyte ploidy classes result from liver regeneration after partial hepatectomy, or from the effect of toxins (especially after repeated administration). The relations in these cases are obviously simple, and unconnected with aging.

Nevertheless, it is thought that the deterioration of some liver functions with age may be partly related to hepatocyte polyploidization. Thus, the retention (in the blood) of bromosulfalein, injected into old rats, is considered to be due to decreased membrane-binding in the liver, owing to the enlargement of the cells with age (De Leeuw-Israel *et al.*, 1972). Although the reason for functional impairment, with age, of the liver and other organs, is not completely clear, it is difficult to object to the above conclusion. In fact, the properties of the polyploid cell which are beneficial to one tissue may have negative consequences for another. But, as can be seen even in the example of bromosulfalein retention, aging may not be the cause but rather it may be a consequence of ontogenetic polyploidization.

One of the main manifestations of tissue aging may be the decrease of the cells' ability to proliferate (Buetow, 1971; Cristofalo, 1974). In the liver, such a change has an effect on regenerative ability.

It is known that liver regeneration after a partial hepatectomy occurs more slowly with age; and in old animals regeneration does not always lead to the complete restoration of the organ. One of the reasons for the lessened ability for regeneration may be the decrease in the number of cells involved in liver growth and in forming the proliferative pool. In young rats, continuous infusion of [^{3}H]thymidine labels practically the entire parenchyma of the regenerating liver; likewise, in adult rats and mice, almost all hepatocytes participate in regeneration. But, in old animals, under the same experimental conditions, a maximum of 77% become labelled in rats (Stöcker *et al.*, 1972) and approximately 70% in mice (Uryvaeva & Faktor, 1974). The hepatocytes of old mice which have lost the ability to divide belong primarily to the high-ploid classes ($4C \times 2$, $8C$, $8C \times 2$, $16C$ and $16C \times 2$). In adult, but not in old animals, the cells of the same high-ploid types (for example, $4C \times 2$) can be stimulated to divide. Consequently, the aging process develops parallel to polyploidy but is not determined by it.

A hypothesis has been put forward with regard to the relationship between polyploidy and aging (Gahan, 1977). Polyploidy in mammals is suggested to be characteristic of slowly renewing or long-lived cells (the liver was taken as a model). If the cells live for a long time, they should have a protective mechanism against the accumulation of errors and damage, which (according to some views) form the essence of aging. The increase in the number of chromosome sets is just such a mechanism. Diploid systems, which have minimal protection, possess cells which 'wear out' quickly. Gahan assumes the genetic resilience of polyploid cells. However, there are already data available which contradict the hypothesis that polyploidy is a stratagem to combat rapid aging. Firstly, the most stable long-lived cells (neurons) have a diploid nucleus. Secondly, the lifespan of diploid and polyploid lines of fibroblasts does not differ *in vitro* (Thompson & Holliday, 1978). Thirdly, among mammals, there is an inverse relationship between ploidy level in the liver and potential species-specific lifespan. As mentioned above, the level of polyploidy in the liver is higher in those species of mammals in which the frequency of spontaneous chromosome aberrations is highest. According to Curtis (1963), these species are characterized precisely by the shortest lifespan.

The most interesting aspects of the relationship between polyploidy and aging are revealed by the analysis of aging of cells in culture. Cells of the normal human and animal tissues cultivated outside the organism are known to have a limited lifespan. Non-proliferating explants (for example, from adult liver and heart) are subject to fairly rapid degeneration and death. Cell cultures capable of mitosis outside the organism (for example, fibroblasts) also have a limited period of existence. In this case, the lifespan of the culture is determined by its proliferative potential (i.e. the number of divisions which may occur in these cells). Following the discovery of this property (Hayflick & Moorhead, 1961), diploid cells in culture were used as a model system for the study of cellular aging.

After reaching the proliferative limit, the rate of growth of a diploid

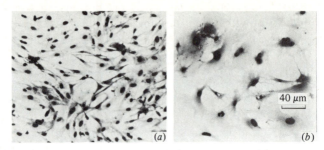

Fig. 7.34. Polyploidization of cultured chick embryo fibroblasts (*a*) young cells, (*b*) old cells of late passage. Both micrographs are at the same magnification (after Kaji & Matsuo, 1979).

culture diminishes rapidly, signifying the onset of senescence; this ends in the death of the cells. Fairly recently it was established that senescence is characterized by the formation of large polyploid and multinucleate (mainly binucleate) cells (Fig. 7.34). This has repeatedly been described in various cultures of diploid cells: chick embryo fibroblast (Kaji & Matsuo, 1978, 1979, 1981), WI-38 and WI-26 cultures of fibroblast-like cells from human embryonic lung (Yanishevsky *et al.*, 1974; Mitsui & Schneider, 1976; Matsumura, Zerrudo & Hayflick, 1979; Matsumura, 1980), IMR-90 culture of human embryo diploid fibroblasts (Kelley, Perdue & Uruchurtu-Valdivia, 1983), MRC-5 line of fetal lung fibroblasts (Thompson & Holliday, 1975), and epithelioid cells from embryonic human liver and kidney (Miller *et al.*, 1977; Matsumura, 1980).

Even the very first generations of cells *in vitro*, in a rapidly proliferating cell culture reveal a relatively high frequency of polyploid cells. These also represent a fraction of slowly cycling cells with a reduced clonogenic capacity; they incorporate [³H]thymidine and consequently do not leave the cell cycle irreversibly (Yanishevsky *et al.*, 1974). With successive passages, the number of polyploid cells increases steadily (Fig. 7.35). When proliferation ceases, the proportion of binucleate cells reaches a maximum of 50% (Matsumura *et al.*, 1979). A detailed study, with the help of flow cytophotometry, has shown the presence of distinct euploid classes of nuclei containing 2, 4, 8, 16, 32, and 64C DNA with a frequency of 41.8, 36.0, 10.9, 4.4, 1.7, and 0.2%, respectively (Kaji & Matsuo, 1979). Individual nuclei were registered, but it was not determined whether they belonged to mononucleate or binucleate cells. A study using DNA-Feulgen cytophotometry revealed six main classes of aging human fibroblasts – mononucleate 2C, 4C and 8C cells and binucleate 2C × 2, 4C × 2 and 8C × 2 cells with each nucleus having an equal DNA content (Matsumura, 1980). Matsumura put forward the hypothesis that the diploid cell which has left the ordinary cell cycle, transfers to another state called 'the cell spiral'.

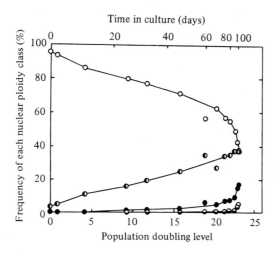

Fig. 7.35. Variation in the percentage of each nuclear ploidy class in cultured chick cells as a function of population doubling level. 2C, O—O; 4C, ◑—◑; 8C, ●—●; 16C, ◒—◒ (after Kaji & Matsuo, 1981).

The spiralling cells increase their level of ploidy by alternating mono-nuclearity and binuclearity, with accompanying DNA synthesis.

It was recently discovered that cultured rat liver cells have a limited proliferative potential. Epithelial-like cells from fetal rat liver undergo a few complete mitoses in the primary culture; then the mitotic cycle is prolonged, mitoses become acytokinetic, and binucleate and polyploid cells are formed (Matsumura *et al.*, 1983).

The similarity between polyploidization of a senescent culture of fibroblasts and ontogenetic polyploidization in the liver of rodents (Chapter 2A) is striking. It is difficult to say how similar are the modes of formation of polyploid cells in each case. Polyploid classes of hepatocytes are formed by polyploidizing mitoses, first acytokinetic, and then with bimitosis of the binucleate cell. Kaji & Matsuo (1981) believe that, in an aging fibroblast culture, the mononucleate 4n cells are unlikely to be formed via binuclearity although, at the end of the life of the cultures, 20–25% of the cells are binucleate. It is also possible that, in contrast to the liver, the endoredupli-cation observed in 12–25% of the polyploid metaphases is responsible for the accumulation of the mononucleate cells with an increased DNA content (Miller *et al.*, 1977). Cell fusion does not play a substantial part of the genesis of polyploid mononucleate and binucleate cells in culture.

The causes of polyploidization in cultivated cells are not clear. It is thought that it may be the result of a gradual reduction of the mitotic function, prior to a complete switch-off (Kaji & Matsuo, 1979). At first, the ability for cytokinesis is lost and then karyokinesis, while the ability to synthesize DNA is retained. In addition, there is yet another feature that underlines the superficial similarity of the cell aging *in vitro* and ontogenetic polyploidization of liver cells. Aging *in vitro* is accompanied by a sharp increase in chromosomal aneuploidy and structural chromosomal changes. The frequency of the latter reaches 30% and more (Saksela & Moorhead, 1963; Thompson & Holliday, 1975; Miller *et al.*, 1977). One interesting circumstance noted by all the authors cited is that chromosomal aberrations are not characteristic of a rapidly proliferating culture, and only develop after 40–50 passages, although the frequency of polyploid cells increases considerably in very early-passage cultures. It is more likely that chromosomal anomalies and polyploidy develop independently and in parallel. It is hardly likely that chromosomal aberrations are the reasons for aging and degeneration of the culture; most probably they arise through deterioration of cell functions which itself forms the basis of the aging process. It is possible that in a diploid culture, just as in murine hepatocytes, the combined polyploid genome saves the cell from death after aberrant mitoses, and prolongs the culture's life during aging.

A comparison of the best studied polyploidization of the hepatocytes in ontogenesis and fibroblast-like cells in an aging culture should be continued. During embryonic and early postnatal life hepatocytes are a diploid population of rapidly proliferating cells. Polyploidization usually occurs in the final two or three cycles of the proliferative period. Polyploid, slowly cycling or non-cycling hepatocytes persist throughout the life of the animal. In fibroblasts, polyploidization of some of the cells begins very early during the first population doublings of the culture. Does it reflect an intrinsic property of fibroblast differentiation manifested *in vitro* rather than the result of sustained growth? In fact, although a systematic study of fibroblasts has not been conducted, data in the literature testify to the tendency of these cells to polyploidize *in vivo*. The death of an aged culture of fibroblasts after their complete polyploidization gives rise to the mistaken impression that they are not viable. But, in actual fact, the culture becomes extinct owing to the absence of cellular reproduction and, hence, the non-replacement of those cells lost during experimental manipulation, the changing of the medium, and so on. It is possible that, in the organism, polyploid fibroblasts are as long-lived as hepatocytes.

There is yet another interesting aspect to the study of aging of a diploid cell culture – the relationship between polyploidization and heteroploid transformation. Aging does not always lead to the cessation of proliferation and the death of the cell. Sometimes the normal diploid karyotype becomes aneuploid and the culture is transformed into a cell line capable of infinite proliferation. In the work of Levan & Biesele (1958), stages in the formation of an aneuploid line from an originally diploid culture were traced with the help of karyotype analysis. Tetraploid euploid metaphases already formed approximately 30% of all metaphases by the fourth passage. By the 16th passage, a single near-tetraploid modal class was formed. This class became stable and the culture underwent malignant transformation.

Today there are still few data on the interrelationship between polyploidization and the heteroploid transformation of cultures. Further research will show whether early polyploidization during the first passages facilitates the formation of aneuploid cells, or whether, on the contrary, these processes reflect two alternative lines of behavior of cell cultures and determine two outcomes – either the loss of the proliferative activity and death, or the acquisition of the ability for indefinite proliferation.

The behavior of human cells in tissue culture is perhaps different from that of mouse cells. The human cell cultures obviously fail to develop into cell strains and usually become senescent, while murine cell cultures undergo heteroploid transformation.

Conclusion

The biological significance of regular polyploidization may lie in advantages in the functioning of the cell, the cell population, or the entire tissue. In many of the cases studied, the polyploid cells are qualitatively the analog of the parent diploid cells. Polyploidy leads to quantitative changes in structure and function, which increase in proportion with the gene dosage (see also Chang *et al.*, 1983). The reason for the proportionality of the polyploid cell mass to their ploidy level is that the cells repeatedly pass through successive mitotic growth cycles in parallel with the nuclear chromosome cycles. Hence, one polyploid cell supercedes several diploid cells, and polyploidization looks like the equivalent of cell multiplication. Cell giantism, which is largely achieved by multiplication of the genome, is evidently beneficial to the functioning of some tissues.

For all that, it is still not clear whether proportionality between gene dosage and functional activity is always retained, especially in high-ploid and polytene cells. Differences between the protein spectrum, the RNA fractions and the activity of the isozymes are known in diploid and polyploid cells. If such research continues, the role of polyploidization in differentiation may be elucidated.

An important feature of polyploid cells is the redundancy of their genetic information. This property makes polyploid cells resistant to damage to the genetic apparatus and determines their advantage over diploid cells.

Polyploidization of cells is involved in the process of evolution of autonomously growing cell populations, such as cancers or cell lines cultivated *in vitro*. Selection of genetically changed cells is continuously taking place in these populations. Polyploid cells may be more suitable than diploid ones for the development of aneuploid variants, and may be optimally adapted to unusual environmental conditions.

8

Control of genome multiplication in ontogenesis

The first cell in metazoan development, the zygote, is diploid. Polyploid cells appear later in many somatic tissues. The emergence of polyploid cells at certain stages of ontogenesis, and also the expression of polyploidy in homologous tissues in organisms of different species, indicate the regularity of the phenomenon. These observations, and also the data on the advantages of polyploid cells over diploid ones, prompt us to suggest that organisms with polyploid cells have been selected and conserved during evolution.

As to reasons for the switch from cellular reproduction to reproduction of the genomes alone, two possibilities may be discussed: one concerns changes in the genetic regulation of the mitotic cycle, the other concerns the competition that exists between mitotic and other functions within the cell.

A Genetic control of the mitotic cycle and its incomplete variants

The mitotic cycle is a complex integrated process that includes events which ensure cellular growth, reproduction of the genetic material and cell division (Mitchison, 1971; Prescott, 1976; Epifanova, Terskikh & Polunovsky, 1984). It is common knowledge that initiation of DNA synthesis and entry of the cell into mitosis, as well as expression of its separate phases, require the synthesis of special proteins; this is assumed to be directed by special 'mitotic' genes. For example, a G_2-phase cell contains at least nine proteins specific to this part of the cell cycle (Al-Bader, Orengo & Rao, 1978).

Most of the data on genetic regulation of the mitotic cycle have followed from the isolation of temperature-sensitive (ts) recessive mutants of yeasts, fungi, protozoans and mammalian cell lines. The most important results were obtained by Hartwell's group during the study of ts-mutants of *Saccharomyces cerevisiae* (Hartwell *et al.*, 1973; Johnston, Pringle & Hartwell, 1977; Hartwell, 1978). Approximately 150 ts-mutants affecting the cell division cycle have been isolated. Genetic analysis has shown more than 30 genes to be active at certain phases in the cycle. In Fig. 8.1 a

234

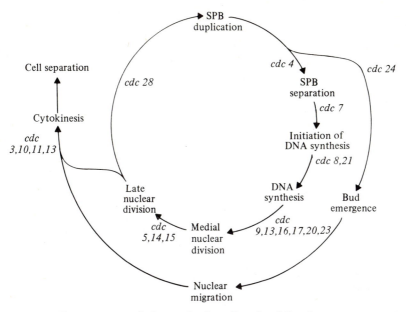

Fig. 8.1. The gene-controlled steps in the cell cycle of *Saccharomyces cerevisiae*. Each gene (numbered) has a particular time in the cycle at which it is expressed. *cdc*, Cell division control gene; SPB, spindle pole body (after Hartwell, 1978).

diagram is given of the results of Hartwell's work (1978), in which the gene-controlled stages in the *S. cerevisiae* cycle have been summarized. Based on the behavior and morphology of cells at a nonpermissive temperature, genes were discovered which regulate the initiation of DNA synthesis, DNA replication, cellular budding, nuclear division, cell division, and cell-wall separation. The observations corroborate the old suggestion that events in the cycle are independent of each other. For example, a gene was discovered whose expression coincides with bud emergence. DNA synthesis and nuclear division are not inhibited in these mutants and so, at the restrictive temperature, large multinucleate cells are formed. Cell growth occurs relatively independently of the mitotic cycle itself, and so mutants blocked at different phases of the cell cycle continue to grow.

In the infusorion *Tetrahymena pyriformis* 14 recessive ts mutations affecting the cell cycle have been found (Frankel, Jenkins & De Bault, 1976). An analysis of these mutants revealed genetic markers for macronuclear DNA synthesis, macronuclear division, cytokinesis, micronuclear division, and oral development.

Isolation of the cell-cycle specific ts mutants of mammalian cells have been achieved (for references, see Lloyd, Poole & Edwards, 1982). Cell-cycle-defective mutants, in which the nonpermissive temperature causes arrest and specific interruptions of mitosis and, as a result, the accumulation of mononucleate and multinucleate polyploid cells, have been isolated.

Defective prophases, with the formation of unusual chromosome structures have been described (Wang, 1976). The mitotic cells, arrested in metaphase at a non-optimal temperature, were similar to the well known colchicine mitoses (Wang, 1974). A recently isolated ts-mutant of a Syrian hamster cell line exhibits abnormal mitoses because of non-separation of the centrioles, and the consequent formation of monopolar pro-metaphases. Such cells do not divide, but they are capable of repeating cycles of chromosome reproduction and abnormal mitosis. As a result, cells of a high ploidy level are formed (Wang *et al.*, 1983).

Sometimes, the post-metaphase processes occur incorrectly. The nuclear membrane may form around odd groups of chromosomes and where cell cleavage is absent, these groups form the nuclei of mono-, bi-, or multinucleate cells (Wissinger & Wang, 1978). Mutant lines are known in which the non-optimal temperature affects cytokinesis, although other types of cell movement and other phases of the cell cycle and mitosis are not disturbed. Ts mutants blocked in cytokinesis, were isolated from cultures of Syrian hamster cells (Smith & Wigglesworth, 1972), Chinese hamster ovary cells (Hatzfeld & Buttin, 1975; Thompson & Lindl, 1976; Hori, 1980) and murine leukemic cells (Shiomi & Sato, 1978). A characteristic feature of such cell lines is the accumulation of binucleate and multinucleate cells. It is thought that ts-cell lines contain mutated genes that control the production of proteins for the spindle, centrioles, pericentriolar material (and other mitosis-related proteins). As a result, structural defects occur because these proteins become disorganized or unorientated under restrictive conditions.

One of the recently isolated ts mutants of the hamster cell line has a genetic defect in the processing of ribosomal RNA. This defect is mainly reflected in the inability to divide, although it is not clear which phase of mitosis is affected. At a nonpermissive temperature, the number of cells in the population does not increase but large, 4C cells accumulate (Mora, Darżynkiewicz & Baserga, 1980).

Recently, a novel class of mutants of Chinese hamster ovary cells has been described (Cabral, 1983; Cabral *et al.*, 1983). These mutants require the microtubule-stabilizing drug (taxol) for cell division. One of these mutant cell lines, when cultured in the absence of taxol, forms only a rudimentary mitotic spindle. Mitosis progresses up to pro-metaphase and cytokinesis does not occur. However, taxol-deprived mutant cells are able to continue DNA synthesis. Flow cytometry and karyotyping show the cells to have increased nuclear DNA content and chromosome number, up to at least an octaploid level.

Mutant forms of animals are also known which are specifically defective in mitosis and accumulate polyploid cells in developing tissues. For example, a temperature-sensitive mutation in the nematode *Caenorhabditis elegans* causes failure of nuclear and cell divisions in precursor cells for the ventral nerve cord (Albertson, Sulston & White, 1978; Sulston & Horvitz, 1981). Prophase appears to be normal but, in metaphase, the chromosomes did not become attached to the spindle. After metaphase, the nucleus returns to an interphase structure. Three cycles of reproduction (as in wild-type animals) occur, in spite of the metaphase block and failure to produce daughter nuclei. Some embryonic lethal mutants continue DNA replication and nuclear (but not cellular) division at a restrictive temperature. Very large nuclei with a high DNA content could have arisen by fusion in the ensuing cycles of binucleate cells (Hecht *et al.*, 1982).

Binuclearity of erythrocytes has been described in the turkey, caused by a mutation in an autosomal recessive gene (Bloom, Buss & Strother, 1970; Bloom & Searle, 1973; Searle & Bloom, 1979). This gene controls the structure and function of the mitotic apparatus. In homozygotes, disruption of mitosis in the polychromatophil erythroblasts appears even during embryonic development. This leads to the accumulation in the peripheral blood of mature erythrocytes with two or more nuclei (40% of the population) and also of mononucleate polyploid cells. The chromosomes in such cells are not separated. It is interesting that such anomalies in turkey blood are cytologically similar to some types of congenital human dyserythropoietic anemias (also inherited through autosomal recessive genes). These illnesses are characterized by giant multinucleate cell forms, which correspond to the normoblast (with respect to maturity). No abnormalities of mitosis were observed in early precursor stages (Heimpel, 1976).

As can be seen, polyploidizing mitoses such as those observed during the normal development of some tissues are the result of mutations in the genes regulating mitosis. The examples cited in this section indicate that the change-over to incomplete mitotic cycles and polyploidization may be in accordance with the realization of a genetic program. This seems probable for the polytene cycle, which may actually arise through the inactivation of mitotic genes. However, in many cases of polyploidy, there is no direct evidence for the genetic regulation of the incomplete mitosis. We will return to examine the control of polyploidy at the end of this chapter, but first we shall take note of other possible explanations for an incomplete cycle.

B Modification of the mitotic cycle as a result of competition between intracellular metabolic processes

Elsewhere we have substantiated the existence of a competitive relation between proliferation and differentiation as a cause of blocked mitoses (Brodsky & Uryvaeva, 1970, 1977). But, which polyploid cell types result from genetic control or from the aforementioned competition is not yet clear. We cannot rule out the latter possibility, at least in those cases where polyploidy is a rare and functionally dispensable phenomenon. However, competition between proliferation and differentiation may play a general role in the regulation of cell functions, especially in multifunctional systems. This is reminiscent of the old embryological tenet of the antagonism between cell division and the expression of differentiated functions. But whereas antagonism was formerly regarded as the complete repression of one process by another ('a cell either divides or differentiates'), recent works have shown certain mutual interrelations where functions are balanced and there is the occurrence of quantitative influences of one function upon the other.

The peculiarities of differentiation of polyploid cell populations

Even if analysis is restricted to mammals, quite a large number of mature and completely differentiated cells are found to retain the capacity for mitosis (Table 8.1). It is precisely in such cell lines that polyploidization affects virtually all or most of the cells during their development. Striking examples are hepatocytes, heart muscle cells, melanocytes, cells of the bladder epithelium, and megakaryocytes. In other tissues, proliferation frequently does not cease with differentiation, but here polyploidization is not regular and probably does not have any particular functional significance for the tissue. It is sufficient to recall the occurrence of binucleate cells in the epidermis (Zhinkin, Brodsky & Lebedeva, 1961), mast cells (Allen, 1962), chondrocytes, antibody-synthesizing cells (Russell & Diener, 1970), polyploid and binucleate fibroblasts (Brodsky & Khrushchov, 1962) and polyploid cells of the adrenal gland, thyroid and other endocrine glands (which are regarded as cases of facultative polyploidy) (Gilbert & Pfitzer, 1977). Polyploidy or multinuclearity was recently noted in Sertoli cells of aged human testis (Schulze & Schulze, 1981), in pericryptal fibroblasts of murine colon (Neal & Potten, 1981) and in aortic smooth muscle cells of hypertensive rats (Owens et al., 1981; Owens & Schwartz, 1982). The possibility of suppressing proliferation and accumulating G_2 cells is achieved in the erythroid line where it ensures a rapid acceleration of differentiation when an emergency arises (see Chapter 2J).

Division of the cell population into fixed categories is a shortcoming of Table 8.1. This limitation is a fairly conventional one, as the assignment of the tissue to one category or another depends, to a considerable extent, on the criteria of differentiation selected. It is easy to imagine that, were other criteria chosen, the distribution of examples might be quite different. Thus, myofibrils and the ability to contract emerge in skeletal muscles only after the cessation of proliferation in the myoblasts, their fusion and the formation of a common cytoplasm. But small quantities of contractile

Table 8.1. *Relationships between proliferation and differentiation during tissue development in mammals*

Categories	Cell type	References
Arrest of growth and irreversible exit from the cell cycle with the beginning of terminal differentiation	Skeletal myocytes Neurons of the central nervous system Retinal ganglionic cells	Stockdale & Holtzer, 1961 Fujita, 1964 Jacobson, 1968
Mitotic ability in differentiating cells with its loss during terminal differentiation	Erythroid cells Megakaryocytes Neurons of the peripheral nervous system Chondroblasts Epithelium of the lens β-cells of the islets of Langerhans in the pancreas Keratinocytes	Cameron & Jeter, 1971 Odell, Jackson & Reiter, 1968 Cohen, 1974 Lasher, 1971 Modak, Morris & Yamada, 1968 Von Denffer, 1970 Potten, 1974
Ability of completely differentiated cells to undergo mitosis	Hepatocytes Cardiomyocytes Retinal pigment cells Acinar cells of lacrymal glands Secretory cells of the mammary glands Follicular thyroid cells Thyroid calcitocytes Mucosal cells of the large intestine Exocrine cells of the pancreas Duodenal goblet cells Cells of the smooth muscles Transitional epithelium of the bladder Antibody-synthesizing cells Mast cells	Tsanev, 1975 Rumyantsev, 1981 Stroeva & Mitashov, 1983 Desaive, 1967 Franke & Keenan, 1979 Zeligs & Wollman, 1979 Gorbunova, 1976 Chang & Leblond, 1971 Pictet *et al.*, 1972 Troughton & Trier, 1969 Cobb & Bennett, 1970 Hicks, 1976 Russell & Diener, 1970 Enerbäck & Rundquist, 1981

protein (myosin) are found in dividing myoblasts long before fusion occurs (Cameron & Jeter, 1971).

Differentiation of the neurons in the central nervous system is closely correlated with the loss of proliferative activity. The formation of the dendritic processes, the accumulation of neurofibrils and the establishment of synapses occur only after mitoses have ceased. It is not completely clear what determines these correlations. It cannot be ruled out that the formation of the axon and other processes make impossible the changes in cell shape needed for mitosis. However, in neuroblastoma cells *in vitro*, induction of axon growth after the introduction of cAMP into the medium and also the intensive synthesis of the neurospecific protein 14-3-2 occurs before the cessation of cell division (Prasad, 1975). DNA synthesis and mitosis have been induced in culture of motor neurons isolated from the spinal cord and which possessed signs of complete differentiation (Cone & Cone, 1976). The accumulation of specific catecholamine granules and the formation of neurofibrils in cells synthesizing DNA and entering into mitosis is quite normal in the neuroblasts of the peripheral nervous system, unlike those of the central nervous system (Cohen, 1974).

The cessation of division as maturity is reached, is evidently determined in many cell lineages by the nature of terminal differentiation itself; this represents not only a functional stage in these lineages, but also imminent cell death. This phase of development in cells of the erythroid lineage and of the lens fibres is accompanied by inactivation of the genome and the destruction of the nucleus. The tissue-specific products sometimes fill the cytoplasm (for example, in mature mast and adipose cells) and, in some cases, this product practically replaces all the cellular structures (e.g. in keratinocytes and in cells of the crystalline lens).

The idea that the accumulation of a specific product may influence the ability of the cell to undergo mitosis is discussed in morphological studies (see, for example, Goss, 1966). However, in the pigmented epithelium of the eye, mitoses occur even where there is a considerable cellular melanin content (Marshak *et al.*, 1972). Similarly, mitoses occur in heart myocytes after accumulation of a large quantity of myofibrils (see Rumyantsev, 1981).

The interrelated processes of growth and specialization are influenced not only by the nature of terminal differentiation, but also by other aspects of the cell lineage. Each lineage is an integrated system of stem cells, intermediate, maturing and terminal (functioning) forms, sometimes described as a 'differon' (Vogel, Niewisch & Matioli, 1969). The composition and activity of these cell groups are determined by the rate of cell division (ensuring the growth, or renewal, of the tissue) and also by the lifespan of the differentiated-cell compartment. This, in turn, depends on the physiology of the tissues. Differentiated cells retain the capacity for mitosis in those tissues whose cells live for a long time and there is no cell renewal or where renewal occurs very slowly. These tissues do not have a stem cell reserve, and the main source of increase in mass is the reproduction of differentiated cells. Examples include hepatocytes, heart myocytes and the pigment epithelium. Mitoses are also observed in differentiated cells, both in slowly renewing tissues with a cambial reserve (such as the exocrine glands), and in the bladder transitional epithelium.

There is a sharper separation of proliferative and specialized functions between the various cells (and, consequently, between the different subdivisions of the differon) in tissues where rapid cell renewal occurs. Thus, in the erythroid cell lineage (erythron) two main parts can be distinguished (Rifkind, 1974). In one of these, hemoglobin is synthesized and the required number of cells is produced. The other consists of the circulating erythrocytes, and is therefore the functional compartment; it includes cells which are no longer capable of division or of tissue-specific syntheses.

In summing up the above, it should be emphasized that the question of whether cell division and differentiation are compatible or incompatible is still a point for discussion not only as a general question of cell physiology, but as a specific characteristic of one cell lineage or another. The accumulation of new facts and also the improvement of observation techniques have shaken the initial concept of the incompatibility of these processes.

There can no longer be any doubt that the cell can indeed carry out both specific and proliferative syntheses simultaneously. However, although the cell expresses tissue functions and replicates simultaneously, these processes are indeed antagonistic. Competition between proliferation and differentiation lies at the basis of this antagonism.

Limitations to cellular metabolism

Cell differentiation may be characterized not only by the production of distinct proteins but also by the constraints exerted on their production. In addition to the tissue-specific proteins (which are not essential to the life of the cell), other significant proteins are synthesized in differentiated cells. These proteins are the structural components of the cells and the enzymes necessary for maintaining vital activity. In differentiated cells capable of mitosis, special division proteins are also synthesized without which the cell cannot progress through the cell cycle, cannot reduplicate DNA, and cannot form a division apparatus. In addition chromosomal proteins are synthesized during the mitotic cycle and the entire cell mass doubles. Thus, the total production of proteins in the mitotic cycle of the differentiated cell is quite large. Energy losses during DNA replication and protein synthesis in the cell cycle are also considerable. For example, during the hepatocyte S-phase, the glycogen content falls by approximately one third from its level in the $G_1(G_0)$-phase (Shalakhmetova *et al.*, 1981 *b*).

Since the rate of polypeptide-chain synthesis is fairly constant in animal cells (Palmiter, 1975), the protein production is largely determined by the number of ribosomes. According to cytophotometric data, the quantity of ribosomal RNA varies in different functional states of nerve cells (up to 1.5–2-fold) and, in gland cells, slightly more (Brodsky, 1975). If the percentage of ribosomes contained in polysomes does not change (polysomes of differentiated cells of mammals usually contain as much as 80% of the overall number of ribosomes), such changes of 50-100% in the total quantity of RNA also indicate possible changes in the rate of protein synthesis. Stimulation of protein synthesis often causes new ribosomes to be formed in a quantity over and above that required for normal tissue function. Even in these cases, however, the requirements of differentiation and division may not be completely satisfied. Thus, the injection of estradiol into cockerels or male frogs stimulates the synthesis of huge quantities of proteins (precursors of vitellogenin) in the liver. The determining factor in intensifying synthesis is the formation of new ribosomes. The number of polysomes increases several-fold (Bast *et al.*, 1977), a situation which is unlike that in other differentiated cells. However, this great increase in the protein-synthesizing machinery is evidently insufficient to maintain normal liver functions. The concentration of the serum proteins falls sharply after induction of vitellogenin synthesis (Follet, Nichols & Redshaw, 1968).

So, each type of differentiated cell is characterized by a certain limit in the production of both tissue-specific and other cell proteins. Limiting factors, in addition to the number of ribosomes, may be the amount of tRNA, the size of precursor pools, and ATP levels.

Competition between cell functions

It may be imagined that the substances or structures common to different metabolic pathways (ATP, NADPH, precursors, ribosomes) may be the limiting factors with regard to cell functions. In other words, the expression of a particular cellular function may depend on the intensity of other functions. Examples may be cited of such an interdependence and special attention should be paid to manifestations of antagonism between cell division and differentiation. These interactions lie at the root of such competitive processes.

The concentration of uridine triphosphate (UTP) in the medium is a limiting factor in the expression of the differentiated phenotype of cartilaginous cells *in vitro* (Marzullo & Lash, 1970). Under optimal conditions, the cells divide and synthesize chondroitin sulfate. Otherwise only cell division occurs, and the cells change their phenotype. The specific enzymes for the transformation of glucosamines into chondroitin sulfate have been discovered in both cases; i.e. only the phenotype was

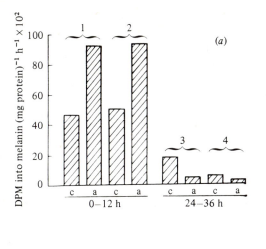

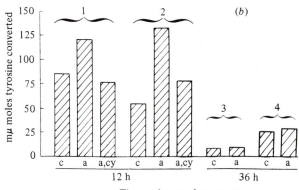

Time and type of treatment

Fig. 8.2. (*a*) Melanin synthesis expressed as tyrosinase-dependent incorporation of [^{14}C]tyrosine into the acid-insoluble cell fraction in the retinal pigment cells of chick embryo *in vitro*, and the effect of actinomycin D on melanin synthesis. (*b*) The level of tyrosinase. c, Control cultures; a, cultures with 3 μg ml^{-1} actinomycin D; a, cy, actinomycin D and cycloheximide 5 μg ml^{-1} (time of the treatment represented below). Each experiment (numbers above the histograms) repeated twice (after Whittaker, 1968*a*).

changed, the genes for tissue function were not blocked. The formation of chondroitin sulfate depends on the concentration of the UTP, which is necessary for the ultimate step in the chondroitin-sulfate synthesis, and also for the preparation of the cell for mitosis. In suboptimal conditions, mitosis is preferred.

Another factor which limits the intensity of protein synthesis is the quantity of ribosomes. Mitoses may be stimulated in primary cultures of pigment cells from the chicken retina. Even before the first division takes place, the cells become devoid of pigment. As explained by Whittaker

(1968*a*,*b*, 1970), the loss of color is the result of the insufficient restoration of melanin, because of a reduced rate of synthesis. Early in the culture period, the activity of tyrosinase, the key enzyme in melanin synthesis, decreases; it falls approximately threefold after 12 h and 30-fold after 36 h. By contrast, the total quantity of cellular proteins increases by more than tenfold during this period. It is noteworthy, however, that tyrosinase activity is maintained for at least 36 h (Fig. 8.2). This is considerably in excess of the lifetime of the tyrosinase mRNA. Thus, stimulation of mitosis does not entirely block the tyrosinase gene. According to Whittaker, the reduction in the rate of tyrosinase synthesis coincides with the saturation of the protein-synthesizing apparatus with new mRNAs, which did not exist before the explantation of the tissues. Actinomycin D, introduced into the medium during the first 12 h of cultivation (i.e. during the synthesis of these new – probably mitotic – mRNAs), resulted in increased tyrosinase activity and melanin synthesis (Fig. 8.2). Actinomycin D plus cycloheximide does not lead to such an effect. By 24–26 h, when mitoses are already proceeding in the cells, actinomycin D does not influence tyrosinase activity and melanin synthesis. Whittaker suggests that mRNAs for tyrosinase (tissue-specific) and mitotic proteins compete for the translation apparatus. Such an explanation may also be suitable for the increased rate of melanogenesis in pigment cells of the chick retina *in vivo*, after injection of fluorodeoxyuridine and inhibition of DNA synthesis (Zimmermann, 1975).

Stimulation of mitoses in the embryonic cells of the crystalline lens *in vitro* is accompanied by an approximately twofold drop in the rate of the synthesis of crystallins, although the quantity of mRNA does not decrease. In the different functional states of the cell, the quantity of tissue-specific proteins differs fivefold per molecule of mRNA (Beebe & Piatigorsky, 1977). The authors suggest either a change in the activity of the mRNA for crystallin or an interaction of this mRNA with other mRNAs, the quantity of which increases sharply in dividing cells.

Data on the competition by different mRNAs for translation have been obtained in experiments with mRNA injected into oocytes. Globin mRNA is translated after injection into the *Xenopus laevis* oocyte and globin is synthesized (Laskey *et al.*, 1977). At the same time, the incorporation of amino acids in the endogenous proteins of the oocyte drops considerably (Fig. 8.3); the total level of protein synthesis does not change. The authors conclude that there is competition by different mRNAs for active ribosomes.

In oocytes of *Xenopus laevis*, only 1–2% of the ribosomes occur in polysomes (Woodland, 1974). The number of polysomes does not increase when new mRNA appears. Consequently, in oocytes (just as in differentiated cells), competitive relations are the result of the cell's limited metabolic capacity. The precise degree of limitation is characteristic of each cell type.

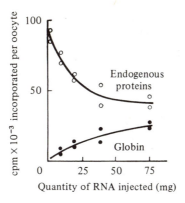

Fig. 8.3. The effect of injected globin mRNA on the incorporation of [¹⁴C]amino acids into oocyte proteins. Synthesis of endogenous proteins is inhibited owing to competition for translation between the injected globin mRNA and endogenous mRNAs (after Laskey *et al.*, 1977).

Data on the stability of total protein production during changes in the protein complement support the competitive hypothesis. Thus, in chick embryonic hepatocytes *in vitro*, the synthesis of albumin slows down simultaneously with the acceleration of synthesis of two other proteins, while the total level of protein synthesis remains constant (Grieninger & Granick, 1975). Protein production during the lag phase of growth and in stationary-phase culture of kidney cells also remains constant. Prior to mitosis, while replication-related components are being synthesized, the rates of lactate dehydrogenase, esterase, and catalase syntheses were observed to fall sharply (Ruddle & Rapola, 1970).

More precise data on the post-transcriptional limitations of protein synthesis are provided by experiments where estradiol was injected into cockerels (Bast *et al.*, 1977). This induces the synthesis of huge quantities of vitellogenin in the liver. Although ribosomes had already been formed in large numbers, the rate of synthesis of serum proteins hardly decreased. It is known that, after the stimulation of vitellogenin synthesis and the increased transcription of the corresponding mRNA, the relative content of active albumin mRNA decreases twofold (Williams, Wang & Klett, 1978). The level of albumin synthesis in the liver diminishes by approximately the same extent, and the concentration of albumin in the blood falls. As can be seen, the situation is reminiscent of Whittaker's description (see above) of the limitation of tyrosinase synthesis owing to competition by different mRNAs for the translation apparatus.

Liver cells are multifunctional; mammalian hepatocytes synthesize no fewer than 20 tissue-specific proteins, while many (and possibly all) of these proteins can be synthesized within the same cell (Tsanev, 1975; Salganik, 1979). The microsomal oxidizing enzymes, essential to the liver's drug-metabolizing function, are also relevant to our review of competing

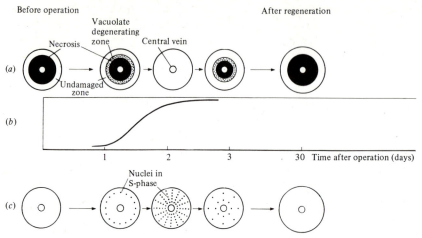

Fig. 8.4. Reciprocal interrelationship between drug-metabolizing and mitotic functions in regenerating mouse liver. Adult mice were subjected to partial hepatectomy and then exposed to carbon tetrachloride at different times after the operation. (*a*) Localization of carbon tetrachloride-induced necroses in the liver lobule. (*b*) Cumulative labelling with [³H]thymidine. (*c*) Localization of DNA synthesis in the liver lobule. Necroses do not appear during the period of active proliferation (after Uryvaeva & Faktor, 1976*a*).

cellular functions. Cholesterol-7α-hydroxylase is a key enzyme in fatty-acid synthesis, and is one of the microsomal hydroxylases. The activity of this enzyme is inversely proportional to the rate of cellular proliferation in the liver. Enzyme activity is low in the growing liver of young animals; in adults there is a circadian rhythm in enzyme activity which is out of phase with mitotic activity (Barbason, Van Cantfort & Houbrechts, 1974). After partial hepatectomy, hydroxylase activity falls sharply. According to the authors, these observations indicate a direct interrelation of the genes for mitosis and for tissue-specific functions, which prevents their simultaneous activity; the expression of some genes is accompanied by the repression of others (Van Cantfort & Barbason, 1972).

A temporary decline in the activity of drug-metabolizing enzymes results in the resistance of the regenerating liver to hepatotoxins such as carbon tetrachloride and paracetamol (Uryvaeva & Faktor, 1976*a*,*b*). Carbon tetrachloride itself, as well as many other hepatotoxins, is non-toxic for hepatocytes; its toxic action appears only after metabolic conversion. Hence, the well known centrolobular liver necrosis induced by carbon tetrachloride poisoning may be regarded as a morphological sign of normal activity of the drug-metabolizing enzyme system (see Gerhard, Schultze & Maurer, 1972).

Judging from variation in the area of necrosis induced by carbon tetrachloride in regenerating mouse liver, it is clear that there is suppression of the drug-metabolizing function after stimulation of mitosis (Fig. 8.4). The loss of this specific function was compared with the mitotic rate and in terms of tissue architecture within the liver lobule. It is doubtful that the interaction of these two functions occurs at the gene level. The synthesis of new types of mRNA, which is to be regarded as genome reprograming (Markov *et al.*, 1975), takes place after the simultaneous stimulation of mitosis in all the zones of the liver lobule (Rabes & Brändle, 1968). The loss of the drug-metabolizing function, however, occurs asynchronously in the cells of a competent zone of the lobule. It begins at the periphery and progresses towards the center, similar to the wave-like propagation of DNA synthesis and mitosis through the lobule (Fabrikant, 1968).

Examples of competitive relations in liver function are mentioned by Salganik (1979). Thus, phenobarbital induces the intensification of the synthesis of drug-metabolizing oxidases. Cortisol stimulates the synthesis of tyrosine aminotransferase. If cortisol and phenobarbital were injected simultaneously for several days, then the activity of the tyrosine amino-transferase increased, although to a lesser extent than under the injection of cortisol alone. The activity of the oxidases studied did not increase. These and similar data may be interpreted as being the result of the competitive interrelations between the inducers and the receptors for inducers, as well as between the biochemical substances which play a part in the realization of tissue function.

The activity of microsomal enzymes decreases during stimulation of liver growth by various methods: partial hepatectomy, injection of growth hormone, or even the implantation of a tumor which produces this hormone (Henderson & Kersten, 1971, Wilson & Spelsberg, 1976). It has been hypothesized that specialized functions of the hepatocytes, including drug metabolism, decrease owing to the diversion of the resources for cell metabolism to the provision of the requirements for tissue growth (Wilson & Frohman, 1974).

The reciprocal interrelationships of growth-related and special liver functions were discovered in primary monolayer cultures derived from normal adult rat liver. Hepatocytes retained differentiated liver functions *in vitro*. The transition of hepatocytes from a resting state to proliferation is accompanied by the inhibition of specialized functions (measured as enzyme activity): glutathione S-transferase B, alcohol dehydrogenase, and so on. After cessation of the growth cycle, however, these activities reappear; the curves defining their modulations turn upwards (Leffert *et al.*, 1978; Koch & Leffert, 1980; Brown *et al.*, 1983). The cell density conditions influence the patterns of the cell functions, too. At low cell density, hepatocyte proliferation and growth-related functions are stimulated, at high cell density proliferation is inhibited and another program of liver function is stimulated (namely, tissue-specific functions; see Nakamura *et al.*, 1983). According to these authors, the regulation of and inversion of the different cell syntheses is accomplished at the transcriptional level.

The morphological effects caused by metabolic inhibitors on cycling cells may resemble the incomplete course of mitosis and, consequently, polyploidization in normal development. If mitosis depends on the adequate provision of materials generated during interphase, the effects of inhibitors of transcription and translation must reduce mitosis and favor polyploidy. Thus, after arrest of protein synthesis by chloramphenicol or streptonigrin in transformed lymphocytes *in vitro*, tetraploid mitoses and metaphase plates with diplochromosomes arise (Nasjleti & Spencer, 1967, 1968). Base analogs such as 8-azaguanine switch the cells of the pea root meristem over to endoreduplication (Nuti Ronchi, Avanzi & D'Amato, 1965). Actinomycin D and puromycin have the same effect on cultured animal cells (Sutou & Arai, 1975). Nagl (1970c, 1978) demonstrated the arrest of the middle and final phases of mitosis with actinomycin D in the meristematic root cells of *Allium* and a corresponding accumulation of 4C cells.

The somewhat biased selection of examples in this section can be explained by our wish to draw attention to the competitive nature of the interactions between cellular functions and the possible epigenetic control in the expression of proliferation and differentiation events.

Conclusion

It has been repeatedly stressed that polyploidization and polytenization develop as regular and inheritable processes of ontogenesis. In this chapter, two alternative control mechanisms are examined. The polyploidization of somatic cells may be fixed by inherited changes in the system of genes controlling mitosis. This point of view is substantiated by examples of mutant cell lines *in vitro* and also by organisms in which mitotic defects are inherited and hence are evidence of mutation. The other control mechanism is the influence of cellular metabolism on the mitotic processes. This suggests that polyploidy is more an epigenetic event although, ultimately, an inherited one.

We believe that only one example can be cited of a possible gene-independent inhibition of mitosis occurring at the time of increase of other (non-mitotic) cell functions. This is the compensatory reaction in the hemopoietic system of the pigeon in response to acute hemolytic anemia, described in Chapter 2I. In order to accelerate differentiation, several mitotic cycles in the cells of the erythroid series are eliminated and the last mitotic cycle is inhibited after a phase of DNA synthesis. The maturing cells live for a long time as a G_2-population with a 4C DNA content. Inhibition of the mitotic function here may be regarded as the result of a sharp increase in the rate of the processes of differentiation and the synthesis of hemoglobin.

Evidently, polyploidization can be regarded as an unprogramed event in those cases where the number of polyploid cells in the tissue is small or varies greatly. Examples of such facultative polyploidy were cited in this chapter.

It is difficult to determine ways of proving one or other of the controlling mechanisms in the ontogenetic, non-experimental expression of polyploidy. Possibly, one way is to study the behavior of polyploid cell populations during their growth outside the organism or in ectopic transplants. It is known that polytene cells will develop in such abnormal conditions. Thus, cells of the salivary gland of *Drosophila* larvae which have been transplanted to the body cavity of the adult fly still become polytene, but a higher level of polyteny is attained here than in normal development (Chapter 3A). Also, cells of the rat trophoblast become giant when cultivated *in vitro* or when transplanted under the kidney capsule or in the liver (Chapter 2G). In these cases, the polytene cycles remain just the same as *in situ*. It is more probable that polyteny rather than polyploidy develops as a result of changes in the genetic programing of mitosis.

9

General conclusions

This chapter is not just an ordinary summary: for, if read in order, it is the concluding remarks to each chapter, which will give a brief impression of the contents of the book. Here, in this chapter, we should like to pay further attention to some of the more general, theoretical problems that remain unresolved in the field of somatic polyploidy. Particularly, we wish to examine whether there is some fundamental distinction between the multiplied genome and the diploid genome and also between polyploidy and polyteny; furthermore, what is the significance of these phenomena for development.

The usual composition of the genome in eucaryotes is the diploid chromosome set. This state is maintained through many cell generations by mitosis. Diploidy provides for two fundamental aspects of cell reproduction: (i) reproduction of the cell mass and of all the cellular components and (ii) their even distribution between the daughter cells. In the incomplete mitotic cycle, cell growth and the reproduction of DNA and chromosomes occur unchanged, but the division of the cell into two daughter cells does not take place. In general, the formation of a multiplied cell genome is accompanied by many processes of the mitotic cycle, but the second component, the partitioning of cellular material, is partly eliminated.

It is convenient to discern two main variants of the incomplete mitotic cycle – the endocycle and polyploidizing mitosis. The endocycle, in which mitosis is completely excluded, is responsible for the appearance of endoreduplicated diplochromosomes, of giant polytene chromosomes and also of giant polytene nuclei with the usual interphase nuclear structure. In a cell which has undergone an endocycle, the sister chromosomes remain undivided and connected with each other to varying extents. On the other hand, polyploidizing mitosis is an ordinary mitotic process which has been interrupted at some stage. Acytokinetic mitosis, c-mitosis, and 'restitution mitosis' (mitosis with nuclear fusion during telophase) are well known forms of polyploidizing mitosis. The sister chromosomes definitely separate during polyploidizing mitosis. Endocycles produce only mononucleate cells. But polyploidizing mitoses, especially if recurrent, rearrange the

250

nuclear material in a single cell and tissues come to contain a mixed cell population of mono-, bi-, and multinucleate cells.

The terms 'polyploidizing mitosis' and 'mitotic polyploidization' we use (they were introduced by us in our review in *International Review of Cytology*, vol. 50, 1977) for the purpose of differentiating the mitotic process from polyteny and cell fusion. Cell fusion, although it produces structures with a multiple genome (binucleate or multinucleate cells) is not connected with the mitotic cycle. The cells involved in fusion need not be sister cells and, thus, may have different genetic properties. We hope that the terms 'endocycle' and 'polyploidizing mitosis' cover practically all the multifarious nuclear phenomena in Metazoa and Protozoa, the essence of which is the formation of cells with a multiple genome.

Whereas polyploidization of somatic cells is characteristic of vertebrates, polytenization is typical of invertebrates and plants. Most of the specialized organs of insect larvae consist exclusively of polytene cells. However, both in insects (and in other invertebrates) and also in plants, tissues have been described in which polyploid cells are typical. On the other hand, polytene populations develop in the extraembryonic tissues of mammals.

It must, nevertheless, be admitted that in multicellular organisms the element established by evolution, and, consequently, the element fundamental to life, is the mononucleate cell with a diploid chromosome set. Why, then, is the development of certain specialized cell lines accompanied by the formation of polyploid or polytene cells? This is not the place to discuss the general tendencies in the evolution of tissues and, in this sense, to examine whether the multiplication of the genome is a progressive or retrogressive evolutionary step, since information on the expression and distribution of the phenomenon is still fragmentary. Moreover, cases are known where the number of polyploid cells varies considerably in one and the same tissue even in related species. There are species of animals and plants in which there are very few polyploid cells. This indicates the distribution of polyploid cells is species-specific and that generalization between species may not be possible. We have attempted to find reasons for the evolutionarily determined degree of multiplication of the cellular genome in tissue-formation by comparing the physiological and genetic properties of these cells with the specific physiological features of the tissue.

Cells with a multiple genome differ from diploid ones in their physical properties. They are larger, and their size depends on the number of nuclear doublings. It is known that the sizes of diploid cells of animals and plants are defined within a fairly narrow range, determined by the quantity of genetic material and by the limit which the latter establishes for a working cytoplasmic volume. Genome multiplication relaxes restrictions on the expansion of the cytoplasmic mass and makes it possible for large cells to be formed. When the need for large and giant cells arose during evolution, modifications of the mechanisms of the mitotic cycle and, only very rarely, fusion of diploid cells were selected to bring them into being.

Although the nuclear cycle that accompanies differentiation of various cell types may be similar, the nuclei themselves may have quite different appearances. Examples are the salivary gland cells of insect larvae which have giant polytene chromosomes, the giant neurons of molluscs with a structure typical of interphase nuclei, the convoluted nuclei of the silk gland in the silkworm or the digestive gland in ascarids. All these giant cells are formed as a result of numerous endocycles. The highly polytene cells are characteristic of rapidly growing and actively functioning tissues. It is possible that the intensification of the protein synthesis is facilitated by the change-over of the cells to the endocycle, in which mitosis is completely excluded.

All known giant cells (in insects, ascarids, plants, and protozoans) have a polytene nature and undergo as many as 10–20 doubling cycles. The cells of somatic tissues in mammals do not usually attain such giant sizes; they frequently contain 4 (and up to a maximum of 64–128) chromosome sets. They usually undergo polyploidizing mitosis and not an endocycle. Evidently the achievement of giant cell sizes is not necessary for the functioning of mammalian tissues and polyploidization has, therefore, another purpose.

We should like to draw attention to one more essential property of the cell with a multiplied genome, namely the redundancy of genetic material. It is known that the duplication of the informative part of any system is an effective way of increasing its reliability. The absolute prevalence in nature of diploid forms testifies to the adequate informational reliability of the diploid chromosome set. It is also obvious that in some types of differentiated cells, especially those whose genome is liable to injury, a diploid set may not be adequate to compensate for induced injuries. The sensitivity of DNA to injury may be a species-specific and tissue-specific feature reflecting a high rate of production of genetically harmful or mutagenic metabolites, or a low level of protective substances, or defects in the DNA repair system, and so forth. Theoretically, cells with a multiple genome may be able to compensate for injury of a recessive nature to genes and chromosomes.

Mutagenic agents often affect mitosis. Evidently, the change-over to the endocycle and the exclusion of mitosis from the cell should stabilize the chromosome complement. The question arises as to why, when faced with the alternative between polyteny and polyploidy, polyteny is chosen less frequently. The drastic enlargement and concomitant decrease in surface area of the cell cannot be favorable for the functioning of many cell types. Each round of DNA replication in the endocycle brings in its wake an increase in the cell size. On the other hand, the cells undergoing poly-ploidizing mitosis do not lose their ability to undergo complete mitosis and can also increase in number. In polyploid cell populations, unlike in polytene ones, there is not only an increase in the size of individual cells but also in their number (when cell reproduction and genome multiplication occur in the life cycle of the same cell).

The change-over to a polyploid mode of existence can frequently be found in autonomous cell populations such as tumor cells and cells growing in culture. This is probably connected with the adaptive possibilities open to the polyploid genome and may form the basis for establishing new genetic combinations. The informational redundancy of the multiple genome gives this scope.

In conclusion, we shall try to define the role played by genome multiplication in developmental events. Usually, this occurs not at an early but at a late stage of development of the cell lineages; it does, in a certain sense, look like polyploid- or polytene-directed cell differentiation. We may ask whether it occurs in a stable or an expanding cell population. If cells with an increased genome are regularly formed in renewing tissues, then this process occurs only in the maturing and mature compartment but not in the precursor cell compartment. If we discuss the phenomenon of genome multiplication in terms of the life of a single cell (from its birth to its death which, in the organism, may be the result of programed cell death), then polyploidization or polytenization occurs in a cell's terminal reproductive cycle. This is quite understandable since the change-over of the population from a diploid to a polyploid or polytene state imposes definite restrictions on morphogenetic processes, cell migration, and on further increases in cell number. Consequently, although the processes of cell genome multiplication appear to be necessary to cell differentiation, they do not affect the processes of activation and transcription of specific genes. Polytenization and polyploidization are phenotypic traits of terminally differentiated cells. They do not determine the direction of differentiation, but facilitate the functional expression of the differentiated cells.

References

Abdel-Hameed, F. (1972). Hemoglobin concentration in normal diploid and intersex triploid chickens: genetic inactivation or canalization? *Science*, **178**, 864–5.

Adrian, E. K. (1971). The metabolic stability of nuclear DNA. In *Cell and Molecular Renewal in the Mammalian Body*, ed. I. L. Cameron & J. D. Thrasher, pp. 25–40. New York & London: Academic Press.

Aizenshtadt, T. B. (1977). Growth of oocytes and vitellogenesis. In *Modern Problems of Oogenesis*, ed. T. A. Detlaf, pp. 5–50. Moscow: Nauka (in Russian).

Aizenshtadt, T. B., Brodsky, V. Ya. & Gazaryan, K. G. (1967). Radioautographic study of RNA and protein syntheses in the gonads of animals with different types of oogenesis. *Tsitologia*, **9**, 397–406 (in Russian).

Aizenshtadt, T. B. & Marshak, T. L. (1969). Nuclear DNA in the nurse cells of some invertebrates: a cytophotometric study. *Arkhiv Anatomii, Histologii, Embryologii*, **56**, 15–21 (in Russian).

Aizenshtadt, T. B. & Marshak, T. L. (1974). A study of oogenesis in hydra: II. *Ontogenez*, **5**, 446–53 (in Russian, English translation in *Soviet J. Dev. Biol.*, **5**, 394–401).

Al-Bader, A. A., Orengo, A. & Rao, P. N. (1978). G_2-phase-specific proteins of HeLa cells. *Proc. Natl. Acad. Sci. USA*, **75**, 6064–8.

Albert, M. D. & Bucher, N. L. R. (1960). Latent injury and repair in rat liver induced to regenerate at intervals after X-radiation. *Cancer Res.* **20**, 1514–22.

Albertson, D. G., Sulston, J. E. & White, J. G. (1978). Cell cycling and DNA replication in a mutant blocked in cell division in the nematode *Caenorhabditis elegans*. *Dev. Biol.*, **63**, 165–78.

Alfert, M. (1954). Composition and structure of giant chromosomes. *Int. Rev. Cytol*, **3**, 131–75.

Alfert, M. & Das, N. K. (1969). Evidence for control of the rate of nuclear DNA synthesis by the nuclear membrane in eucaryotic cells. *Proc. Natl. Acad. Sci. USA*, **63**, 123–8.

Alfert, M. & Geschwind, I. (1958). The development of polysomaty in rat liver. *Exp. Cell Res.*, **15**, 230–2.

Alfert, M. & Siegel, E. P. (1963). Rates of DNA synthesis in diploid and tetraploid mouse liver nuclei. *J. Cell Biol.*, **19**, 3A (abstract).

Allen, A. M. (1962). Mitosis and binucleation in mast cells of the rat. *J. Natl. Cancer Inst.*, **28**, 1125–51.

Altmann, H.-W. (1966). Der Zellersatz insbesondere an den parenchymatösen Organen. *Verhandl. deutsch. Ges. Pathol.*, **50**, 815–53.

Alvarez, M. & Cowden, R. (1966). Karyometric and cytophotometric study of hepatocyte nuclei of frogs exposed to cold and prolonged starvation. *Z. Zellforsch. mikr. Anat.*, **75**, 240–9.

Ammermann, D. (1973). Cell development and differentiation in the ciliate *Stylonychia mytilus*. In *The Cell Cycle in Development and Differentiation*, ed. M. Balls & F. S. Billett, pp. 51–60. Cambridge University Press.

Ammermann, D. & Muenz, A. (1982). DNA and protein content of different hypotrich ciliates. *Europ. J. Cell Biol*, **27**, 22–4.

Ammermann, D., Steinbruck, G., von Berger, L. & Henning, W. (1974). The development of the macronucleus in the ciliated protozoan *Stylonychia mytilus*. *Chromosoma*, **45**, 401–29.

Anderson, J. M. (1954). Studies on the cardiac stomach of the starfish *Asterias forbesi*. *Biol. Bull.*, **107**, 157–73.

Anderson, J. M. (1965). Studies on visceral regeneration in sea-star. III. Regeneration of the cardiac stomach in *Asterias forbesi* (Desor). *Biol. Bull.*, **129**, 454–70.

Andersson, L. C., Berlin, T., Collste, L., Granber-Ohman, G., Gustafsson, H., Tribukait, B. & Wijkström, H. (1980). Newer aspects of research in the field of bladder tumours. In *Bladder Tumors and other Topics in Urological Oncology*, ed. M. Pavone-Macaluso, P. H. Smith & F. Edsmyr, pp. 59–67. New York: Plenum Press.

Andersson, L. C., Lehto, V. P., Stenman, S., Badley, R. A. & Virtanen, G. (1981). Diazepam induces mitotic arrest at prometaphase by inhibiting centriolar separation. *Nature (London)*, **291**, 247–8.

Andreeva, L. F. (1964). A study of DNA synthesis and cell populations kinetics and giant cells and in cell trophoblast of placenta. In *Cell Cycle and Nucleic Acid Metabolism during Differentiation*, ed. L. N. Zhinkin and A. A. Zavarzin pp. 136–47. Leningrad: Nauka (in Russian).

Anisimov, A. P. (1973). DNA content and cell size differences in spermatozoa of *Ascaris suum*. *Tsitologia*, **15**, 1162–5 (in Russian).

Anisimov, A. P. (1974). Polyploidization and nuclear growth in ontogenesis of uterine epithelium of *Ascaris suum* (Nematoda). *Arkhiv Anatomii, Histologii, Embryologii*, **66**, 36–44 (in Russian).

Anisimov, A. P. (1976). Polyploidization and growth of giant nuclei of oesophageal glands during postnatal ontogenesis of *Ascaris suum*. *Tsitologia*, **18**, 445–50 (in Russian).

Anisimov, A. P. (1984). Reduplication and transcription in cycles of cell polyploidization in the mollusc *Succinea putris*. In *Cell Nucleus*, ed. I. B. Zbarsky, Nauka, Pushchino, in press.

Anisimov, A. P. & Tokmakova, N. P. (1973). Proliferation and growth of the intestinal epithelium of *Ascaris suum* (Nematoda) in postnatal ontogenesis. I. Mitotic activity the intestinal epithelium. *Ontogenez*, **4**, 373–82 (in Russian, English translation in the *Soviet J. Dev. Biol.*, **4**, 341–8).

Anisimov, A. P. & Usheva, L. N. (1973). Proliferation and growth of the intestinal epithelium of *Ascaris suum* (Nematoda) during postnatal ontogenesis. II. Increase in the cell size and number as a factor of tissue growth. *Ontogenez*, **4**, 416–20 (in Russian, English translation in the *Soviet J. Dev. Biol.*, **4**, 379–83).

Ansell, J. D., Barlow, P. W. & McLaren, A. (1974). Binucleate and polyploid cells in the decidua of the mouse. *J. Embryol. Exp. Morphol.*, **31**, 223–7.

Arefyeva, A. M., Durova, S. I., Meerson, F. Z. & Brodsky, W. Y. (1982). Growth of rat cardiomyocytes in the course of adaptation to high altitude hypoxia. *Tsitologia*, **24**, 1435–9 (in Russian).

Armstrong, S. W. & Davidson, D. (1982). Difference in protein content of sister nuclei: evidence from binucleate and mononucleate cells. *Can. J. Biochem.*, **60**, 371–8.

Arold, R. & Sandritter, W. (1967). Zytophotometrische Bestimungen des DNS-Gehaltes von Zellkernen des Nebennierenmarkes, der Nebennierenrinde und der Schilddrüse unter verschiedenen experimentellen Bedingungen. *Histochemie*, **10**, 88–97.

Ashburner, M. (1970). Function and structure of polytene chromosomes during insect development. *Adv. Insect Physiol.*, **7**, 1–95.

Astrin, S. M. & Rothberg, P. G. (1983). Oncogenes and cancer. *Cancer Invest.*, **1**, 355–64.

Atkin, N. B. (1976). Prognostic significance of ploidy level in human tumors. I. Carcinoma of the uterus. *J. Natl. Cancer Inst.*, **56**, 909–10.

Atkin, N. B. & Kay, R. (1979). Prognostic significance of modal DNA value and other factors in malignant tumours, based on 1465 cases. *Brit. J. Cancer*, **40**, 210–21.

Atkin, N. B., Mattinson, G. & Baker, M. C. (1966). A comparison of the DNA content and chromosome number of fifty human tumours. *Brit. J. Cancer*, **20**, 87–101.

Auer, G. U., Caspersson, T. O. & Wallgren, A. S. (1980). DNA content and survival in mammary carcinoma. *Analyt. and Quant. Cytol.*, **2**, 161–5.

Avanzi, S., Cionini, P. G. & D'Amato, F. (1970). Cytochemical and autoradiographic analyses on the embryo suspensor cells of *Phaseolus coccineus*. *Caryologia*, **23**, 605–38.

Avery, G. & Hunt, C. (1972). Giant cell formation in ectopic mouse trophoblast. *Exp. Cell Res.*, **74**, 3–8.

Avtandilov, G. G. (1982). Die gesetzmäßige exponentielle Zunahme des durchschnittlichen DNA-Gehaltes der Zellkerne in der Wachstumszone des Epithels bei Hyperplasie und Malignisierung. *Acta Histochem.*, **26** (Suppl.), 67–70.

Avtandilov, G. G., Krasnova, V. G. & Nikitina, N. I. (1975). Microspectrophotometry of nuclear DNA in cells of laryngeal tumours. *Arkhiv Patologii*, **37**, 15–9 (in Russian).

Bachmann, K. (1970). Specific nuclear DNA amounts in toads of the genus *Bufo*. *Chromosoma*, **29**, 365–74.

Bachmann, K. (1972). Genome size in mammals. *Chromosoma*, **37**, 85–93.

Bachmann, K. & Cowden, R. (1965*a*). Quantitative cytophotometric studies on polyploid liver cell nuclei of frog and rat. *Chromosoma*, **17**, 181–93.

Bachmann, K. & Cowden, R. (1965*b*). Quantitative cytophotometric study on isolated liver cell nuclei of bullfrog, *Rana catesbiana*. *Chromosoma*, **17**, 22–34.

Bachmann, K. & Cowden, R. R. (1967). A quantitative cytochemical study of the 'polyploid' liver nuclei of the frog genus *Pseudacris*. *Trans. Amer. Microscop. Soc.*, **86**, 454–63.

Bachmann, K., Goin, O. B. & Goin, C. I. (1966). Hylid frogs: polyploid classes of DNA in liver nuclei. *Science*, **154**, 650–1.

Bachmann, K., Harrington, B. A. & Craig, J. P. (1972). Genome size in birds. *Chromosoma*, **37**, 405–16.

Bachmann, K., Konrad, A., Oeldorf, E. & Hemmer, H. (1978). Genome size in the green toad (*Bufo viridis*) group. *Experientia*, **34**, 331–2.

Bachop, W. E. & Schwartz, F. J. (1974). Quantitative nucleic acid histochemistry of the yolk sac syncytium of oviparous teleosts: Implications for hypothesis of yolk utilization. In *The Early History of Fish*, ed. J. H. S. Blaxter, pp. 345–53. Berlin–New York: Springer Verlag.

Bahr, G. & Weid, G. (1966). Cytochemical determinations of DNA and basic protein in bull spermatozoa. UV spectrophotometry, cytophotometry, microfluorometry. *Acta Cytol.*, **10**, 393–412.

Bakhtadze, G. I. (1981). Protein synthesis in testicular follicle cells of the desert locust. *Tsitologia*, **23**, 51–4 (in Russian).

Bansal, J. & Davidson, D. (1978). Analysis of growth of tetraploid nuclei in roots of *Vicia faba*. *Cell Tiss. Kinet.*, **11**, 193–200.

Barbason, H., Van Cantfort, J. & Houbrechts, N. (1974). Correlation between tissular and division functions in the liver of young rats. *Cell Tissue Kinet.*, **7**, 319–26.

Barigozzi, C. & Semenza, L. (1952). A preliminary note on the biology and chromosome cycle of *Aphiochaeta xanthina* sp. *Amer. Nat.*, **86**, 123–4.

Barka, T. (1967*a*). Stimulation of DNA synthesis in the salivary gland by isoproterenol. *Exp. Cell Res.*, **48**, 53–60.

Barka, T. (1967*b*). Further studies on the stimulation of DNA synthesis in the salivary gland by isoproterenol. *J. Cell Biol.*, **35**, 9A–10A.

Barka, T. (1970). Further studies on the effect of isoproterenol on RNA synthesis of salivary glands. *Exp. Cell. Res.*, **62**, 50–60.

Barlogie, B., Göhde, W., Johnston, A., Smallwood, L., Schumann, J., Drewinko, B. & Freireich, E. I. (1978). Determination of ploidy and proliferative characteristics of human solid tumours by pulse cytophotometry. *Cancer Res.*, **38**, 3333–9.

Barlogie, B., Raber, M. N., Schumann, J., Johnson, T. S., Drewinko, B., Swartzendruber, D. E., Göhde, W., Andreef, M. & Freireich, E. I. (1983). Flow cytometry in clinical cancer research. *Cancer Res.*, **43**, 3982–97.

Barlow, P. W. (1975). The polytene nucleus of the giant hair cell of *Bryonia* anthers. *Protoplasma*, **83**, 339–49.

Barlow, P. W. (1976). The relationship of the dispersion phase of chromocentric nuclei in the mitotic cycle to DNA synthesis. *Protoplasma* **90**, 381–91.

Barlow, P. W. (1977a). An experimental study of cell and nuclear growth and their relation to cell diversification within a plant tissue. *Differentiation*, **8**, 153–7.

Barlow, P. W. (1977b). The time-course of endoreduplication of nuclear DNA in the root cap of *Zea mays*. *Cytobiologie*, **16**, 98–105.

Barlow, P. W. (1978a). The interrelationship of the cycles of chromosome condensation and reduplication in cell growth processes. *The Nucleus*, **21**, 1–11.

Barlow, P. W. (1978b). Endopolyploidy: towards an understanding of its biological significance. *Acta Biotheoret.*, **27**, 1–18.

Barlow, P. W. & Nevin, D. (1976). Quantitative karyology of some species of *Luzula*. *Plant Syst. Evol.*, **125**, 77–86.

Barlow, P. W. & Sherman, M. I. (1972). The biochemistry of differentiation of mouse trophoblast: studies in polyploidy. *J. Embryol. Exp. Morphol.*, **27**, 447–65.

Barlow, P. W. & Sherman, M. I. (1974). Cytological studies on the organization of DNA in giant trophoblast nuclei of the mouse and the rat. *Chromosoma*, **47**, 119–31.

Barr, M. (1959). Sex chromatin and phenotype in man. *Science*, **130**, 679–85.

Baserga, R. (1970). Induction of DNA synthesis by a purified chemical compound. *Fed. Proc.*, **29**, 1443–6.

Baserga, R. (1984). Growth in size and cell DNA replication. *Exp. Cell. Res.*, **151**, 1–5.

Baserga, R., Sasaki, F. & Whitlock, J. P. (1969). The prereplicative phase of isoproterenol-stimulated DNA synthesis. In *Biochemistry of Cell Division*, ed. R. Baserga, pp. 77–90. London: C. C. Thomas.

Bast, R. E., Garfield, S. A., Gehrke, L. & Ilan, J. (1977). Coordination of ribosome content and polysome formation during estradiol stimulation of vitellogenin synthesis in immature male chick livers. *Proc. Natl. Acad. Sci. USA*, **74**, 3133–7.

Baudoin, J. & Ormières, R. (1973). Sur quelques particularités de l'ultrastructuré de *Didymophyes chaudefouri* Ormières, Eugrégarine, Didymophidae. *C.r. Acad. Sci, Paris*, Sér. D., **277**, 73–5.

Bauer, A. (1974). Der Einfluss der Ploidie auf die Strahlenreaktion von Säugerzellen. I. Eine neue tetraploide chinesische Hamster Zellinien. *Humangenetik*, **23**, 289–96.

Bauer, H. (1936). Structure and arrangement of salivary gland chromosomes in *Drosophila* species. *Proc. Natl. Acad. Sci., USA*, **22**, 216–25.

Beams, H. & King, R. (1942). The origin of binucleate and large mononucleate cells in the liver of the rat. *Anat. Rec*, **83**, 281–97.

Becker, F. F., Klein, K. M., Wolman, S. R., Asofsky, R. & Sell, S. (1973). Characterization of primary hepatocellular carcinomas and initial transplant generations. *Cancer Res.*, **33**, 3330–8.

Becker, H. J. (1976). Mitotic recombination. In *The Genetics and Biology of Drosophila*, ed. M. Ashburner & E. Novitski, pp. 1020–87, London & New York: Academic Press.

Beckingham, K. & Thompson, H. (1982). Under-replication of intron[+] rDNA cistrons in polyploid nurse cell nuclei of *Calliphora erythrocephala*. *Chromosoma*, **87**, 177–96.

Beçak, M. L. & Beçak, W. (1974). Studies on polyploid amphibians-karyotype evolution and phylogeny of the genus *Odontophrynus*. *J. Herpetol.*, **8**, 337–41.

Beçak, W. & Goissis, G. (1971). DNA and RNA content in diploid and tetraploid amphibians. *Experientia*, **27**, 345–6.

Bedo, D. G. (1982). Differential sex chromosome replication and dosage compensation in polytene trichogen cells of *Lucilia cuprina* (Diptera: Calliphoridae). *Chromosoma*, **87**, 21–32.

Beebe, D. C. & Piatigorsky, I. (1977). The control of δ-crystallin gene expression during lens development: dissociation of cell elongation, cell division, δ-crystallin synthesis and δ-crystallin mRNA accumulation. *Dev. Biol.*, **59**, 174–82.

Beerman, S. (1977). The diminution of heterochromatic chromosomal segments in *Cyclops* (Crustacea, Copepoda). *Chromosoma*, **60**, 297–344.

Beerman, W. (1952). Chromomerenkonstanz und spezifische Modifikationen der Chromosomenstruktur in der Entwicklung und Organdifferenzierung von *Chironomus tentans*. *Chromosoma*, **5**, 139–98.

Bell, A. G. (1974). Diploid and endoreduplicated cells: measurements of DNA. *Science*, **143**, 139–40.

Benedict, W. F. & Jones, P. A. (1979). Mutagenic, clastogenic and oncogenic effects of 1-β-D-arabinofuranosylcytosine. *Mut. Res.*, **65**, 1–20.

Bennett, M. D. (1970). Natural variation in nuclear characters of meristems in *Vicia faba*. *Chromosoma*, **29**, 317–35.

Bennett, M. D. (1972). Nuclear DNA content and minimum generation time in herbaceous plants. *Proc. R. Soc. London*, **B 181**, 109–35.

Bennett, M. D. (1974). Nuclear characters in plants. In *Basic Mechanisms in Plant Morphogenesis*. Brookhaven Symposia in Biology, **25**, pp. 344–66. New York: Associated Universities Inc.

Bennett, M. D. & Smith, J. B. (1976). Nuclear DNA in angiosperms. *Phil. Trans. R. Soc. London*, **B 274**, 227–74.

Bennett, M. D., Smith, I. B., Ward, J. P., & Finch, R. A. (1982). The relationship between chromosome volume and DNA content in unsquashed metaphase cells of barley *Hordeum vulgare* cv. Tuleen 346. *J. Cell. Sci.*, **56**, 101–11.

Ben-Ze'ev, A. & Raz, A. (1981). Multinucleation and inhibition of cytokinesis in suspended cells: reversal upon reattachment to a substrate. *Cell*, **26**, 107–15.

Berendes, H. D. (1973). Synthetic activity of polytene chromosomes. *Int. Rev. Cytol.*, **35**, 61–116.

Berger, J. D. (1982). Effects of gene dosage on protein synthesis rate in *Paramecium tetraurelia*: implications for regulation of cell mass, DNA content and the cell cycle. *Exp. Cell Res.*, **141**, 261–75.

Berman, E. R., Schwell, H. & Feeney, L. (1974). The retinal pigment epithelium: chemical composition and structure. *Invest. Ophthalmol.*, **13**, 675–87.

Bernocchi, G. (1975). Contenuto in DNA e area nucleare dei neuroni durante l'istogenesi cerebellare del ratto. *Rend. Ist Lomb. Accad. Sci. Lett. (B)*, **109**, 143–61.

Berry, R. J. (1963). Quantitative studies of relationships between tumor cell ploidy and dose response to ionizing radiation *in vivo*. *Radiat. Res.* **18**, 236–45.

Bershadsky, A. D., Gelfant, V. I. & Vasiliev, J. M. (1981). Multinucleation of transformed cells normalizes their spreading on the substratum and their cytoskeleton structure. *Cell Biol. Int. Rep.*, **5**, 143–50.

Bezruchko, S. M., Voshenina, N. I., Gazaryan, K. G., Kulminskaya, A. S. & Kukhtin, V. A. (1969). Radioautographic study of DNA synthesis in giant neurons of mollusc *Tritonia diomedia*. *Biofisica*, **14**, 1052–4 (in Russian, English translation in *Biophysics*, **14**, 1107–11).

Bhadra, S. K., Shaikh, M. A. Q. & Mia, M. M. (1979). Mitotic chromosomal abnormalities in diploid and colchicine-induced tetraploid jute (*Corchorus capsularis* L.) following gamma-rays treatment. *Cytologia*, **44**, 359–64.

Bhattacharya, R. D., van Noorden, C. J. F. & James, J. (1983). A time-dependent distribution pattern of ploidy classes in adult rat liver parenchyma. *Acta Anat.*, **116**, 168–73.

Bier, K. E. (1957). Endomitose und Polytänie in der Nährzellkernen von *Calliphora erythrocephala*. *Chromosoma*, **8**, 493–522.

Bier, K. E. (1959). Quantitative Untersuchungen über die Variabilität der Nährzellkernstruktur und ihre Beeinflussung durch die Temperatur. *Chromosoma*, **10**, 619–53.

Bier, K. E. (1960). Der Karyotyp von *Calliphora erythrocephala* unter besonderer Berücksichtigung der Nährzellkernchromosomen in gebündelten und gepaarten Zustand. *Chromosoma*, **11**, 335–64.

Bier, K. E. (1970). Oogenestypen bei Insekten und Vertebraten, ihre Bedeutung für die Embryogenese und Phylogenese. *Zool. Anz., Suppl.*, **33**, 7–29.

Bird, A. F. (1973). Observations on chromosomes and nucleoli in syncytia induced by *Meloidogyne javanica*. *Physiol. Plant Pathol.*, **3**, 387–91.

Bloom, S. E., Buss, E. G. & Strother, G. K. (1970). Cytological and cytophotometric analysis of binucleated red blood cell mutants in turkeys. *Genetics*, **65**, 51–63.

Bloom, S. E. & Searle, B. M. (1973). Genetic control of red blood cell mitosis in *bn* mutant turkeys. *Genetics*, **74** Suppl. 2, 26 (abstract).

Bocquet-Védrine, J. (1970). Polyploidie et multiplication amitotique des cellules glandulaires cémentaires chez le crustacé cirripède operculé *Balanus crenatus*. *C.r. Acad. Sci., Paris., sér. D*, **270**, 506–8.

Boer, H. H., Groot, C., De Jong-Brink, M., & Cornelisse, C. J. (1977). Polyploidy in the freshwater snail *Lymnaea stagnalis* (Gastropoda, Pulmonata): a cytophotometric analysis of the DNA in neurons and some other cell types. *Neth. J. Zool.*, **27**, 245–52.

Böhm, N. & Noltemeyer, N. (1981a). Development of binuclearity and DNA-polyploidization in the growing mouse liver. *Histochem.*, **72**, 55–61.

Böhm, N. & Noltemeyer, N. (1981b). Excessive reversible phenobarbital induced nuclear DNA-polyploidization in the growing mouse liver. *Histochem.*, **72**, 63–74.

Böhm, N. & Sandritter, W. (1975). DNA in human tumor; a cytophotometric study. *Curr. Top. Pathol.*, **60**, 152–210.

Boivin, A., Vendrely, R. & Vendrely, C. (1948). D'acide désoxyribonucléique du noyau cellulaire dépositaire de caractères héréditaires. *C.r. Acad. Sci., Paris.*, **226**, 1061–3.

Borovyagin, V. L. & Sakharov, D. A. (1968). *Ultrastructure of Giant Neurons of Tritonia.* Moscow: Nauka (in Russian).

Borsook, H., Lingrell, J., Scaro, I. L. & Millette, R. L. (1962). Synthesis of hemoglobin in relation to the maturation of erythroid cells. *Nature (London)*, **196**, 347–50.

Boyle, W. S. (1968). Effect of type and level of polyploidy on radiosensitivity. *J. Hered.*, **59**, 151–4.

Brady, T. (1973). Feulgen cytophotometric determination of the DNA content of the embryo proper and suspensor cells of *Phaseolus coccineus*. *Cell Differentiation*, **2**, 65–75.

Brady, T. & Clutter, M. (1974). Structure and replication of *Phaseolus* polytene chromosomes. *Chromosoma*, **45**, 63–79.

Brasch, K. (1980). Endopolyploidy in vertebrate liver: an evolutionary perspective. *Cell Biol. Int. Rep.*, **4**, 217–26.

Brasch, K. (1982). Age- and ploidy-related changes in rat liver nuclear proteins as assessed by one- and two-dimensional gel electrophoresis. *Can. J. Biochem.*, **60**, 204–14.

Brenner, S., Branch, A., Meredith, S. & Berns, M. W. (1977). The absence of centrioles from spindle poles of rat kangaroo (PtK$_2$) cells undergoing meiotic-like reduction division *in vitro*. *J. Cell Biol.*, **72**, 368–79.

Bridges, C. B. (1936). Salivary chromosome maps with a key to the banding of the chromosomes of *Drosophila melanogaster*. *J. Hered.* **26**, 60–4.

Britten, R. J. & Davidson, E. H. (1969). Gene regulation for higher cells: a theory. *Science*, **165**, 349–57.

Brodsky, V. Ya. (1966). *Functional Cytochemistry*, Moscow: Nauka (in Russian).

Brodsky, V. Ya. (1975). Protein synthesis rhythm. *J. Theoret. Biol.* **55**, 167–200.

Brodsky, V. Ya., Agroskin, L. S., Marshak, T. L., Papajan, G. W., Segal, O. L., Sokolova, G. A. & Yarigin, K. N. (1974). Stability and variation of DNA content in cerebellar cells. *Zhurnal obshchei Biologii*, **35**, 917–25 (in Russian).

Brodsky, V. Ya., Arefyeva, A. M. & Uryvaeva, I. V. (1980*a*). Mitotic polyploidization of mouse heart myocytes during the first postnatal week. *Cell Tissue Res.*, **210**, 133–44.

Brodsky, V. Ya., Bukhvalov, I. B., Nechaeva, N. V., Uryvaeva, I. V. & Chernorotova, T. E. (1971*a*). The DNA content of nuclei of blastula cells and certain differentiated cells of the loach (*Misgurnus fossilis*). *Ontogenez*, **1**, 587–94 (in Russian, English translation in the *Soviet J. Dev. Biol.*, **1**, 430–6).

Brodsky, V. Ya., Faktor, V. M. & Uryvaeva, I. V. (1973). Reproduction of hepatocytes in postnatal development of mouse liver. *Arkhiv Anatomii, Histologii, Embryologii*, **65**, 7–15 (in Russian).

Brodsky, V. Ya. & Krushchov, N. G. (1962). DNA cytophotometry in the course of amitosis. *Doklady Acad. Nauk SSSR*, **147**, 939–42 (in Russian, English translation in *Doklady Biological Sciences*, **147**, 1277–80).

Brodsky, V. Ya., Kudryavtsev, B. N. & Marshak, T. L. (1980*b*). Determination of DNA surplus in the nuclei of some cerebellar Purkinje cells with different cytochemical methods. *Basic Appl. Histochem.*, **24**, 401–8.

Brodsky, V. Ya., Kvinihidze, G. S., Magakyan, Y. A., Uryvaeva, I. V. & Karalova, E. M. (1971*b*). DNA content of the forebrain and the retinal cells of chick embryo. *Tsitologia*, **13**, 522–5 (in Russian).

Brodsky, V. Ya., Marshak, T. L., Mareš, V., Lodin, Z., Fülöp, Z. & Lebedev, E. A. (1979). Constancy and variability in the content of DNA in cerebellar Purkinje cell nuclei. *Histochem.*, **59**, 233–48.

Brodsky, V. Ya., Marshak, T. L., Mikeladze, Z. A., Moskovkin, G. N. & Satdycova, M. K. (1983). DNA synthesis in Purkinje neurons. *Basic Appl. Histochem.*, **28**, 187–94.

Brodsky, V. Ya., Miljutina, N. A., Faktor, V. M. & Uryvaeva, I. V. (1969). A study of DNA synthesis and mitosis in the regenerating mouse liver in order to reveal the G_2-population in hepatocytes. *Doklady Acad. Nauk SSSR*, **189**, 639–42 (in Russian, English translation in *Doklady Biol. Sci.*, **189**, 791–3).

Brodsky, V. Ya., Tsirekidze, N. N. & Arefyeva, A. M. (1984). Mitotic-cyclic and cycle-independent growth of cardiomyocytes. *J. Mol. Cell Cardiol.* (in press).

Brodsky, V. Ya., & Uryvaeva, I. V. (1970). Development and properties of polyploid cell populations in the ontogeny of mammals. *Soviet J. Dev. Biol.*, **1**, 181–95.

Brodsky, V. Ya., & Uryvaeva, I. V. (1974). Somatic polyploidy in tissue development. *Soviet J. Dev. Biol.*, **5**, 594–605.

Brodsky, V. Ya., & Uryvaeva, I. V. (1977). Cell polyploidy: its relation to tissue growth and function. *Int. Rev. Cytol.*, **50**, 275–332.

Broekaert, D. & van Parijs, R. (1975). The origin of wound-induced satellite DNA in pea seedlings. *Cell Differentiation*, **4**, 139–45.

Broekaert, D. & Van Parijs, R. (1978). Histophotometric DNA measurements on Umbelliferae, Solanaceae and Compositae before and after crown gall tumor induction. *Protoplasma*, **93**, 433–42.

Brossard, D. (1978). Microspectrophotometric and ultrastructural analyses of a case of cell differentiation without endopolyploidization: the pith of *Crepis capillaris* (L.) Wallsr. *Protoplasma*, **93**, 369–80.

Brown, J. W., Lad, P. J., Skelly, H., Koch, K. S., Lin, M. & Leffert, H. (1983).

Expression of differentiated function by hepatocytes in primary culture: variable effects of glucagon and insulin on gluconeogenesis during cell growth. *Differentiation*, **25**, 176–84.

Brunori, A. (1971). Synthesis of DNA and mitosis in relation to cell differentiation in the roots of *Vicia faba* and *Latuca sativa*. *Caryologia*, **24**, 209–15.

Bucher, N. (1963). Regeneration of mammalian liver. *Int. Rev. Cytol.*, **15**, 245–300.

Bucher, O. (1959). Die Amitose der tierischen und menschlichen Zelle. *Protoplasmatologia*, **6**, 1–159. Berlin & Vienna: Springer Verlag.

Bucher, O. (1971). Zum Problem der Amitose. In: *Handbuch der allgemeinen Pathologie*: *Der Zellkern*, vol. 1, ed. H.-W. Attmann, pp. 627–99. Berlin: Springer Verlag.

Buetow, D. E. (1971). Cellular content and cellular proliferation changes in the tissues and organs of the aging mammals. In *Cellular and Molecular Renewal in the Mammalian Body*, ed. I. L. Cameron & J. D. Thrasher, pp. 87–106. New York & London: Academic Press.

Butterfass, T. (1973). Control of plastid division by means of nuclear DNA amount. *Protoplasma*, **76**, 167–95.

Byrt, P. (1966). Secretion and synthesis of amylase in the rat parotid gland after isoprenaline. *Nature (London)*, **212**, 1212–15.

Cabral, F. R. (1983). Isolation of Chinese hamster ovary cell mutants requiring the continuous presence of taxol for cell division. *J. Cell Biol.*, **97**, 22–9.

Cabral, F. R., Wible, L., Brenner, S. & Brinkley, B. R. (1983). Taxol-requiring mutant of Chinese hamster ovary cells with impaired mitotic spindle assembly. *J. Cell Biol.*, **97**, 30–9.

Cameron, I. L. & Jeter, J. (1971). Relationship between cell proliferation and cytodifferentiation in embryonic chick tissues. In *Developmental Aspects of the Cell Cycle*, ed. I. Cameron, G. Padilla & A. Zimmerman, pp. 191–222. New York & London: Academic Press.

Cameron, I. L., Pool, M. R. H. & Hoage, T. R. (1979). Low level incorporation of tritiated thymidine into the nuclear DNA of Purkinje neurons of adult mice. *Cell Tissue Kinet.*, **12**, 445–51.

Campbell, G., Weintraub, H., Mayall, B. & Holtzer, H. (1971). Primitive erythropoiesis in early chick embryogenesis. II. Correlation between hemoglobin synthesis and the mitotic history. *J. Cell Biol.*, **50**, 669–81.

Campbell, R. (1973). Endopolyploidy in giant endoderm cells of the coelenterate polyp *Corymorpha palma*. *Trans. Amer. Microscop. Soc.*, **92**, 129–34.

Capanna, E. & Manfredi Romanini, M. G. (1971). Nuclear DNA content and morphology of the karyotype in certain palearctic Microchiroptera. *Caryologia*, **24**, 471–82.

Capesius, I. & Stöhr, M. (1974). Endopolyploidisierung während des Streckungswachstums der Hypokotyle von *Sinapis alba*. *Protoplasma*, **82**, 147–53.

Carlsson, E., Kjörell, U., Thornell, L. E., Lambertsson, A. & Strehler, E. (1982). Differentiation of myofibrils and the intermediate filament system during postnatal development of the rat heart. *Europ. J. Cell Biol.*, **27**, 62–73.

Carrano, A. V. (1973). Chromosome aberrations and radiation-induced cell death. II. Predicted and observed cell survival. *Mutation Res.*, **17**, 355–66.

Carrière, R. (1969). The growth of liver parenchymal nuclei and its endocrine regulation. *Int. Rev. Cytol.*, **25**, 201–77.

Carter, S. B. (1967). Effects of cytochalasins on mammalian cells. *Nature (London)*, **213**, 261–4.

Caspersson, T. (1951). *Cell Growth and Cell Function*. New York & London: Academic Press.

Catarino, F. M. (1965). Salt water, a growth inhibitor causing endopolyploidy. *Portug. Acta Biol.* (ser. A), **9**, 131–52.

Cavallini, A., Cionini, P. G. & d'Amato, F. (1981). Location of Heitz's Zerstäubungsstadium (dispersion phase) in the mitotic cycle of *Phaseolus coccineus* and the concept of angiosperm endomitosis. *Protoplasma*, **109**, 403–14.

Cave, M. D. (1975). Absence of rDNA amplification in the meroistic (telotrophic) ovary of the large milkweed bug *Oncopeltus fasciatus* (Dallas) (Hemiptera: Hygaeidae). *J. Cell Biol.*, **66**, 461–9.

Chacko, S. (1973). DNA synthesis, mitosis and differentiation in cardiac myogenesis. *Dev. Biol.*, **35**, 1–18.

Chadwick, K. H. & Leenhouts, H. P. (1981). The molecular theory of radiation biology. *Monographs in Theoretical and Applied Genetics*, **5**. Berlin & Heidelberg: Springer Verlag.

Chamley, J. & Campbell, G. (1974). Mitosis of contractile smooth muscle cells in tissue culture. *Exp. Cell Res.*, **84**, 105–10.

Chang, P. L., Georgiadis, N., Joubert, G. I. & Davidson, R. G. (1983). Gene dosage effects in human diploid and tetraploid fibroblasts. *Exp. Cell Res.*, **145**, 277–84.

Chang, W. W. L. & Leblond, C. P. (1971). Renewal of the epithelium in the descending colon of the mouse. I. Presence of three cell populations: vacuolar-columnar, mucous and argentaffin. *Amer. J. Anat.*, **131**, 73–100.

Chapman, V. M., Ansell, J. D. & McLaren, A. (1972). Trophoblast giant cell differentiation in the mouse. *Dev. Biol.*, **29**, 48–54.

Chasin, L. A. (1973). The effect of ploidy on chemical mutagenesis in cultured Chinese hamster cells. *J. Cell. Physiol.*, **82**, 299–307.

Chen, T. T. (1940). A further study on polyploidy in *Paramecium*. *J. Hered.* **31**, 249–51.

Christensen, B. (1966). Cytophotometric studies on the DNA content in diploid and polyploid Enchytraeidae. *Chromosoma*, **18**, 305–15.

Cionini, P. G., Cavallini, A., Baroncelli, S., Lercari, B. & D'Amato, F. D. (1983). Diploidy and chromosome endoreduplication in the development of epidermal cell lines in the first foliage leaf of Durum wheat (*Triticum durum* Desf.). *Protoplasma*, **118**, 36–43.

Clutter, M., Brady, T., Walbot, V. & Sussex, I. (1974). Macromolecular synthesis during plant embryogeny: cellular rates of RNA synthesis in diploid and polyploid cells in bean embryos. *J. Cell Biol.*, **63**, 1097–102.

Cobb, J. L. S. & Bennett, T. (1970). An ultrastructural study of mitotic division in differentiated gastric smooth muscle cells. *Z. Zellforsch. mikr. Anat.*, **108**, 177 – 89.

Coggeshall, R. E., Yaksta, B. A. & Swartz, F. J. (1970). A cytophotometrical analysis of the DNA in the nucleus of the giant cell, R-2, in *Aplysia*. *Chromosoma*, **32**, 205–12.

Cohen, A. M. (1974). DNA synthesis and cell division in differentiating avian adrenergic neuroblasts. In *Dynamics of Degeneration and Growth in Neurons*, ed. K. Fux, L. Olson & Y. Zotterman, pp. 359–70. Oxford: Pergamon Press.

Cohen, L. H. & Gotchel, B. V. (1971). Histones of polytene and non-polytene nuclei of *Drosophila melanogaster*. *J. Biol. Chem.*, **246**, 1841–8.

Cohn, N. & Van Duijn, P. (1967). Constancy of DNA content in adrenal medulla nuclei of cold-treated rats. *J. Cell Biol.*, **33**, 349–54.

Collins, J. M. (1972). Amplification of ribosomal RNA cistrons in regenerating lens of *Triturus*. *Biochemistry*, **11**, 1259–63.

Collins, J. M. (1978). RNA synthesis in rat liver cells with different DNA contents. *J. Biol. Chem.*, **253**, 5769–73.

Collins, S. & Groudine (1982). Amplification of endogenous *myc*-related DNA sequences in a human myeloid leukaemia cell line. *Nature (London)*, **298**, 679 – 81.

Comings, D. E. (1972). Evidence for ancient tetraploidy and conservation of linkage groups in mammalian chromosomes. *Nature (London)*, **238**, 455–7.

Cone, C. D. & Cone, Ch. M. (1976). Induction of mitosis in mature neurons in central nervous system by sustained depolarization. *Science*, **192**, 155–8.

Cooper, G. M. (1982). Cellular transforming genes. *Science*, **217**, 801–6.

Coulombre, A. I. (1956). The role of intraocular pressure in the development of the chick eye. 1. Control of eye size. *J. Exp. Zool.*, **133**, 211–26.

Cowden, R. R. (1972). Cytological and cytochemical examination of the neuronal nuclei of the central nervous systems of pulmonate gastropods and some other molluscs. *Trans. Amer. Microscop. Soc.*, **91**,130–43.

Cowell, J. K. (1980). Consistent chromosome abnormalities associated with mouse bladder epithelial cell lines transformed *in vitro. J. Natl. Cancer Inst.*, **65**, 955–9.

Cowell, J. K. & Wigley, C. B. (1980). Changes in DNA content during *in vitro* transformation of mouse salivary gland epithelium. *J. Natl. Cancer Inst.*, **64**, 1443–9.

Cowell, J. K. & Wigley, C. B. (1982). Chromosome changes associated with the progression of cell lines from preneoplastic to tumorigenic phenotype during transformation of mouse salivary gland epithelium *in vitro. J. Natl. Cancer Inst.*, **69**, 425–33.

Cristofalo, V. J. (1974). Aging. In *Concepts of Development*, ed. J. Lash & J. R. Whittaker, pp. 429–47. Stamford: Sinauer.

Curtis, H. J. (1963). Biological mechanisms underlying the aging process. *Science*, **141**, 686–94.

Cuvelier, C. A. & Roels, H. J. (1979). Cytophotometric studies of the nuclear DNA content in cartilaginous tumours. *Cancer*, **44**, 1363–74.

Dalla Favera, R., Wong-Staal, F. & Gallo, R. C. (1982). One gene amplification in promyelocytic leukemia cell line HL-60 and primary leukaemic cells of the same patient. *Nature (London)*, **299**, 61–3.

D'Amato, F. (1952). Polyploidy in the differentiation of tissues and cells in plants; a critical examination of the literature. *Caryologia*, **4**, 311–58.

D'Amato, F. (1977). *Nuclear Cytology in Relation to Development.* Cambridge University Press.

Daneholt, B. & Edström, J. E. (1967). The content of deoxyribonucleic acid in individual polytene chromosomes of *Chironomus tentans. Cytogenetics*, **6**, 350–6.

Darrow, J. M. & Clever, U. (1970). Chromosome activity and cell function in polytenic cells. *Dev. Biol.*, **21**, 331–48.

Davidson, E. H. & Britten, R. J. (1979). Regulation of gene expression: possible role of repetitive sequences. *Science*, **204**, 1052–9.

Davies, D. R. (1976). DNA and RNA contents in relation to cell and seed weight in *Pisum sativum. Plant Sci. Letts.*, **7**, 17–25.

De, N. (1961). Polyploidy and radiosensitive behaviour of human malignant cells *in vivo. Brit. J. Cancer*, **15**, 54–61.

Defendi, V. & Manson, L. A. (1963). Analysis of the life-cycle in mammalian cells. *Nature (London)*, **198**, 359–61.

De Flierdt, K. (1975). No multistrandedness in mitotic chromosomes of *Drosophila melanogaster. Chromosoma*, **50**, 431–4.

De Leeuw-Israel, F. R., van Bezooijen, C. F. A. & Hollander, C. F. (1972). Ploidy as a possible explanation for the variation in liver function during the life span of the rat. *Z. Alternsforsch.*, **26**, 29–38.

De Leval, M. (1964). Dosages cytophotométriques d'ADN dans les mégacaryocytes normaux de cobaye. *C.r. Soc. Biol. (Paris)*, **158**, 2198–201.

Del Pino, E. M. & Humphries, A. A. (1978). Multiple nuclei during early oogenesis in *Flectonopus pygmaeus* and other marsupial frogs. *Biol. Bull.*, **154**, 198–212.

Dennhöfer, L. (1981). Cytophotometric DNA determinations and autoradiographic studies in salivary gland nuclei from larvae with different karyotypes in *Drosophila melanogaster. Chromosoma*, **86**, 123–47.

Desaive, C. (1965). Détermination par cytophotométrie de la teneur en acides désoxyribonucléiques des noyaux de la glande de Loewenthal du rat albino mâle. *C.r. Acad. Sci., Paris*, **260**, 315–17.

Desaive, C. (1967). Etude cytophotométrique des acides désoxyribonucléiques dans les noyaux de la glande de Loewenthal du rat albinos mâle: à differents ages on après modifications de l'état endocrinien. *Arch. Biol.*, **78**, 521–73.

Deschênes, J., Valet, J.-P. & Marceau, N. (1981). The relationship between cell volume, ploidy and functional activity in differentiating hepatocytes. *Cell Biophysics*, **3**, 321–34.

Deumling, B. & Nagl, W. (1978). DNA characterization, satellite DNA localization and nuclear organization in *Tropaeolum majus*. *Cytobiologie*, **16**, 412–20.

Dickerman, L. H. & Goldman, R. D. (1974). A rapid method for production of binucleate cells. *Exp. Cell Res.*, **82**, 433–6.

Digernes, V. (1981). The cellular DNA content and profile of dimethyl nitrosamine induced murine lung tumors in relation to their histopathology. *Carcinogenesis*, **2**, 195–8.

Dolnick, B. J., Berenson, R. J., Bertino, J. R., Kaufman, R. J., Nunberg, J. H. & Schimke, R. T. (1979). Correlation of dihydrofolate reductase elevation with gene amplification in a homogeneously staining chromosomal region. *J. Cell Biol.*, **83**, 394–402.

Domon, M., Okano, T., Sasaki, T. & Nakamura, T. (1978). Regression of the mouse parotid gland induced with isoproterenol. *Cell Tiss. Res.*, **11**, 567–71.

Dorgan, W. I. & Schultz, R. L. (1971). An *in vitro* study of programmed death in rat placental giant cells. *J. Exp. Zool.*, **178**, 497–512.

Dreher, R., Keller, H. U., Hess, M. W., Roos, B. H. & Cottier, M. D. (1978). Early appearance and mitotic activity of multinucleated giant cells in mice after combined injection of talc and prednisolone acetate. *Lab. Invest.*, **38**, 149–56.

Dubovaya, T. K., Kushch, A. A. & Brodsky, V. Ya. (1977). Peculiarities of regeneration in nonpolyploid liver. *Byulleten exper. Biologii Meditsini*, **84**, 221–4 (in Russian, English translation in *Bull. Exp. Biol. Med.*, **84**, 1183–6).

Dupont, H., Dupont, M.-A., Bricaud, H. & Bosseau, M.-R. (1983). Megakaryocytes separation in homogeneous classes by unit gravity sedimentation: physico-chemical, ultrastructural and cytophotometric characterizations. *Biol. Cell*, **49**, 137–44.

Eglitis, M. A. & Wiley, L. M. (1981). Tetraploidy and early development: effects on developmental timing and embryonic metabolism. *J. Embryol. Exp. Morph.*, **66**, 91–108.

Ehria, M. & Swartz, F. (1972). Polyploid β-cells in the human islet of Langerhans. *Anat. Rec.*, **172**, 305–6.

Eisenstein, R. & Wied, G. L. (1970). Myocardial DNA and protein in maturing and hypertrophied human hearts. *Proc. Soc. Exp. Biol. Med.*, **133**, 176–9.

Endow, S. A. & Gall, J. G. (1975). Differential replication of satellite DNA in polyploid tissues of *Drosophila virilis*. *Chromosoma*, **50**, 175–92.

Enerbäck, L. & Rundquist, I. (1981). DNA distribution of mast cell populations in growing rats. *Histochem.*, **71**, 521–31.

Enesko, H. E. & Samborsky, J. (1983). Liver polyploidy: influence of age and of dietary restriction. *Exp. Gerontol.*, **18**, 79–87.

Engelmann, G. L., Richardson, A., Katz, A. & Fierer, J. A. (1981). Age-related changes in isolated rat hepatocytes: comparison of size, morphology, binucleation, and protein content. *Mech. Ageing Dev.*, **16**, 385–95.

Enzenberg, U. (1961). Beiträge zur Karyologie des Endosperms. *Österr. Bot. Z.*, **108**, 245–85.

Epifanova, O. I., Terskikh, V. V. & Polunovsky, V. A. (1984). *The Resting Cells*. New York & London: Academic Press (in press).

Epstein, C. J. (1967). Cell size, nuclear content and the development of polyploidy in the mammalian liver. *Proc. Natl. Acad. Sci. USA*, **57**, 327–34.

Evans, G. M., Rees, H., Snell, C. L. & Sun, S. (1972). The relationship between nuclear DNA amount and the duration of the mitotic cycle. *Chromosomes Today*, **3**, 24–31.

Evans, I. H. (1976). Polyploidization in the rat liver: the role of binucleate cells. *Cytobios*, **16**, 115–24.

Evans, L. S. & Van't Hof, J. (1975). Is polyploidy necessary for tissue differentiation in higher plants? *Amer. J. Bot.*, **62**, 1060–4.

Evans, L. S. & Verville, K. M. (1978). Is the predominant period of cell arrest and presence of endoreduplicated cells coincident with production of secondary vascular tissues in intact and cultured roots of *Raphanus sativus? Amer. J. Bot.*, **65**, 70–4.

Evgen'ev, M. B. & Polianskaya, G. G. (1976). The pattern of polytene chromosome synapsis in *Drosophila* species and interspecific hybrids. *Chromosoma*, **57**, 285–95.

Fabrikant, J. (1967). The effect of prior continuous irradiation on the G_2, M and S phases of proliferating parenchymal cells in the regenerating liver. *Radiat. Res.*, **31**, 304–14.

Fabrikant, J. (1968). The kinetics of cellular proliferation in regenerating liver. *J. Cell Biol.*, **36**, 557–67.

Faktor, V. M., Malyutin, V. F., Li, S. E. & Brodsky, V. Ya. (1979). Delay of polyploidization of rat hepatocytes caused by liver growth inhibition. *Tsitologia* **21**, 397–400 (in Russian).

Faktor, V. M., Poltoranina, V. S. & Uryvaeva, I. V. (1982). Change of the hepatocyte proliferation pattern in the mouse liver and adenomatous nodes in carcinogenesis induced with CCl_4. *Byulleten exper. Biologii Meditsini*, **93**, 79–82 (in Russian).

Faktor, V. M. & Uryvaeva, I. V. (1972). Mitotic cycle of diploid and polyploid cells in regenerating mouse liver. *Tsitologia*, **14**, 868–72 (in Russian).

Faktor, V. M. & Uryvaeva, I. V. (1975). Progression of polyploidy in the mouse liver by repeated hepatectomy. *Tsitologia*, **17**, 909–15 (in Russian).

Faktor, V. M. & Uryvaeva, I. V. (1983). Variants of murine urethane-induced hepatomas by their DNA distribution profiles. *Abstracts of XII Meeting of the European Study Group for Cell Proliferation*, p. 124. Budapest.

Faktor, V. M., Uryvaeva, I. V., Sokolova, A. S., Chernov, V. A., & Brodsky, V. Ya. (1980). Kinetics of cellular proliferation in regenerating mouse liver pretreated with the alkylating drug Dipin. *Virchows Arch. B.*, **33**, 187–97.

Fallon, M. D., Teitelbaum, S. L., Kahn, A. J. (1983). Multinucleation enhances macrophage-mediated bone resorption. *Lab. Invest.*, **49**, 159–64.

Farsund, T. (1975). Cell kinetics of mouse urinary bladder epithelium. I. Circadian and age variations in cell proliferation and nuclear DNA content. *Virchows Arch. B*, **18**, 35–49.

Farsund, T. (1976). Cell kinetics of mouse urinary bladder epithelium. II. Changes in proliferation and nuclear DNA content during necrosis, regeneration, and hyperplasia caused by a single dose of cyclophosphamide. *Virchows Arch. B.*, **21**, 279–98.

Farsund, T. & Dahl, E. (1978). Cell kinetics of mouse urinary bladder epithelium. III. A histological and ultrastructural study of bladder epithelium during regeneration after a single dose of cyclophosphamide with special reference to the mechanism by which polyploid cells are formed. *Virchows Arch. B.*, **26**, 215–23.

Farsund, T. & Iversen, O. H. (1978). Cell kinetics of mouse urinary bladder epithelium. VII. Changes in proliferation and nuclear DNA content after repeated doses of dibutylnitroamine. *Virchows Arch. B*, **27**, 119–35.

Fenaux, R. (1971). La couche oikoplastique de l'appendiculaire *Oikopleura albicans* (Tunicata). *Z. Morphol. Tiere.*, **69**, 184–200.

Firket, H. & Hopper, A. F. (1970). Abnormal cycle phase duration in polyploid cells or cells with irregular mitosis in a HeLa cell population. *J. Cell Biol.*, **44**, 678–81.

Fischman, D. A. (1979). Multinucleation: comparative aspects and functional implications. In *Muscle Regeneration*, ed. A. Mauro, R. Bischoff, B. M. Carlson, S. A. Shafig, J. Konigsberg & B. Lipton, pp. 9–12. New York: Raven Press.

Flavell, R. B., Bennett, M. D., Smith, J. B. & Smith, D. B. (1974). Genome size and the proportion of repeated nucleotide sequence DNA in plants. *Biochem. Genet.*, **12**, 257–69.

Fleroff, N. (1936). Studien über den Bau und die funktionelle Struktur des Harnblasenepithels des Nagetiere. *Z. Zellforsch. mikr. Anat.*, **24**, 360–92.

Floyd, A. & Swartz, F. (1969). The uterine epithelium of *Ascaris lumbricoides* as a model system for the study of polyploidy. *Exp. Cell Res.*, **56**, 275–80.

Follet, B. K., Nichols, T. I. & Redshaw, M. R. (1968). The vitellogenic response in the South African clawed toad (*Xenopus laevis*). *J. Cell Physiol.*, **72** suppl., 91–102.

Fontaine, J. C. & Swartz, F. (1972). Fluctuation of Feulgen- and diphenylamine-DNA in peripheral leukocytes. *J. Cell Physiol*, **80**, 281–90.

Fontana, F. (1974). Cytophotometric studies on the nuclear DNA content in somatic tissues of *Schistocerca gregaria* Forskål (Acrididae: Orthoptera). *Caryologia*, **27**, 73–82.

Formenti, D. (1975). Nuclear DNA content in some species of the genus *Cercopithecus* (Primates: Cercopithecidae). *Genetica*, **45**, 307–13.

Fournier, K. E. & Pardee, A. B. (1975). Cell cycle studies of mononucleate and cytochalasin-B-induced binucleate fibroblasts. *Proc. Natl. Acad. Sci. USA*, **72**, 869–73.

Fox, D. P. (1969). DNA values in somatic tissues of *Dermestes* (Dermestidae: Coleoptera). 1. Abdominal fat body and testis wall of the adult. *Chromosoma*, **28**, 445–56.

Fox, D. P. (1970). A non-doubling DNA series in somatic tissues of the locusts *Schistocerca gregaria* (Forskål) and *Locusta migratoria* Linn. *Chromosoma*, **29**, 446–61.

Franke, W. W. & Keenan, Th. W. (1979). Mitosis in milk secreting epithelial cells of mammary gland: an ultrastructural study. *Differentiation*, **13**, 81–8.

Frankel, J., Jenkins, L. M. & De Bault, L. E. (1976). Causal relations among cell cycle processes in *Tetrahymena pyriformis*: an analysis employing temperature-sensitive mutants. *J. Cell Biol.*, **71**, 242–60.

Frederiks, W. M., Slob, A. & Schröder, M. (1980). Histochemical determination of histone and nonhistone protein contents in rat liver nuclei. *Histochem.*, **68**, 49–53.

Frenkel, M. A. (1978). DNA content in micronuclei and chromatin bodies of divided macronucleus in ciliate *Colpoda steini*. *Tsitologia*, **20**, 465–9 (in Russian).

Frolova, S. L. (1936). Nuclear structure in salivary gland of some *Drosophila* species. *Biologicheskii Zhurnal*, **5**, 271–92 (in Russian).

Fujita, S. (1964). Analysis of neuron differentiation in the central nervous system by tritiated thymidine autoradiography. *J. Comp. Neurol.*, **122**, 311–27.

Gage, L. P. (1974a). The *Bombyx mori* genome: analysis by DNA reassociation kinetics. *Chromosoma*, **45**, 27–42.

Gage, L. P. (1974b). Polyploidization of the silk gland of *Bombyx mori*. *J. Mol. Biol.*, **86**, 97–108.

Gahan, P. B. (1977). Increased levels of euploidy as a strategy against rapid ageing in diploid mammalian systems: an hypothesis. *Exp. Gerontol*, **12**, 133–6.

Gahan, P. B. & Middleton, J. (1982). Hepatocyte euploidization is a typical mammalian physiological specialization. *Comp. Biochem. Physiol.*, **71A**, 345–8.

Gambarini, A. G. & Lara, F. J. S. (1974). Under-replication of ribosomal cistrons in polytene chromosomes of *Rhynchosciara*. *J. Cell Biol.*, **62**, 215–22.

Garber, A. T. & Brasch, K. (1979). Age- and ploidy-related changes in the non-histone proteins of rat liver nuclei. *Exp. Cell Res.*, **120**, 412–17.

Garcia, A. (1964). Feulgen-DNA values in megakaryocytes. *J. Cell Biol.*, **20**, 342–5.

Gay, H., Das, C. C., Forward, K. & Kaufmann, B. (1970). DNA content of mitotically-active condensed chromosomes of *Drosophila melanogaster*. *Chromosoma*, **32**, 213–23.

Gazaryan, K. G. (1982). Genome activity and gene expression in avian erythroid cells. *Int. Rev. Cytol.*, **74**, 95–126.

Gearhart, J. D. & Mintz, B. (1972). Glucose phosphate isomerase subunit-reassociation tests for maternal-fetal and fetal–fetal cell fusion in the mouse placenta. *Dev. Biol.*, **29**, 53–64.

Geitler, L. (1934). *Grundriß der Cytologie*. Berlin: Borntraeger.

Geitler, L. (1938). Über den Bau des Ruhekerns mit besonderer Berücksichtigung der Heteropteren und Dipteren. *Biol. Zentralbl.*, **58**, 152–79.

Geitler, L. (1939). Die Entstehung der polyploiden Somakerne der Heteropteren durch Chromosomenteilung ohne Kernteilung. *Chromosoma*, **1**, 1–22.

Geitler, L. (1944). Zur Kentniss des Kern- und Chromosomenbau der Heuschrecken und Wanzen. *Chromosoma*, **2**, 531–43.

Geitler, L. (1953). Endomitose und endomitotische Polyploidisierung. *Protoplasmatologia*, **6c**, 1–89.

Gelfant, S. (1963*a*). A new theory on the mechanism of cell division. In *Cell Growth and Cell Division*, ed. R. J. C. Harris, pp. 229–59. New York & London: Academic Press.

Gelfant, S. (1963*b*). Inhibition of cell division: a critical and experimental analysis. *Int. Rev. Cytol.*, **14**, 1–39.

Gelfant, S. (1966). Patterns of cell division: the demonstration of discrete cell population. In *Methods in Cell Physiology*, ed. D. M. Prescott, vol. 2, pp. 359–95. New York & London: Academic Press.

Gentcheff, G. & Gustafsson, A. (1939). The double chromosome reproduction in *Spinacia* and its causes. *Hereditas*, **25**, 349–58.

Gerhard, H., Schultze, B. & Maurer, W. (1971). Zellklassen und Ploidiestufen in isolierten Leberzellen der Maus. *Exp. Cell Res.*, **69**, 223–6.

Gerhard, H., Schultze, B. & Maurer, W. (1972). Wirkung einer zweiten CCl_4-Intoxikation auf die CCl_4-geschädigte Leber der Maus. *Virchows Arch. B*, **10**, 184–99.

Gerhard, H., Schultze, B. & Maurer, W. (1973). Quantitative Untersuchungen über Wachstum und Polyploidisierung bei der Regeneration der CCl_4-Leber der Maus. *Virchows Arch. B*, **14**, 345–59.

Gerzelli, G. & Barni, S. (1976). Changes in liver cells ploidy of young rats following isoprenaline treatment. *Cell Tiss. Kinet.*, **9**, 267–72.

Geschwind, I., Alfert, M. & Schooley, C. (1958). Liver regeneration and hepatic polyploidy in the hypophysectomized rat. *Exp. Cell Res.*, **15**, 232–5.

Geschwind, I., Alfert, M. & Schooley, C. (1960). The effects of thyroxin and growth hormone on liver polyploidy. *Biol. Bull.*, **118**, 66–9.

Geyer-Duszyǹska, I. (1959). Experimental research on chromosome elimination in Cecidomyiidae (Diptera). *J. Exp. Zool.*, **141**, 174–8.

Ghosh, S., Paweletz, N. & Ghosh, I. (1978). Mitotic asynchrony of multinucleate cells in tissue culture. *Chromosoma*, **65**, 293–300.

Gilbert, P. & Pfitzer, P. (1977). Facultative polyploidy in endocrine tissues. *Virchows Arch. B*, **25**, 233–42.

Giménez-Martín, G., López-Sáez, J. F., Moreno, P., González-Fernández, A. (1968). On the triggering of mitosis and the division cycle of polynucleate cells. *Chromosoma*, **25**, 282–96.

Gläss, E. (1957). Das Problem der Genomsonderung in der Mitosen unbehandelter Rattenlebern. *Chromosoma*, **8**, 468–92.

Gledhill, B., Gledhill, M., Rigler, R. & Ringertz, N. (1966). Changes in deoxyribonucleoprotein during spermiogenesis in the bull. *Exp. Cell Res.*, **41**, 652–65.

Gómez-Lechón, M. J., Barbera, E., Gil, R. & Baguena, J. (1981). Evolutive changes of ploidy and polynucleation in adult rat hepatocytes in culture. *Cell Mol. Biol.*, **27**, 695–701.

González-Fernández, A., López-Sáez, J. F. & Giménez-Martín, G. (1966). Duration of the division cycle in binucleate and mononucleate cells. *Exp. Cell Res.*, **43**, 255–67.

Goode, D. (1975). Mitosis of embryonic heart muscle cells in vitro. *Cytobiologie*, **11**, 203–29.

Gopalan, H. N. B. & Das, C. M. S. (1972). Polyteny and salivary gland secretion in the melon fly *Dacus cucurbitae*. *Experientia*, **28**, 684–6.

Gorbunova, M. P. (1976). Mitosis in C-cells of thyroid gland: electron microscopy. *Tsitologia*, **18**, 219–21 (in Russian).

Goss, R. (1966). Hypertrophy versus hyperplasia. *Science*, **153**, 1615–20.

Goyanes-Villaescusa, V. (1969). Cycles of reduplication in megakaryocyte nuclei. *Cell Tiss. Kinet.*, **2**, 165–8.

Gräbner, W. & Pfitzer, P. (1974). Number of nuclei in isolated myocardial cells of pigs. *Virchows Arch. B.*, **15**, 279–94.

Grafl, I. (1939). Kernwachstum durch Chromosomenvermehrung als regelmässiger Vorgang bei pflanzlichen Gewebe Differenzierung. *Chromosoma*, **1**, 265–75.

Grafl, I. (1941). Über das Wachstum der Antipodenkerne von *Caltha palustris*. *Chromosoma*, **2**, 1–11.

Grell, K. G. (1973). *Protozoology*. Berlin & Heidelberg: Springer Verlag.

Grell, M. (1946a). Cytological studies in *Culex*. I. Somatic reduction divisions. *Genetics*, **31**, 60–76.

Grell, M. (1946b). Cytological studies in *Culex*. II Diploid and meiotic divisions. *Genetics*, **31**, 76–94.

Grieninger, G. & Granick, S. (1975). Synthesis and differentiation of plasma proteins in cultured embryonic chicken liver cells: a system for study of regulation of protein synthesis. *Proc. Natl. Acad. Sci. USA*, **72**, 5007–11.

Grif, V. G. (1980). Changes of DNA content in plants under low temperature: a cytophotometric study. *Tsitologia*, **22**, 1185–92 (in Russian).

Griffith, J. K. & Anderson, O. D. (1979). Synthesis and turnover of hdRNA and mRNA in the salivary gland of the blowfly *Calliphora erythrocephala*. *Dev. Biol.*, **69**, 480–95.

Grime, J. P. & Mowforth, M. A. (1982). Variation in genome size – an ecological interpretation. *Nature (London)*, **299**, 151–3.

Grisham, G. W. (1969). Cellular proliferation in the liver. In *Normal and Malignant Cell Growth*, ed. J. M. Fry, M. L. Griem & W. H. Kirsten, pp. 28–43. Berlin & Heidelberg: Springer Verlag.

Grosset, L. & Odartchenko, N. (1975a). Relationship between cell cycle duration, S-period and nuclear DNA content in erythroblasts of four vertebrate species. *Cell Tiss. Kinet.* **8**, 81–90.

Grosset, L. & Odartchenko, N. (1975b). Duration of mitosis and separate mitotic phases compared to nuclear DNA content in erythroblasts of four vertebrates. *Cell Tiss. Kinet.*, **8**, 91–6.

Grygoryeva, T. A., Kushch, A. A., Astakhova, A. M. & Dubovaya, T. K. (1970). Binucleation and ploidy of the liver cells. *Doklady MOIP (ser. Biol.)*, 28–30 (in Russian).

Hadorn, E., Ruch, F. & Staub, M. (1964). Zum DNS-Gehalt in Speicheldrüsenkernen mit 'übergrossen Riesenchromosomen' von *Drosophila melanogaster*. *Experientia*, **20**, 566–7.

Hägele, K. (1976). Prolongation of replication time after doubling of the DNA content of polytene chromosomes bands of *Chironomus*. *Chromosoma*, **55**, 253–8.

Harris, M. (1971a). Polyploid series of mammalian cells. *Exp. Cell Res.*, **66**, 329–36.

Harris, M. (1971b). Mutation rates in cells at different ploidy levels. *J. Cell Physiol.*, **78**, 177–84.

Hartwell, L. H. (1978). Cell division from a genetic perspective. *J. Cell Biol.*, **77**, 627–37.

Hartwell, L. H., Mortimer, R. K., Culotti, J. & Culotti, M. (1973). Genetic control of the cell division cycle in yeast. V. Genetic analysis of *cdc* mutants. *Genetics*, **74**, 267–86.

Hatch, F. T., Bodner, A. J., Mazrimas J. A. & Moore, D. H. (1976). Satellite DNA and cytogenetic evolution. *Chromosoma*, **58**, 155–68.

Hatzfeld, J. & Buttin, G. (1975). Temperature-sensitive cell cycle mutants: a Chinese hamster cell line with a reversible block in cytokinesis. *Cell*, **5**, 123–9.

Hayflick, L. (1981). Cell death *in vitro*. In *Cell Death in Biology and Pathology*, ed. I. D. Bowen & R. A. Lockshin, pp. 243–85. New York: Chapman & Hall.

Hayflick, L. & Moorhead, P. E. (1961). The serial cultivation of human diploid cell strains. *Exp. Cell Res.*, **25**, 585–621.

Hecht, R. M., Wall, S. M., Schomer, D. F., Oro, J. A. & Bartel, A. H. (1982). DNA replication may be uncoupled from nuclear and cellular division in temperature-sensitive embryonic lethal mutants of *Caenorhabditis elegans*. *Dev. Biol.*, **94**, 183–91.

Heimpel, H. (1976). Congenital dyserythropoeitic anaemia type I: clinical and experimental aspects. In *Congenital Disorders of Erythropoiesis*, CIBA Symposium **37**, pp. 135–49. Amsterdam: Elsevier.

Heitz, E. (1928). Das Heterochromatin der Moose. *Jahrb. wiss. Bot.*, **69**, 762–818.

Henderson, P.Th. & Kersten, K. I. (1971). Alteration of drug metabolism during liver regeneration. *Arch. Internat. Pharmacodynam.* **189**, 373–5.

Henderson, S. A. (1967). The salivary gland chromosomes of *Dasyneura crataegi* (Diptera: Cecidomyiidae). *Chromosoma*, **23**, 38–58.

Herreros, B. & Giannelli, F. (1967). Spatial distribution of old and new chromatid sub-units and frequency of chromatic exchanges in induced human lymphocyte endoreduplication. *Nature (London)*, **216**, 286–8.

Hertwig, G. (1928). Allgemeine mikroskopische Anatomie der lebenden Masse. In *Handbuch der Mikroskopischen Anatomie des Menschen.* ed. von. W. Möllendorff, **1**, 218–55.

Hervás, J. P. (1975). Mitotic activity of endopolyploid root cells in *Allium cepa*. *Experientia*, **31**, 1143–4.

Hervás, J. P., López-Sáez, J. F. & Giménez-Martín, G. (1982). Multinucleate plant cells. II. Requirements for nuclear DNA synthesis. *Exp. Cell Res.*, **139**, 341–50.

Hesse, M. (1968). Karyologische Anatomie von Zoocecidien und ihre Kernstrukturen. *Österr. Bot. Z.*, **115**, 34–83.

Hesse, M. (1969). Anatomische und Karyologische Untersuchungen an der Galle von *Mayetiola poae* auf *Poa nemoralis*. *Österr. Bot. Z.*, **117**, 411–25.

Hesse, M. (1971). Häufigkeit und Mechanismen der durch gallbildende Organismen ausgelösten somatischen Polyploidisierung. *Österr. Bot. Z.*, **119**, 454–63.

Hicks, R. M. (1975). The mammalian urinary bladder: an accommodating organ. *Biol. Rev.*, **50**, 215–45.

Hicks, R. M. (1976). Changes in differentiation of the urinary bladder during benign and neoplastic hyperplasia. In *Progress in Differentiation Research*, ed. N. Müller-Bérat, pp. 339–53. Amsterdam: Elsevier–North-Holland.

Hoehn, H., Sprague, C. A., Martin, G. M. (1973). Effects of cytochalasin B on cultivated human diploid fibroblast and its use for the isolation of tetraploid clones. *Exp. Cell Res.*, **76**, 170–4.

Hollenberg, C. P. (1976). Proportional representation of rDNA and Balbiani ring DNA in polytene chromosomes of *Chironomus tentans*. *Chromosoma*, **57**, 185–97.

Holtzer, H., Weintraub, H., Mayne, R. & Mochan, B. (1972). The cell cycle, cell lineages and cell differentiation. *Curr. Top. Dev. Biol.*, **7**, 229–54.

Hori, T. (1980). Polyploidization and multinucleation in a temperature-sensitive mutant of CHO cells. *Europ. J. Cell Biol.*, **22**, 494 (abstract).

Hsu, T. C. (1961). Chromosomal evolution in cell population. *Int. Rev. Cytol.*, **12**, 69–161.

Hsu, T. C. & Moorhead, P. S. (1956). Chromosome anomalies in human neoplasms with special reference to the mechanisms of polyploidization and aneuploidization in the HeLa strain. *Ann. New York Acad. Sci.*, **63**, 1083–94.

Hunt, C. V. & Avery, G. B. (1971). Increased levels of deoxyribonucleic acid during trophoblast giant cell formation in mice. *J. Reprod. Fertil.*, **25**, 85–91.

Huskins, C. L. (1948). Segregation and reduction in somatic tissues. I. Initial observations on *Allium cepa*. *J. Hered.*, **39**, 311–25.

Ilgren, E. B. (1980). Polyploidization of extraembryonic tissues during mouse embryogenesis. *J. Embryol. Exp. Morphol.*, **59**, 103–11.

Ilgren, E. B. (1981*a*). On the control of the trophoblastic giant-cell transformation in the mouse: homotypic cellular interactions and polyploidy. *J. Embryol. Exp. Morphol.*, **62**, 183–202.

Ilgren, E. B. (1981*b*). The initiation and control of trophoblastic growth in the mouse: binucleation and polyploidy. *Placenta*, **2**, 317–32.

Inamdar, N. (1958). Development of polyploidy in mouse liver. *J. Morphol.*, **103**, 65–90.

Ingle, J. & Timmis, J. N. (1975). A role for differential replication of DNA in development. In *Modification of the Information Content of Plant Cells*, ed. R. Markham, pp. 37–52. Amsterdam: Elsevier–North-Holland.

Inui, N., Takayama, S. & Kuwabara, N. (1971). DNA measurement on cell nucleus of normal liver, adenoma and hepatoma in mice: histologic features. *J. Natl. Cancer. Inst.*, **47**, 47–55.

Ishii, K. (1980). Multinucleation in mouse fibroblasts cultured in methocel medium. *J. Cell Physiol.*, **103**, 105–8.

Ivanov, V. B. (1974). *Cellular Basis of Plant Growth*. Moscow: Nauka (in Russian).

Ivanov, V. B. (1978). DNA content in the nucleus and rate of development in plants. *Soviet J. Dev. Biol.*, **9**, 28–40.

Jacobson, M. (1968). Cessation of DNA synthesis in retinal ganglion cells correlated with the time of specification of their central connections. *Dev. Biol.*, **17**, 219–32.

Jacobj, W. (1925). Über das rhythmische Wachstum der Zellen durch Verdopplung ihres Volumens. *Wilhelm Roux' Arch. Entwicklungsmech. Org.*, **106**, 125–92.

Jakobson, A., Bichel, P. & Sell, A. (1979). Correlation of DNA distribution and cytological differentiation of human cervical carcinoma. *Virchows Arch. B*, **31**, 75–9.

James, J., Schopman, M. & Delfgaauw, P. (1966). The nuclear pattern of the parenchymal cells of the liver after partial hepatectomy. *Exp. Cell Res.*, **42**, 375–9.

James, J., Tas, J., Bosch, K. S., de Meere, A. I. P. & Schuyt, H. Cl. (1979). Growth patterns of rat hepatocytes during postnatal development. *Europ. J. Cell Biol.*, **19**, 222–6.

Jataganas, X., Gahrton, G. & Thorell, B. (1970). DNA, RNA and hemoglobin during erythroblast maturation. *Exp. Cell Res.*, **62**, 254–61.

Jehle, E. & Kiefer, G. (1979). Incorporation of uridine into polyploid and binuclear cells of mouse liver. *Europ. J. Cell Biol.*, **19**, 89–91.

Jimbow, K., Roth, S. I., Fitzpatrick, T. B. & Szabo, G. (1975). Mitotic activity in non-neoplastic melanocytes *in vivo* as determined by histochemical, autoradiographic and electron microscope studies. *J. Cell Biol.*, **66**, 663–70.

Johnston, G. C., Pringle, J. R. & Hartwell, L. H. (1977). Coordination of growth with cell division in the yeast *Saccharomyces cerevisiae*. *Exp. Cell Res.*, **105**, 79–98.

Johnston, I. R. & Mathias, A. P. (1972). The biochemical properties of nuclei fractionated by zonal centrifugation. In *Subcellular Components*, ed. G. D. Burnie, pp. 53–75. London: Butterworths.

Jollie, W. P. (1960). The persistence of trophoblast on extrauterine tissues in rat. *Amer. J. Anat.*, **106**, 109–15.

Jones, R. N. & Brown, L. M. (1976). Chromosome evolution and DNA variation in *Crepis*. *Heredity*, **36**, 91–104.

Jost, E. & Mameli, M. (1970). The nuclear content of DNA in three different strains of *Artemia salina* (Phyllopoda, Branchiopodidae). *Experientia*, **26**, 795–6.

Jype, P., Bhargava, P. & Tasker, A. (1965). Some aspects of the chemical and cellular compositions of adult rat liver. *Exp. Cell Res.*, **40**, 233–51.

Kaczanowski, A. (1968). Mitosis and polyploidy of nuclei of *Opalina ranarum*. *Experientia*, **24**, 846–7.

Kafatos, F. C. (1972). The cocoonase zymogen cells of silk moths: a model of terminal cell differentiation for specific protein synthesis. *Curr. Top. Dev. Biol.*, **7**, 125–91.

Kaji, K. & Matsuo, M. (1978). Ageing of chick embryo fibroblasts *in vitro*. II. Relationship between cell proliferation and increased multinuclear cells. *Mech. Ageing Develop.*, **8**, 233–9.

Kaji, K. & Matsuo, M. (1979). Ageing of chick embryo fibroblasts *in vitro*. III. Polyploid cell accumulation. *Exp. Cell. Res.*, **119**, 231–6.

Kaji, K. & Matsuo, M. (1981). Ageing of chick embryo fibroblasts *in vitro*. V. Time course studies on polyploid nucleus accumulation. *Exp. Cell Res.*, **131**, 410–12.

Kaminskaya, E. V., Stepanyan, L. I. & Vakhtin, Y. B. (1981). DNA content variability of the clonogenic cell progenitors in the rat rabdomiosarcoma. *Tsitologia*, **23**, 811–17 (in Russian).

Karpovskaya, E. V. & Belyaeva, E. S. (1973). DNA synthesis duration in cells of diploid and tetraploid rye (*Secale cereale* L.). *Tsitologia*, **15**, 104–7 (in Russian).

Kasten, F. H. (1972). Rat myocardial cells *in vitro*: mitosis and differentiated properties. *In Vitro*, **8**, 128–49.

Kasten, F. H., Kudryavtsev, B. N. & Rumyantsev, P. P. (1982). DNA content of isolated rat heart cells during postnatal development. *J. Mol. Cell. Cardiol.*, **14** (suppl. 1), 43–4.

Katzberg, A., Farmer, B. & Harris, R. (1977). The predominance of binucleation in isolated rat heart myocytes. *Amer. J. Anat.*, **149**, 489–500.

Kausch, A. P. & Horner, H. T. (1984). Increased nuclear DNA content in raphide crystal idioblasts during development in *Vanilla planifolia* (Orchidaceae). *Europ. J. Cell Biol.*, **33**, 7–12.

Kelley, R. O., Perdue, B. D. & Uruchurtu-Valdivia, R. (1983). Isolation by flow sorting of cytokinetic and morphological heterogeneity in late-passage cultures of human diploid fibroblasts (IMR-90). *Anat. Rec.*, **206**, 329–39.

Keyl, H. G. (1965). Duplicationen von Untereinheiten der Chromosomalen DNS während der Evolution von *Chironomus thummi*. *Chromosoma*, **17**, 139–80.

Khrushchov, N. G., Lange, M. A. & Satdykova, G. P. (1978). The giant foreign body cells from foci of aseptic inflammation: An electron microscopic and radioautographic study. *Arkhiv Anatomii, Histologii, Embryologii*, **75**, 43–9 (in Russian).

Kiknadze, I. I. (1972). *Functional Organization of Chromosomes*. Leningrad: Nauka (in Russian).

Kiknadze, I. I., Bakhtadze, G. I. & Istomina, A. G. (1975a). DNA reduplication in the endomitotic cells of *Schistocerca gregaria*. *Tsitologia*, **17**, 509–17 (in Russian).

Kiknadze, I. I. & Istomina, A. G. (1980). Endomitosis in grasshoppers. I. Nuclear morphology and synthesis of DNA and RNA in the endopolyploid cells of the inner parietal layer of the testicular follicle. *Europ. J. Cell Biol.*, **21**, 122–33.

Kiknadze, I. I., Kolesnikov, N. N. & Lopatin, O. E. (1975b). *Chironomus thummi* Kieff in laboratory culture. In *Methods of Developmental Biology*, ed. T. A. Detlaf, pp. 95–127. Moscow: Nauka (in Russian).

Kiknadze, I. I. & Tuturova, K. F. (1970). On the transcription capacity of endomitotic cells in the testicle of *Chrysochraon dispar*. *Tsitologia*, **12**, 844–53 (in Russian).

Kiknadze, I. I. & Vysotskaya, L. V. (1979). Nuclear DNA content in locust species with different numbers of chromosomes. *Tsitologia*, **12**, 1100–7 (in Russian).

Kiknadze, I. I., Valeeva, F. S., Vlasova, I. E., Panova, T. M., Sebeleva, T. E. & Kolesnikov, N. N. (1979). Puffing and special function of salivary gland cells of *Chironomus thummi*. 1. Quantitative changes of protein and glycoproteins in the developing larval salivary gland. *Ontogenez*, **10**, 161–72 (in Russian, English translation in *Soviet J. Dev. Biol.*, **10**, 141–50).

Kimler, B. F., Leeper, D. B. & Schneiderman, M. H. (1981). Radiation-induced division delay in Chinese hamster ovary fibroblast and carcinoma cells: dose effect and ploidy. *Radiat. Res.*, **85**, 270–80.

King, R. C. (1970). *Ovarian Development in Drosophila melanogaster*. New York & London: Academic Press.

King, R. C., Riley, S. F., Cassidy, J. D. & White, P. E. (1981). Giant polytene chromosomes from the ovaries of a *Drosophila* mutant. *Science*, **212**, 441–2.

Kinosita, R., Ohno, S. & Nakasawa, M. (1959). On differentiation of the thrombocytic series of cells. *Proc. Amer. Assoc. Canc. Res.*, **3**, 33 (abstract).

Kinsella, A. R. (1980). The role of promoters in two-step carcinogenesis: a hypothesis. In *Progress in Environmental Mutagenesis*, ed. M. Alačevic, pp. 261–73. Amsterdam & New York: Elsevier.

Kirpichnikov, W. S. (1982). *The Genetic Bases of Fish Selection*. New York & Berlin: Springer-Verlag.

Klein, F. A., Herr, H. W., Sogani, P. C., Whitmore, W. F. & Melamed, M. M. (1982). Detection and follow-up of carcinoma of the urinary bladder by flow cytometry. *Cancer*, **50**, 389–95.

Klimenko, V. V. (1971). Changes in weight of diploid, triploid and tetraploid forms of silkworm *Bombyx mori* L. during larval development. *Ontogenez*, **2**, 626–31 (in Russian, English translation in *Soviet J. Dev. Biol.*, **2**, 500–4.

Klinge, O. (1968). Altersabhängige Beeinträchtigung der Zellverteilung in der regenerierenden Rattenleber. *Virchows Arch. B.*, **1**, 342–5.

Klinge, O. (1970). Karyokinese und Kernmuster im Herzmuskel wachsender Ratten. *Virchows Arch. B.*, **6**, 208–19.

Klinge, O. (1971). Das Kernmuster im postnatalen Rattenherzen als Funktion der mitotischen Aktivität. *Verhandl. Deutsch. Ges. Pathol.*, **55**, 458–64.

Klinge, O. (1973). Kernveränderungen und Kernteilungsstörungen der Altersleber. *Gerontologia*, **19**, 314–29.

Klinger, H. P. & Schwarzacher, H. G. (1960). The sex chromatin and heterochromatic bodies in human diploid and polyploid nuclei. *J. Biochem. Biophys. Cytol.*, **8**, 345–64.

Kloetzel, J. A. (1970). Compartmentalization of the developing macronucleus following conjugation in *Stylonychia* and *Euplotes*. *J. Cell Biol.*, **47**, 395–407.

Knudson, A. G. (1971). Mutation and cancer: statistical study of retinoblastoma. *Proc. Natl. Acad. Sci., USA*, **68**, 820–3.

Koch, K. C. & Leffert, H. L. (1979). Increased sodium ion influx is necessary to initiate rat hepatocyte proliferation. *Cell*, **18**, 153–63.

Koch, K. S. & Leffert, H. L. (1980). Growth regulation of adult rat hepatocytes in primary culture. *Ann. NY Acad. Sci.*, **349**, 111–27.

Kogan, M. E., Belov, L. N. & Leontyeva, T. A. (1976). Estimation of cell quantity in different organs and tissues after alkaline dissociation. *Arkhiv patologii*, **38**, 77–80 (in Russian).

Koike, J., Suzuki, J., Nagata, A., Furuta, S. & Nagata, T. (1982). Studies on DNA content of hepatocytes in cirrhosis and hepatoma by means of microspectrophotometry and radioautography. *Histochem.*, **73**, 549–62.

Koltzoff, M. K. (1934). The structure of the chromosomes in the salivary glands of *Drosophila*. *Science*, **80**, 312–13.

Komarov, S. A. (1976). The changes of DNA content in the nuclei of intersegmental abdominal muscle fibres of *Bombyx mori* larvae during the intermolting period. *Tsitologia*, **18**, 458–63 (in Russian).

Korecky, B. & Rakusan, K. (1978). Normal and hypertrophic growth of the rat heart. *Amer. J. Physiol.*, **234**, 123–8.

Korecky, B., Sweet, S. & Rakusan, K. (1979). Number of nuclei in mammalian cardiac myocytes. *Can. J. Physiol. Pharmacol.*, **57**, 1122–9.

Kovaleva, V. G. & Raikov, I. B. (1978). Diminution and resynthesis of DNA during development and senescence of the 'diploid' macronuclei of the ciliate *Trachelonema sulcata*. *Chromosoma*, **67**, 177–92.

Krause, E. (1960). Untersuchungen über die Neurosekretion im Schlundring von *Helix pomatia*. *Z. Zellforsch. mikr. Anat.*, **51**, 748–76.

Kreicbergs, A., Zetterberg, A. & Söderberg, G. (1980). The prognostic significance of nuclear DNA content in chrondrosarcoma. *Analyt. Quant. Cytol.*, **2**, 272–9.

Krishan, A. & Ray-Chandhuri, R. (1964). Asynchrony of nuclear development in cytochalasin-induced multinucleate cells. *J. Cell Biol.*, **43**, 618–21.

Kudryavtsev, B. N., Kudryavtseva, M. V. & Shalakhmetova, T. M. (1979). Cytophotometry of glycogen in rat hepatocytes of different ploidy. *Tsitologia*, **21**, 218–21 (in Russian).

Kudryavtsev, B. N., Kudryavtseva, M. V., Zavadskaya, E. E., Smirnova, S. A. & Skorina, A. D. (1982). Polyploidy in the human liver: normal and hepatitis. *Tsitologia*, **24**, 436–44 (in Russian).

Kudryavtsev, B. N., Tomilin, N. B., Aprelikova, O. N., Maitesyan, E. S. & Turoverova, L. B. (1984). Activity of uracyl-DNA-glycosylase, the enzyme of DNA reparation, in mammalian species differing in lifetime and liver cell ploidy. *Tsitologia*, **26**, 83–90 (in Russian).

Kuhlman, D. (1969). Bestimmung des DNA-Gehaltes in Zellkernen des Nervengewebes von *Helix pomatia* and *Planorbis corneus*. *Experientia*, **25**, 848–9.

Kühn, A. (1971). *Lectures on Developmental Physiology*. Berlin & Heidelberg: Springer-Verlag.

Kuhn, H., Pfitzer, P. & Stoepel, K. (1974). DNA content and DNA synthesis in the myocardium of rats after induced renal hypertension. *Cardiovasc. Res.*, **8**, 86–91.

Kühnel, W. (1972). Morphologische Untersuchungen an der Glandula infraorbitalis buccalis von Kaninchen. *Z. Zellforsch. mikr. Anat.*, **123**, 55–65.

Kulminskaya, A. S., Brodsky, V. Ya. & Gazaryan, K. G. (1978). Mechanism of genome inactivation in bird erythrocytes. IV. New data on the mechanism of cell differentiation during erythropoiesis. *Ontogenez*, **9**, 601–8 (in Russian, English translation in *Soviet J. Dev. Biol.*, **9**, 523–30.

Kushch, A. A., Kolesnikov, V. A., Nijazmatov, A. A., Tolmachev, V. S. & Zelenin, A. V. (1976). The chromatin of mouse hepatocytes in early period after partial hepatectomy: quantitative cytochemistry. *Tsitologia*, **18**, 490–3 (in Russian).

Küster, E. (1956). *Die Pflanzenzelle*. Jena: Fischer Verlag.

Kusyk, Ch. J., Seski, J. C., Medlin, W. V. & Edwards, C. L. (1981). Progressive chromosome changes associated with different sites of one ovarian carcinoma. *J. Natl. Cancer Inst.*, **66**, 1021–4.

Laird, C. D. (1973). DNA of *Drosophila* chromosomes. *Ann. Rev. Genet.*, **7**, 177–204.

Lakhotia, S. C. (1984). Replication in Drosophila chromosomes. XII. Reconfirmation of underreplication of heterochromatin in polytene nuclei by cytofluorometry. *Chromosoma*, **89**, 63–7.

Lamb, M. J. (1982). The DNA content of polytene nuclei in midgut and Malpighian tubule cells of adult *Drosophila melanogaster*. *Wilhelm Roux Arch. Dev. Biol.*, **191**, 381–4.

Lamppa, G. K., Honda, S. & Bendich, A. J. (1984). The relationship between ribosomal repeat length and genome size in *Vicia*. *Chromosoma*, **89**, 1–7.

Landgren, C. R. (1976). Patterns of mitosis and differentiation in cells derived from pea root protoplasts. *Amer. J. Bot.*, **63**, 473–80.

Lapham, L. (1968). Tetraploid DNA content of Purkinje neurons of human cerebellar cortex. *Science*, **159**, 310–12.

Lasek, R. & Dower, W. (1971). *Aplysia californica*: analysis of nuclear DNA in individual nuclei of giant neurons. *Science*, **172**, 278–80.

Lasher, R. (1971). Studies on cellular proliferation and chondrogenesis. In *Developmental Aspects of the Cell Cycle*, ed. I. Cameron, G. Padilla & A. Zimmerman, pp. 223–41. New York & London: Academic Press.

Laskey, R. A., Mills, A. D., Gurdon, J. B. & Partington, G. A. (1977). Protein synthesis in oocytes of *Xenopus laevis* is not regulated by the supply of messenger RNA. *Cell*, **11**, 345–51.

Lau, Y.-F. (1983). Studies on mammalian chromosome replication. III. Organization and replication of diplochromosomes. *Exp. Cell Res.*, **146**, 445–50.

Lebedeva, G. S. & Diment, A. V. (1982). DNA reassociation kinetics in liver nuclei of different ploidy. *Tsitologia*, **24**, 219–23 (in Russian).

Le Bouton, A. V. (1976). DNA synthesis and cell proliferation in the simple liver acinus of 10 to 20-day-old rats: evidence for cell fusion. *Anat. Rec.*, **184**, 679–87.

Lee, L. & Yunis, J. J. (1971). A developmental study of constitutive heterochromatin in *Microtus agrestis. Chromosoma*, **32**, 237–50.

Leffert, J., Moran, T., Sell, S., Skelly, H., Ibsen, K., Mueller, M. & Arias, I. (1978). Growth state-dependent phenotypes of adult hepatocytes in primary monolayer culture. *Proc. Natl. Acad. Sci. USA*, **75**, 1834–8.

Le Rumeur, E. L., Beaumont, C., Guguen-Guillouzo, Ch. & Guillouzo, A. (1982). Protein synthesis in cultured adult hepatocytes is a function of their degree of ploidy. *Biol. Cell*, **45**, 71–8.

Le Rumeur, E. L., Beaumont, C., Guillouzo, Ch., Rissel, M., Bourel, M. & Guillouzo, A. (1981). All normal rat hepatocytes produce albumin at a rate related to their degree of ploidy. *Biochem. Biophys. Res. Commun.* **101**, 1038–46.

Le Rumeur, E., Guguen-Guillouzo, Ch., Beaumont, C., Saunier, A. & Guillouzo, A. (1983). Albumin secretion and protein synthesis by cultured diploid and tetraploid rat hepatocytes separated by elutriation. *Exp. Cell Res.*, **147**, 247–54.

Leuchtenberger, C., Helweg-Larsen, H. & Murmanis, L. (1954). Relationship between hereditary pituitary dwarfism and the formation of multiple DNA classes in mice. *Lab. Invest.*, **3**, 245–60.

Levan, A. (1939). Cytological phenomena connected with the root swelling caused by growth substances. *Hereditas*, **25**, 87–96.

Levan, A. (1956). The significance of polyploidy for the evolution of mouse tumours. *Exp. Cell Res.*, **11**, 613–29.

Levan, A. & Biesele, J. J. (1958). Role of chromosomes in carcinogenesis, as studied in serial tissue culture of mammalian cells. *Ann. New York Acad. Sci.*, **71**, 1022–53.

Levan, A. & Hauschka, T. S. (1953). Endomitotic reduplication mechanisms in ascites tumors of the mouse. *J. Natl. Cancer. Inst.* **14**, 1–43.

Levi, P. E., Cooper, E. H., Anderson, C. U., Path, M. P. & Williams, R. E. (1969*b*). Analysis of DNA content, nuclear size and cell proliferation of transitional cell carcinoma in man. *Cancer* **23**, 1074–85.

Levi, P. E., Cowen, D. M. & Cooper, E. H. (1969*a*). Induction of cell proliferation in the mouse bladder by 4-ethylsulphonylnaphthalene-1-sulphonamide. *Cell Tiss. Kinet.*, **2**, 249–62.

Li, C. C., Karnovsky, M. J., Lin, P. S. & Lin, E. C. C. (1978). The selection of a stable rat hepatoma variant with concomitant increase in ploidy and permeability to glycerol. *J. Cell. Physiol.*, **94**, 197–204.

Libbenga, K. R. & Torrey, J. G. (1973). Hormone induced endoreduplication prior to mitosis in cultured pea root cortex cells. *Amer. J. Bot.*, **60**, 293–9.

Lima-de-Faria, A., Pero, R., Avanzi, S., Durante, M., Ståle, U., D'Amato, F. & Granström, H. (1975). Relation between rRNA genes and the DNA satellites of *Phaseolus coccineus. Hereditas* **79**, 5–20.

Linden, W. A. (1982). Clinical applications of flow cytometry. In *Cell Growth*, ed. C. Nicolini, pp. 735–44. New York: Plenum Press.

Liozner, L. D. & Markelova, I. V. (1971). Mitotic cycle of hepatocytes in regenerating liver. *Byulleten exper. Biologii Meditsini*, **71**, 99–103 (in Russian).

Lipps, H. J., Gruissem, W. & Prescott, D. M. (1982). Higher order DNA structure in macronuclear chromatin of the hypotrichous ciliate *Oxytricha nova*. *Proc. Natl. Acad. Sci. USA*, **79**, 2495–9.

Lloyd, D., Poole, R. K. & Edwards, S. W. (1982). *The Cell Division Cycle. Temporal Organization and Control of Cellular Growth and Reproduction*. London: Academic Press.

Lohmann, K. (1972). Untersuchungen zur Frage der DNS-Konstanz in der Embryonalentwicklung. *Wilhelm Roux' Arch. Entwicklungmech. Org.*, **169**, 1–40.

Long, M. W. & Williams, N. (1981). Immature megakaryocytes in the mouse: morphology and quantitation by acetylcholinesterase staining. *Blood*, **58**, 1032–9.

Lyapunova, E. A., Ginatulina, L. K., Korablev, V. P., Ginatulin, A. A. & Vorontsov, N. N. (1980). The comparison of intragenic divergence in DNA and heterochromatin content in ground squirrels of the genus *Citellus*. In *Animal Genetics and Evolution*, ed. N. N. Vorontsov & J. M. Van Brink, pp. 229–39. The Hague: W. Junk Publishing House.

McBurney, M. W. & Whitmore, G. F. (1974). Selection for temperature-sensitive mutants of diploid and tetraploid mammalian cells. *J. Cell. Physiol.*, **83**, 69–74.

Mack, L. (1977). Nukleinsäuren- und Hämoglobin-Gehalt bei Eisenmangelanämie, sideroachrestischer und perniciöser Anämie. Ph.D. thesis, University of Tübingen.

Magakyan, J. A., Karalova, E. M. & Hachatryan, M. G. (1979). The oogenesis of the Ararat cochineal. III. Nucleic acid synthesis in the nuclei of auxiliary cells during differentiation and accomplishment of specific functions. *Tsitologia*, **21**, 548–57 (in Russian).

Mahowald, A. P., Caulton, J. H. & Edwards, M. K. (1979). Loss of centrioles and polyploidization in follicle cells of *Drosophila melanogaster*. *Exp. Cell Res.*, **118**, 404–10.

Maljuk, V. I. (1970). *Physiological Regeneration of the Vascular Wall*. Kiev: Naukova Dumka (in Russian).

Manfredi Romanini, M. G. (1974). The DNA content and the evolution of vertebrates. In *Cytotaxonomy and Vertebrate Evolution*, ed. A. B. Chiarelli & E. Capanna, pp. 39–81. London & New York: Academic Press.

Manfredi Romanini, M. G., Fraschini, A. & Bernocchi, G. (1972). DNA content and nuclear area in the neurons of the cerebral ganglion in *Helix pomatia*. *Ann. Histochim.*, **18**, 49–58.

Manfredi Romanini, M. G., Minazza, M. & Capanna, E. (1971). Nuclear DNA content in lymphocytes from *Mus musculus* and *Mus poschiavinus*. *Boll. Zool.*, **38**, 321–6.

Manteuffel, R., Muntz, K., Puckel, M. & Scholz, G. (1976). Phase-dependent changes of DNA, RNA and protein accumulation during ontogenesis of broad bean seeds (*Vicia faba*). *Biochem. Physiol. Pflanzen*, **169**, 595–605.

Marceau, N., Deschênes, J. & Valet, J. P. (1982). Effect of hepatocyte proliferation and neoplastic transformation on alpha-foetoprotein and albumin production per cell: influence of cell specialization and cell size. *Oncodev. Biol. Med.*, **3**, 49–63.

Mariano, M. & Spector, W. G. (1974). The formation and properties of macrophage polykaryons (inflammatory giant cells). *J. Pathol.*, **113**, 1–19.

Mark, G. & Strasser, F. (1966). Pacemaker activity and mitosis in cultures of newborn rat heart ventricle cells. *Exp. Cell Res.*, **44**, 217–33.

Markov, G., Dessev, G., Russev, G. & Tsanev, R. (1975). Effects of γ-irradiation on biosynthesis of different types of ribonucleic acids in normal and regenerating rat liver. *Biochem. J.*, **146**, 41–51.

Marshak, T. L. (1974). Role of mitosis in the formation of binucleate cells in the pigment epithelium of the retina and in the liver of rats. *Ontogenez*, **5**, 192–7 (in Russian, English translation in *Soviet J. Dev. Biol.*, **5**, 166–71).

Marshak, T. L., Gorbunova, M. P. & Stroeva, O. G. (1972). Proliferation of retinal pigment epithelium cells in rats. *Tsitologia*, **14**, 1113–19 (in Russian).

Marshak, T. L., Mareš, V., Lisý, V., Kudryavtsev, B. N., Lodin, Z. & Brodsky, V. Ya. (1980). Constancy and variability of DNA content in Purkinje cells: re-evaluation of data on ontogeny and cytological analysis of this phenomenon. In *Ontogenesis of the Brain*, ed. S. Trojan & F. Stastny, vol. 3, pp. 523–30. Prague: Artia.

Marshak, T. L. & Stroeva, O. G. (1973). Cytophotometric investigation of the DNA content of the retinal pigment epithelium cells in postnatal development of rats: a cytophotometric study. *Ontogenez*, **4**, 516–20 (in Russian, English translation in *Soviet J. Dev. Biol.*, **4**, 472–5).

Marshak, T. L. & Stroeva, O. G. (1974). Cell ploidy of retinal pigment epithelium in hereditary microphthalmic rats. *Archiv Anatomii, Histologii, Embryologii*, **66**, 94–8 (in Russian).

Marshak, T. L., Stroeva, O. G. & Brodsky, V. Ya. (1976). Specialization of mononucleate and binucleate cells of the retinal pigment epithelium in early postnatal development of rats. *Zhurnal obshchei Biologii*, **37**, 608–14 (in Russian).

Martin, B. F. (1972). Cell replacement and differentiation in transitional epithelium: a histological and autoradiographic study of the guinea-pig bladder and ureter. *J. Anat.*, **112**, 433–55.

Marzullo, G. & Lash, J. W. (1970). Control of phenotypic expression in cultured chondrocytes: investigations on the mechanism. *Dev. Biol.*, **22**, 638–54.

Matthews, J. L., Martin, J. H., Race, G. L., & Collins, E. J. (1967). Giant-cell centrioles. *Science*, **155**, 1423–4.

Matsumura, T. (1980). Multinucleation and polyploidization of aging human cells in culture. In *Aging Phenomena: Relationships among Different Levels of Organisation*, ed. K. Oota, T. Makinodan, M. Iriki & L. S. Baker, pp. 31–8. New York: Plenum Press.

Matsumura, T., Masuda, K., Murakami, Y. & Konishi, R. (1983). Family trees representing the finitely proliferative nature of cultured rat liver cells. *Cell Struct. & Func.*, **8**, 293–306.

Matsumura, T., Zerrudo, Z. & Hayflick, L. (1979). Senescent human diploid cells in culture: survival, DNA synthesis and morphology. *J. Gerontol.*, **34**, 328–34.

Matthysse, A. G. & Torrey, J. G. (1967). DNA synthesis in relation to polyploid mitoses in excised pea root segments cultured *in vitro*. *Exp. Cell Res.*, **48**, 484–99.

Matuszewski, B. (1965). Transition from polyteny to polyploidy in salivary glands of Cecidomyiidae. *Chromosoma*, **16**, 22–34.

Maurer, W., Gerhard, H. & Schultze, B. (1973). Quantitatives Modell der Regeneration der CCl₄ Leber der Maus. *Virchows Archiv. B.*, **14**, 361–71.

Mazia, D. (1961). Mitosis and physiology of cell division. In *The Cell*, ed. J. Brachet & A. E. Mirsky, vol. 3, pp. 77–412. New York & London: Academic Press.

Meinders-Groeneveld, J. & James, J. (1971). Some quantitative data regarding the nucleoli in cell nuclei from rat liver of different ploidy classes. *Z. Zellforsch. mikr. Anat.*, **114**, 165–74.

Mello, M. L. S. (1971). Nuclear behavior in the Malpighian tubes of *Triatoma infestans* (Hemiptera). *Cytologia*, **36**, 42–9.

Mendecki, J., Dillmann, W. H., Wolley, M. C., Offenheimer, J. H. & Koss, L. G. (1978). Effect of thyroid hormone on the ploidy of rat liver nuclei as determined by flow cytometry. *Proc. Soc. Exp. Biol. Med.*, **158**, 63–7.

Mermod, J. J., Jacobs-Lorena, M. & Crippa, M. (1977). Changes in rate of RNA synthesis and ribosomal gene number during oogenesis of *Drosophila melanogaster*. *Dev. Biol.*, **57**, 393–402.

Merriam, R. W. & Ris, H. (1954). Size and DNA content of nuclei in various tissues of male, female and worker honey bees. *Chromosoma*, **6**, 522–38.

Mezger-Freed, L. (1977). Haploid vertebrate cell cultures. In *Growth, Nutrition, and Metabolism of Cell Culture*, ed. G. H. Rothblat & V. J. Cristofalo, vol. 3, pp. 57–82. New York & London: Academic Press.

Middleton, J. & Gahan, P. B. (1979). A quantitative cytochemical study of acid phosphatases in rat liver parenchymal cells of different ploidy values. *Histochem. J.*, **11**, 649–59.

Middleton, J. & Gahan, P. B. (1982). A quantitative cytochemical study of acid phosphatases in hepatocytes of different ploidy classes from adult rats. *Exp. Gerontol.*, **17**, 267–72.

Mikeladze, Z. A. & Brodsky, V. Ya. (1980). Non-symmetric histogram of RNA content in nucleoli as possible sign of the ribosome DNA amplification in the cerebellar Purkinje cells of rats. *Byulleten exper. Biologii Meditsini*, **90**, 485–7 (in Russian, English translation in *Bull. Exp. Biol. Med.*, **90**, 1445–7).

Miljutina, N. A., Kogan, E. M. & Nekhljudova, M. A. (1978). Pulsed pattern of proliferation in liver parenchyma of young mice. *Virchows Arch. B.*, **29**, 179–90.

Miller, R. C., Nichols, W. W., Pottash, J. & Aronson, M. M. (1977). *In vitro* aging: cytogenetic comparison of diploid human fibroblast and epithelioid cell lines. *Exp. Cell Res.*, **110**, 63–73.

Millerd, A. & Whitfeld, P. R. (1973). DNA and RNA synthesis during the expansion phase of cotyledon development in *Vicia faba* L. *Plant Physiol.*, **51**, 1005–10.

Mirsky, A. E. & Ris, H. (1951). The desoxyribonucleic acid content of animal cells and its evolutionary significance. *J. Gen. Physiol.*, **34**, 451–62.

Mitchison, J. M. (1971). *The Biology of the Cell Cycle*. Cambridge University Press.

Mitsui, J. & Schneider, E. L. (1976). Increased nuclear sizes in senescent human diploid fibroblast cultures. *Exp. Cell Res.*, **100**, 147–52.

Mittwoch, U., Lele, K. P. & Webster, W. S. (1965). Deoxyribonucleic acid synthesis in cultured human cells and its bearing on the concepts of endoreduplication and polyploidy. *Nature (London)*, **208**, 242–4.

Modak, S., Morris, G. & Yamada, T. (1968). DNA synthesis and mitotic activity during early development of chick lens. *Dev. Biol.*, **17**, 545–61.

Moorhead, P. S. & Hsu, T. C. (1956). Cytologic studies of HeLa, a strain of human cervical carcinoma. *J. Natl. Cancer. Inst.*, **16**, 1047–66.

Mora, M., Darżynkiewicz, Z. & Baserga, R. (1980). DNA synthesis and cell division in a mammalian cell mutant temperature sensitive for the processing of ribosomal RNA. *Exp. Cell Res.*, **125**, 241–9.

Morescalchi, A. (1974). Amphibia. In *Cytotaxonomy and Vertebrate Evolution*, ed. A. B. Chiarelli & E. Capanna, pp. 233–348. London & New York: Academic Press.

Moriwaki, K. & Imai, H. T. (1978). Mechanism of ploidy shift from diploidy to tetraploidy in MSPC-1 mouse myeloma. *Exp. Cell Res.*, **111**, 483–9.

Morselt, A. F. W. & Frederiks, W. N. (1974). Microphotometry of rat liver nuclear protein and RNA before and after partial hepatectomy. *Histochem.*, **41**, 111–18.

Morselt, A. F. W. & Wijgerden, H. G. (1975). Microphotometry of rat liver nucleoproteins during the cell cycle and comparison of diploid nuclei in the G_2 period with tetraploid nuclei. *Histochem.*, **44**, 87–93.

Moskvin-Tarkhanov, M. I. & Onishchenko, G. E. (1978). Centrioles in megakaryocytes of the mouse bone marrow. *Tsitologia*, **20**, 1436–7 (in Russian).

Moubayed, P. A. & Pfitzer, P. (1975). DNS-Gehalt und Zahl der Kerne in den myokardialen Zellen atrophischer Herzen. *Verhandl. deutsch. Ges. Pathol.*, **59**, 338–40.

Müller, D. (1972). Quantitative cytochemical evaluation of hemoglobin synthesis in erythroblasts. In *Synthese, Structur und Function des Hämoglobins*, ed. H. Martin & L. Nowicki, pp. 47–52. Munich: Lehmans.

Müller, D., Lauterbach, H., Pouillon, H. G. & Hahn, E. (1973). Synthese von Hämoglobin, RNS und Proteinen in der normalen Erythropoese. *Acta Haematol.*, **50**, 340–9.

Murin, A. (1976), Polyploidy and mitotic cycle. *Nucleus*, **19**, 192–4.

Murray, B. G. & Williams, C. A. (1973). Polyploidy and flavonoid synthesis in *Briza media* L. *Nature* (*London*), **243**, 87–8.

Nadal, C. (1970). Polyploidie hépatique du rat: mode de formation des cellules binucléées. *J. Microscop.*, **9**, 611–18.

Nadal, C. & Heyman-Blanchet, T. (1965). Mode de formation des cellules binucléées dans le foie du rat. *C.r. Acad. sci., Paris*, **260**, 1763–6.

Nadal, C. & Zajdela, F. (1966). Polyploidie somatique dans le foie du rat. I. Le rôle des cellules binucléées dans la genèse des cellules polyploides. *Exp. Cell Res.*, **42**, 99–116.

Nagl, W. (1962). Über Endopolyploidie, Restitutionkernbildung und Kernstructuren im Suspensor von Angiospermen und einer Gymnosperme. *Österr. Bot. Z.*, **109**, 431–94.

Nagl, W. (1970*a*). Inhibition of polytene chromosome formation in *Phaseolus* by polyploid mitoses. *Cytologia*, **35**, 252–8.

Nagl, W. (1970*b*). Karyologische Anatomie des Endosperms von *Phaseolus coccineus*. *Österr. Bot. Z.*, **118**, 566–71.

Nagl, W. (1970*c*). Differential inhibition by actinomycin D and histone F_1 of mitosis and endomitosis in *Allium carinatum*. *Z. Pflanzenphysiol.*, **63**, 316–26.

Nagl, W. (1972*a*). Molecular and structural aspects of endomitotic chromosome cycle in angiosperms. *Chromosomes Today*, **3**, 17–23.

Nagl, W. (1972*b*). Giant sex chromatin in endopolyploid trophoblast nuclei in the rat. *Experientia*, **28**, 217–18.

Nagl, W. (1973). The mitotic and endomitotic nuclear cycle in *Allium carinatum* IV. ^{3}H-uridine incorporation. *Chromosoma*, **44**, 203–12.

Nagl, W. (1977). Localization of amplified DNA in nuclei of the orchid *Cymbidium* by *in situ* hybridization. *Experientia*, **33**, 1040–1.

Nagl, W. (1978). *Endopolyploidy and Polyteny in Differentiation and Evolution*. Amsterdam: Elsevier–North-Holland.

Nagl, W. (1981). Polytene chromosomes of plants. *Internat. Rev. Cytol.*, **73**, 21–53.

Nagl, W. (1983). Heterochromatin elimination in the orchid *Dendrobium*. *Protoplasma*, **118**, 234–7.

Nagl, W., Peschke, C. & van Gyseghem, R. (1976). Heterochromatin underreduplication in *Tropaeolum* embryogenesis. *Naturwissenschaften*, **63**, 198–9.

Nair, K. K., Chen, Th. T. & Wyatt, G. R. (1981). Juvenile hormone-stimulated polyploidy in adult locust fat body. *Dev. Biol.*, **81**, 356–60.

Nakamura, T., Yoshimoto, K., Nakayama, Y., Tomita, Y. & Ichihara, A. (1983). Reciprocal modulation of growth and differentiated functions of mature rat hepatocytes in primary culture by cell–cell contact and cell membranes. *Proc. Natl. Acad. Sci. USA*, **80**, 7229–33.

Nakanishi, Y. H., Kato, H. & Utsumi, S. (1969). Polytene chromosomes in silk gland cells of the silkworm *Bombyx mori*. *Experientia*, **25**, 384–5.

Nakanishi, K. & Fujita, S. (1977). Molecular mechanism of polyploidization and binucleate formation of the hepatocyte. *Cell Struct. Func.* **2**, 261–5.

Nakeff, A. (1980). Flow cytometric analysis of megakaryocytopoiesis strategy for defining mitotic and endoreduplicative events in the commited progenitor compartment. *Blood Cells*, **6**, 665–78.

Naora, H. (1957). Microspectrophotometry of cells during postnatal growth. *J. Biophys. Biochem. Cytol.*, **3**, 949–75.

Narayan, R. K. J. & Rees, H. (1976). Nuclear DNA variation in *Lathyrus*. *Chromosoma*, **54**, 141–54.

Narushevichus, E. V. & Kviklite, A. V. (1968). A rule of size distribution of nerve cells and its interpretation. *Tsitologia*, **10**, 900–2 (in Russian).

Nasjleti, C. E. & Spencer, H. H. (1967). Chromosome polyploidization in human leukocyte cultures treated with streptonigrin and cyclophosphamide. *Cancer*, **20**, 31–5.

Nasjleti, C. E. & Spencer, H. H. (1968). The effects of chloramphenicol on mitosis of phytohemagglutinin stimulated human leukocytes. *Exp. Cell Res.*, **53**, 11–17.

Neal, J. V. & Potten, C. S. (1981). Polyploidy in the murine colonic pericryptal fibroblast sheath. *Cell Tiss. Kinet.*, **14**, 527–36.

Neyfakh, A. A. & Timofeeva, M. J. (1977). *Molecular Biology of Developmental Processes*. Moscow: Nauka. (In Russian).

Neyfakh, A. A. & Timofeeva, M. Y. (1984). *Molecular Biology of Development*. Part II. Moscow: Plenum Press (in press).

Nowell, P. C. (1976). The clonal evolution of tumor cell population. *Science*, **194**, 23–8.

Nuti Ronchi, V., Avanzi, S. & D'Amato, F. (1965). Chromosome endoreduplication (endopolyploidy) in pea root meristems induced by 8-azaguanine. *Caryologia*, **18**, 599–617.

Oberling, Ch. & Bernhard, W. (1961). The morphology of the cancer cells. In *The Cell*, ed. J. Brachet & A. E. Mirsky, vol. 5, pp. 405–96. New York & London: Academic Press.

Oberpriller, J. O., Ferrans, V. J. & Carroll, R. J. (1983). Changes in DNA content, number of nuclei and cellular dimensions of young rat atrial myocytes in response to left coronary artery ligation. *J. Molec. Cell Cardiol.*, **15**, 31–42.

Odell, T. T. (1972). Regulation of the megakaryocyte-platelet system. In *Regulation of Organ and Tissue Growth*, ed. R. J. Goss, pp. 187–95. New York & London: Academic Press.

Odell, T. T., Burch, E. A., Jackson, C. W. & Friday, T. J. (1969). Megakaryocytopoiesis in mice. *Cell Tiss. Kinet.*, **2**, 363–7.

Odell, T. T. & Jackson, C. W. (1968). Polyploidy and maturation of rat megakaryocytes. *Blood*, **32**, 102–10.

Odell, T. T., Jackson, C. W. & Friday, T. J. (1970). Megakaryocytopoiesis in rats with special reference to polyploidy. *Blood*, **35**, 775–82.

Odell, T. T., Jackson, C. W. & Gosslee, D. G. (1965). Maturation of rat megakaryocytes studied by microspectrophotometric measurement of DNA. *Proc. Soc. Exp. Biol. Med.*, **119**, 1194–9.

Odell, T. T., Jackson, C. W. & Reiter, R. S. (1968). Generation cycle of rat megakaryocytes. *Exp. Cell Res.*, **53**, 321–8.

Ohno, S. (1970). *Evolution by Gene Duplication*. Berlin & Heidelberg: Springer Verlag.

Ohno, S., Kaplan, W. & Kinosita, R. (1959). Formation of the sex chromatin by a single X-chromosome in liver cells of *Rattus norvegicus*. *Exp. Cell Res.*, **18**, 415–18.

Olive, P. L., Leonard, J. C. & Durand, R. E. (1982). Development of tetraploidy in V79 spheroids. *In vitro*, **18**, 700–14.

Olszewski, W., Darżynkiewicz, Z., Rosen, P. P., Schwartz, M. & Melamed, M. M. (1981). Flow cytometry of breast carcinoma: Relation of DNA ploidy level to histology and estrogen receptor. *Cancer*, **48**, 980–4.

Onishchenko, A. M., Mikichur, N. I. & Maksimovsky, L. F. (1976). Nuclear organization and differentiation of the salivary glands in *Drosophila virilis* larvae. 1. Changes of DNA and RNA contents in different structures of cells at stage of the secretion formation. *Ontogenez*, **7**, 76–81 (in Russian, English translation in *Soviet J. Dev. Biol.*, **7**, 60–4).

Onishchenko, G. E. (1978). Correspondence in centrioles number and hepatocyte ploidy in the mouse liver. *Tsitologia*, **20**, 395–9 (in Russian).

Onishchenko, G. E., Bystrevskaya, V. B. & Chentsov, Yu. S. (1979). Multipolar mitoses in tissue culture cells under normal conditions and after induction with 2-mercaptoethanol. *Doklady Akad. Nauk SSSR*, **248**, 1443–6 (in Russian, English translation in *Doklady Biol. Sci.*, **248**, 1071–4).

Onishchenko, G. E. & Chentsov, Yu.S. (1977). Centrioles in polyploid cells. In *The Second Soviet Symposium on Cell Polyploidy*, 89–94. Yerevan: Armenian Academy of Science Publishing Office.

Onishchenko, G. E. & Volkova, L. V. (1983). The centriolar complex in the somatic hybrid cells (mouse × Chinese hamster). *Tsitologia*, **25**, 508–15 (in Russian).

Ordahl, C. P. (1977). Reassociation kinetics of polyploid hepatocyte DNA. *Biochim. Biophys. Acta*, **474**, 17–29.

Orly, J. & Sato, G. (1979). Fibronectin mediates cytokinesis and growth of rat follicular cells in serum-free medium. *Cell*, **17**, 295–305.

Owens, G. K., Rabinovitch, P. S. & Schwartz, S. M. (1981). Smooth muscle cell hypertrophy versus hyperplasia in hypertension. *Proc. Natl. Acad. Sci. USA*, **78**, 7759–63.

Owens, G. K. & Schwartz, S. M. (1982). Alterations in vascular smooth muscle mass in the spontaneously hypertensive rat: role of cellular hypertrophy, hyperploidy, hyperplasia. *Circ. Res.*, **51**, 280–90.

Painter, T. S. & Reindorp, E. C. (1939). Endomitosis in the nurse cells of the ovary of *D. melanogaster*. *Chromosoma*, **1**, 276–83.

Palanker, P. A. & Swartz, F. J. (1971). Nuclear differentiation in *Ascaris* and some unusual cytological characteristics of giant highly polyploid nucleus. *Anat. Rec.*, **169**, 395.

Palmiter, R. (1975). Quantitation of parameters that determine the rate of ovalbumin synthesis. *Cell*, **4**, 189–97.

Panova, I. G. & Stroeva, O. G. (1978). Influence of experimental arrest of eye growth on the proliferation and polyploidization of cells in the pigment epithelium of the rat retina. *Ontogenez*, **9**, 179–83 (in Russian, English translation in *Soviet J. Dev. Biol.* **9**, 146–50).

Papadimitriou, J. M. & Shellam, G. R. (1981). A cytophotometric measurement of DNA in murine hepatocytic nuclei during cytomegalovirus infection. *Histochem.* **72**, 481–7.

Pasternak, I. & Barrell, R. (1976). Quantitation of nuclear DNA in *Ascaris lumbricoides*: DNA constancy and chromatin diminution. *Genet. Res.*, **27**, 339–48.

Patek, E., Johannisson, E., Krauer, F. & Riottou, G. (1980). Microfluorometric grading of mammary tumors: a pilot study. *Analyt. Quant. Cytol.*, **2**, 264–71.

Paulus, J. (1968). Cytophotometric measurements of DNA in thrombopoietic megakaryocytes. *Exp. Cell Res.*, **53**, 310–13.

Pearson, G. G., Timmis, J. N. & Ingle, J. (1974). The differential replication of DNA during plant development. *Chromosoma*, **45**, 281–94.

Pearson, M. J. (1974). Polyteny and the functional significance of the polytene cell cycle. *J. Cell Sci.*, **15**, 457–79.

Pelliciari, C. & Prosperi, E. (1980). Quantitative cytochemical study of chromatin during megakaryocyte maturation. *Basic Appl. Histochem.*, **24**, 409–21.

Pegington, C. & Rees, H. (1970). Chromosome weights and measures in the Triticinae. *Heredity*, **25**, 195–205.

Penington, D. G. & Olsen, T. E. (1970). Megakaryocytes in states of platelet production: cell numbers, size and DNA content. *Brit. J. Haematol.*, **18**, 447–63.

Pera, F. (1975). Arrangement of spindle apparatus in mitoses of different ploidy. *Exp. Cell Res.*, **92**, 419–27.

Pera, F. & Detzer, K. (1975). Heteropyknosis of the chromosomes in liver cells of different ploidy: a nuclear image study. *Beitr. Pathol.*, **156**, 145–54.

Pera, F. & Rainer, B. (1973). Studies of multipolar mitoses in euploid tissue cultures.

I. Somatic reduction to exactly haploid and triploid chromosome sets. *Chromosoma*, **42**, 71–86.

Pera, F. & Schwarzacher, H. G. (1968). Formation and division of binucleated cells in kidney cell cultures of *Microtus agrestis*. *Human Genetics*, **6**, 158–62.

Pera, F. & Schwarzacher, H. G. (1969). Die Verteilung der Chromosomen auf die Tochterzellkerne multipolarer Mitosen in euploiden Gewebekulturen von *Microtus agrestis*. *Chromosoma*, **26**, 337–54.

Perdrix-Gillot, S. (1974). Endoreduplication de la chromatine dans la glande sericigène de *Bombyx mori*. *Annales Embryol. Morphogen.*, **7**, 445–58.

Perdrix-Gillot, S. (1979). DNA synthesis and endomitoses in the giant nuclei of the silkgland of *Bombyx mori*. *Biochimie*, **61**, 171–204.

Perry, L. & Swartz, F. (1967). Evidence for a subpopulation of cells with an extended G_2 period in normal adult mouse liver. *Exp. Cell Res.*, **48**, 155–7.

Pfitzer, P. (1971*a*). Der DNS-Gehalt der Zellkerne im Herzmuskel des Truthahns. *Virchows Arch. B.*, **8**, 175–8.

Pfitzer, P. (1971*b*). Polyploide Zellkerne im Herzmuskel des Schweins. *Virchows Arch. B*, **9**, 180–6.

Pfitzer, P. (1980). Amitosis: a historical misinterpretation. *Pathol. Res. Pract.*, **167**, 292–300.

Pfitzer, P., Knieriem, H. J. & Schulte, H. (1977). Karyological and electron-microscopic studies in myocardial cells of primates after experimentally induced cardiac hypertrophy. *J. Med. Primatol.*, **6**, 349–59.

Philippens, R. M. H., Rover, S. & Abbecht, J. (1981). Circadian rhythmic variations of the relative number of binucleated liver cells in rats. *Prog. Clin. Biol. Res.*, **159c**, 99–108.

Pictet, R., Clark, W., Williams, K. & Rutter, W. (1972). An ultrastructural analysis of the developing embryonic pancreas. *Dev. Biol.*, **29**, 436–67.

Polishchuk, A. M. (1967). Content and kinetics of cell population in the regenerating rat liver early after partial hepatectomy. *Tsitologia*, **9**, 652–7 (in Russian).

Pollinger, G. S. (1973). Growth and DNA synthesis in embryonic chick heart cells *in vivo* and *in vitro*. *Exp. Cell Res.*, **76**, 253–62.

Polyansky, G. I. (1965). Peculiarities of Protozoa organization as cell-organisms. In *Handbook of Cytology*, ed. A. S. Troshin, pp. 409–38. Leningrad: Nauka (in Russian).

Polyansky, G. I. (1972). Evolution of protozoans and morphophysiological rules of evolution process. In *Rules of Progressive Evolution*, ed. K. M. Zavadsky, pp. 286–93. Leningrad: Nauka (in Russian).

Potten, C. S. (1974). The epidermal proliferative unit: the possible role of the central basal cell. *Cell Tiss Kinet.*, **7**, 77–83.

Poulson, D. F. (1950). Histogenesis, organogenesis and differentiation in the embryo of *Drosophila melanogaster*. In *Biology of Drosophila*, ed. M. Demerec, pp. 168–274. New York: Wiley & Sons.

Prasad, N. (1975). Differentiation of neuroblastoma cells in culture. *Biol. Rev.*, **50**, 129–67.

Prasad, K. N. & Cutler, R. G. (1976). Percent satellite DNA as a function of tissue and age of mice. *Biochim. Biophys. Acta*, **418**, 1–23.

Prescott, D. M. (1976). *Reproduction of Eukaryotic Cells*. New York & London: Academic Press.

Prescott, D. M. (1982). Initiation of DNA synthesis and progression through the S-period. In *Cell Growth*, ed. C. Nicolini, pp. 355–64. New York: Plenum Press.

Prescott, D. M., Liskay, R. M. & Stancel, G. M. (1982). The cell life cycle and the G_1 period. In *Cell Growth*, ed. C. Nicolini, pp. 305–14. New York: Plenum Press.

Prokofyeva-Belgovskaya, A. A. (1937). Concerning chromosome structure in the salivary gland of *Drosophila melanogaster*. *Izvestiya Akad. Nauk SSSR*, **2**, 393–9 (in Russian).

Przybalski, R. I. & Chlebowski, G. S. (1972). DNA synthesis, mitosis and fusion of myocardial cells. *J. Morphol.*, **137**, 417–31.

Půža, H. (1969). Sources of mistakes on the interpretation of the appearances of amitotic division. *Folia Morphol.*, **17**, 66–7.

Pylilo, I. V. (1975). Mitoses in the regenerating comb row and binucleate cells of the Ctenophora. *Ontogenez*, **6**, 187–9 (in Russian, English translation in *Soviet J. Dev. Biol.*, **6**, 152–5).

Rabes, H. & Brändle, H. (1968). Beziehungen zwischen RNS- und DNS-Synthese bei der Leberzellproliferation nach partieller Hepatektomie. *Virchows Arch. B*, **1**, 317–22.

Radley, J. M. (1967). Changes in ploidy in the rat submaxillary gland induced by isoprenaline. *Exp. Cell Res.*, **48**, 679–81.

Raghuvanshi, S. S. (1978). Some problems of polyploids and their radiosensitivity. *Nucleus*, **21**, 132–4.

Raikov, I. B. (1972). Nuclear phenomena during conjugation and autogamy in ciliates. In *Research in Protozoology*, vol. 4, ed. T. T. Chen, pp. 147–284. Oxford: Pergamon Press.

Raikov, I. B. (1982). *The Protozoan Nucleus: Morphology and Evolution*. Vienna: Springer-Verlag.

Raikov, I. B. & Ammermann, D. (1976). Recent data in studying of the ciliate macronucleus. In *Karyology and Genetics of Protozoa*, ed. G. I. Polyansky, pp. 64–90. Leningrad: Nauka (in Russian).

Ranek, L. (1976). The nuclear dry weight of liver cells from patients with virus hepatitis and from control. *Acta Cytol.* **20**, 58–61.

Ranek, L. (1976). Cytophotometric studies of the DNA, nucleic acid and protein content of human liver cell nuclei. *Acta Cytol.*, **20**, 151–7.

Rasch, E. M. (1970). DNA cytophotometry of salivary gland nuclei and other tissue system in Dipteran larvae. In *Introduction to Quantitative Cytochemistry*, vol. 2, ed. G. L. Wied & G. F. Bahr, pp. 357–97. New York & London: Academic Press.

Rasch, E. M. (1974). The DNA content of sperm and hemocyte nuclei of silkworm *Bombyx mori* L. *Chromosoma*, **45**, 1–26.

Rasch, E. M., Barr, H. J. & Rasch, R. W. (1971). The DNA content of sperm of *Drosophila melanogaster*. *Chromosoma*, **33**, 1–18.

Rasch, E. M., Cassidy, J. D. & King, R. C. (1971). Evidence for dosage compensation in parthenogenetic hymenoptera. *Chromosoma*, **59**, 323–40.

Rees, H. (1964). The question of polyploidy in the Salmonidae. *Chromosoma*, **15**, 275–9.

Rees, H. (1972). DNA in higher plants. *Brookhaven Symposia in Biology*, **23**, 394–418.

Rees, R. W., Fox, D. P. & Maher, E. P. (1976). DNA content, reiteration and satellites in *Dermestes*. In *Current Chromosome Research*, ed. K. Jones & P. E. Brandham, pp. 33–41. Amsterdam: Elsevier–North-Holland.

Renkawitz-Pohl, R. & Kunz, W. (1975). Underreplication of satellite DNAs in polyploid ovarian tissue of *Drosophila virilis*. *Chromosoma*, **49**, 375–82.

Revesz, L. & Norman, U. (1960). Chromosome ploidy and radiosensitivity of tumours. *Nature (London)*, **187**, 861–2.

Ribbert, D. & Bier, K. (1969). Multiple nucleoli and enhanced nucleolar activity in the nurse cells of the insect ovary. *Chromosoma*, **27**, 178–97.

Richter, R. & Halle, W. (1983). Membrane potential of vascular mono- and multinuclear endothelial cells cultured *in vitro*. *Experientia*, **39**, 55–6.

Riede, J. & Renz, M. (1983). Study on the somatic pairing of polytene chromosomes. *Chromosoma*, **88**, 116–23.

Rifkind, R. A. (1974). Erythroid cell differentiation. In *Concepts of Development*, ed. J. Lash & J. R. Whittaker, pp. 149–62. Stamford: Sinauer.

Rigler, R. (1962). Untersuchungen an isolierten Mäuseleberzellen nebst Angabe eines Isolierverfahrens zur Bestimmung ihres Trockengewichtes. *Exp. Cell Res.*, **28**, 260–8.

Rizzoni, M. & Palitti, F. (1973). Regulatory mechanism of cell division. I. Colchicine-induced endoreduplication. *Exp. Cell Res.*, **77**, 450–8.

Rizzoni, M., Palitti, F. & Perticone, P. (1974). Euploid segregation through multipolar mitosis in mammalian cell cultures. *Chromosoma*, **45**, 151–62.

Roberts, B. (1975). The volume and DNA content of extrachromosomal inclusions in the dorsal foot-pad nuclei of *Tricholioproctia impatiens* (Sarcophagidae, Diptera). *Chromosoma*, **52**, 49–58.

Roberts, B., Whitten, J. M. & Gilbert, L. (1974). DNA synthesis patterns in the giant foot-pad nuclei of *Sarcophaga bullata* (Sarcophagidae, Diptera). *Chromosoma*, **47**, 193–201.

Rodman, T. C. (1967). Control of polytenic replication in Dipteran larvae. I. Increased number of cycles in a mutant strain of *Drosophila melanogaster*. *J. Cell Physiol.*, **70**, 179–86.

Romen, W., Rüter, A., Saito, K., Harms, H. & Aus, H. M. (1980). Relationship of ploidy and chromatin condensation in the rat liver, moreover a comparison of the nuclear texture in sections and touch preparations. *Histochem.*, **67**, 249–56.

Romer, F. (1972). Histologia, Histochemie, Polyploidie und Feinstruktur der Oenocyten von *Gryllus bimaculatus* (Saltatoria). *Cytobiologie*, **6**, 195–213.

Roozemond, R. C. (1976). Ultramicrochemical determination of nucleic acids in individual cells using the Zeiss UMSP-1. *Histochem. J.*, **8**, 625–38.

Röper, W. (1976). Nuclear fusion and irregular cytokinesis in binucleate and tetraploid cells of *Vicia faba* after caffeine treatment. *Experientia*, **32**, 1260–2.

Roszell, J. A., Fredi, J. L. & Irving, Ch.C. (1978). The development of polyploidy in two classes of rat liver nuclei. *Biochim. Biophys. Acta*, **519**, 306–16.

Roth, G. E. & Moritz, K. B. (1981). Restriction enzyme analysis of the germ line limited DNA of *Ascaris suum*. *Chromosoma*, **83**, 169–90.

Rowley, J. D. (1983). Consistent chromosome abnormalities in human leukemia and lymphoma. *Cancer Invest.*, **1**, 267–80.

Ruddle, F. H. & Rapola, J. (1970). Changes in lactate dehydrogenase and esterase-specific activities, isozymic patterns and cellular distribution during the growth cycle of PK cells *in vitro*. *Exp. Cell Res.*, **59**, 399–412.

Rudkin, G. T. (1972). Replication in polytene chromosomes. In *Results and Problems in Cell Differentiation*, vol. 4, ed. W. Beerman, pp. 59–85. Berlin & Heidelberg: Springer-Verlag.

Rudkin, G. T. (1973). Cyclic synthesis of DNA in polytene chromosomes of Diptera. In *The Cell Cycle in Development and Differentiation*, ed. M. Balls & F. S. Billett, pp. 279–92. Cambridge University Press.

Rumyantsev, P. P. (1963). A morphological and autoradiographical study of the peculiarities of differentiation rate, DNA synthesis and nuclear division in the embryonal and postnatal histogenesis of cardiac muscle of white rats. *Folia Histochem. Cytochem.*, **1**, 463–71.

Rumyantsev, P. P. (1966). Autoradiographic study on the synthesis of DNA, RNA and proteins in normal cardiac muscle cells and these changes by experimental injury. *Folia Histochem. Cytochem.*, **4**, 397–424.

Rumyantsev, P. P. (1972). Electron microscope study of the myofibril partial disintegration and recovery in the mitotically dividing cardiac muscle cells. *Z. Zellforsch. mikr. Anat.*, **129**, 471–99.

Rumyantsev, P. P. (1979). Some comparative aspects of myocardial regeneration. In *Muscle Regeneration*, ed. A. Mauro, pp. 335–55, New York: Raven Press.

Rumyantsev, P. P. (1981). New comparative aspects of myocardial regeneration with special reference to cardiomyocyte proliferative behavior. In *Mechanisms of Growth Control*, ed. R. O. Becker, pp. 311–42. Springfield: C. C. Thomas Publ.

Rumyantsev, P. P. & Kassem, A. M. (1976). Cumulative indices of DNA synthesizing myocytes in different compartments of the working myocardium and conductive system of the rat's heart muscle following extensive left ventricle infarction. *Virchows Arch. B*, **20**, 329–42.

Russell, P. J. & Diener, E. (1970). The early antibody-forming response to *Salmonella* antigens. A study of morphology and kinetics *in vivo* and *in vitro*. *Immunology*, **19**, 651–67.

Ruthman, A. (1964). Zellwachstum und RNS-Synthese im Ei-Nährzellverband von *Ophryotrocha puerilis*. *Z. Zellforsch. mikr. Anat.*, **63**, 816–29.

Sabin, A. B. (1981). Suppression of malignancy in human cancer cells: issues and challenges. *Proc. Natl. Acad. Sci. USA*, **78**, 7929–33.

Sager, R. & Kitchin, R. (1975). Selective silencing of eucaryotic DNA. *Science*, **189**, 426–47.

Sakharov, D. A. (1965). Functional organization of giant neurons of molluscs. *Uspekhi Sovremennoi Biologii*, **60**, 365–83 (in Russian).

Sakharov, V. V. (1946). Somatic reduction as a reason for original mosaicism in the tetraploid buckwheat. *Doklady Akad. Nauk SSSR*, **52**, 349–52 (in Russian).

Sakharov, V. V., Mansurova, V. V., Platonova, R. N. & Shcherbakov, V. K. (1960). Detection of physiological protection of autotetraploids of buckwheat against ionizing radiation. *Biofisika*, **5**, 558–65 (in Russian, English translation in *Biophysics*, **5**, 632–41).

Saksela, E. & Moorhead, P. S. (1963). Aneuploidy in the degenerative phase of serial cultivation of human cell strains. *Proc. Natl. Acad. Sci. USA*, **50**, 390–5.

Salganik, R. I. (1979). Some patterns of genetic induction of protein synthesis in animal cells. *The Cell Nucleus*, vol. 7, ed. H. Busch, pp. 327–67. New York & London: Academic Press.

Salomon, D. S. & Sherman, M. I. (1975). The biogenesis of progesterone by cultured mouse midgestation trophoblast cells. *Dev. Biol.*, **47**, 394–406.

Sandberg, A. A. (1977). Chromosome markers and progression in bladder cancer. *Cancer Res.*, **37**, 2950–6.

Sandberg, A. A. (1983). A chromosomal hypothesis of oncogenesis. *Cancer Genetics & Cytogenet.*, **8**, 277–86.

Sandritter, W. & Adler, C. P. (1978). Polyploidization of heart muscle nuclei as a prerequisite for heart growth and numerical hyperplasia in heart hypertrophy. *Recent Advan. Stud. Cardiac Struct. Metabol.*, **12**, 115–27.

Sandritter, W., Nováková, V., Pilny, J. & Kiefer, G. (1967). Cytophotometrische Messungen des Nukleinsäure- und Proteingehaltes von Ganglienzellen der Ratte während der postnatalen Entwicklung und im Alter. *Z. Zellforsch. mikr. Anat.*, **80**, 145–52.

Sandritter, W. & Scomazzoni, G. (1964). Deoxyribonucleic acid content (Feulgen photometry) and dry weight (interference microscopy) of normal and hypertrophic heart muscle fibers. *Nature (London)*, **202**, 100–1.

Sapp, I. P. (1976). An ultrastructural study of nuclear and centriolar configurations in multinucleated giant cells. *Lab. Invest.*, **34**, 109–14.

Sarto, G. E., Stubblefield, P. A. & Therman, E. (1982). Endomitosis in human trophoblast. *Human Genet.*, **62**, 228–32.

Sauaia, H. & Alves, M. A. R. (1969). Non-replication of DNA in heterochromatin of the ovary nurse cells of Sciarids. *Exp. Cell Res.*, **55**, 127–9.

Saxholm, H. J. K. & Digernes, V. (1980). Progressive loss of DNA and lowering of the chromosomal mode in chemically transformed C3H/10T 1/2 cells during development of their oncogenic potential. *Cancer Res.*, **40**, 4254–60.

Scharpé, A. & van Parijs, R. (1973). The formation of polyploid cells in ripening cotyledons of *Pisum sativum* in relation to ribosome and protein synthesis. *J. Exp. Bot.*, **24**, 216–22.

Schimke, R. T. (1982). Studies on gene duplications and amplifications – an historical perspective. In *Gene Amplification*, ed. R. T. Schimke, pp. 1–6. Cold Spring Harbor, New York: Cold Spring Harbor Laboratory.

Schimke, R. T., Alt, F. W., Kellems, R. E., Kaufman, R., Bertino, J. R. (1978). Amplification of dihydrofolate reductase genes in methotrexate resistant cultured mouse cells. *Cold Spring Harbor Symp. Quant. Biol.*, **42**, 649–57.

Schmidtke, J. & Epplen, J. T. (1980). Sequence reorganization of animal nuclear DNA. *Human Genet.*, **55**, 1–18.

Schneider, R. & Pfitzer, P. (1973). Die Zahl der Kerne in isolierten Zellen des menschlichen Myokards. *Virchows Arch. B*, **12**, 238–58.

Schneyer, C. A. (1973). A growth suppressive influence of l-isoproterenol on postnatally developing parotid gland of rat. *Proc. Soc. Exp. Biol. Med.*, **143**, 899–904.

Schneyer, C. A., Finley, W. & Finley, S. (1967). Increased chromosome number of rat parotid cells after isoproterenol. *Proc. Soc. Exp. Biol. Med.*, **125**, 722–8.

Schneyer, C. A. & Shackleford, J. M. (1963). Accelerated development of salivary glands of early postnatal rats following isoproterenol. *Proc. Soc. Exp. Biol. Med.*, **112**, 320–4.

Schulte-Hermann, R., Hoffmann, V., Landgraff, H. (1980). Adaptive responses of rat liver to the gestagen and anti-androgen cyproterone acetate and other inducers. III. Cytological changes. *Chem.–Biol. Interact.*, **31**, 301–11.

Schultze, B., Gerhard, H., Schump, E. & Maurer, W. (1973). Autoradiographische Untersuchungen über die Proliferation der Hepatocyten bei der Regeneration der CCl_4-Leber der Maus. *Virchows Arch. B*, **14**, 329–43.

Schulze, W. & Schulze, C. (1981). Multinucleate sertoli cells in aged human testis. *Cell Tiss. Res.*, **217**, 259–66.

Schwarzacher, H. G. (1968). Zur Frage der Entstehung polyploider Zellen. *Anat. Anz.*, **121**, 209–10.

Schwarzacher, H. G. & Klinger, H. (1963). Die Entstehung mehrkerniger Zellen durch Amitose im Amnionepithel des Menschen und die Aufteilung des chromosomalen Materials auf deren einzelne Zellkerne. *Z. Zellforsch. mikr. Anat.*, **60**, 741–54.

Schwarzacher, H. G. & Schnedl, W. (1965). Endoreduplication in human fibroblast cultures. *Cytogenetics*, **4**, 1–18.

Schwarzacher, H. G. & Schnedl, W. (1966). Position of labelled chromatids in diplochromosomes of endoreduplicated cells after uptake of tritiated thymidine. *Nature (London)*, **209**, 107–8.

Schweizer, D. & Nagl, W. (1976). Heterochromatin diversity in *Cymbidium* and its relationship to differential DNA replication. *Exp. Cell Res.*, **98**, 411–23.

Searle, B. M. & Bloom, S. E. (1979). Influence of the *bn* gene on mitosis of immature red blood cells in turkeys. *J. Hered.*, **70**, 155–60.

Shalakhmetova, T. M., Kudryavtseva, M. V. & Kudryavtsev, B. N. (1981*a*). Protein content of hepatocytes of different ploidy in postnatal development of rats. *Tsitologia*, **23**, 674–80 (in Russian).

Shalakhmetova, T. M., Kudryavtseva, M. V., Zavadskaya, E. E., Komarova, N. I., Komarov, S. A. & Kudryavtsev, B. N. (1981*b*). Glycogen content of hepatocytes synthesizing and not synthesizing DNA in rats of different ages. *Tsitologia*, **23**, 539–44 (in Russian).

Sherman, M. I., McLaren, A. & Walker, P. M. B. (1972). Mechanism of accumulation of DNA in giant cell of mouse trophoblast. *Nature New Biol.*, **27**, 81–4.

Shima, A. (1980). Aging of hepatocytes. In *Aging Phenomenon*, ed. K. Oota, T. Makinodan, M. Iriki & L. S. Baker, New York & London: Plenum Press.

Shiomi, T. & Sato, A. (1978). Cell cycle studies on a temperature-sensitive cell division mutant mammalian cells. *Cell Struct. Funct.* **3**, 95–102.

Sidorov, B. N. & Sokolov, N. N. (1965). Spindle block as cause of polymorphic nuclei formation in the polyploid cells. *Tsitologia*, **7**, 645–9 (in Russian).

Sieger, M., Pera, F. & Schwarzacher, H. G. (1970). Genetic inactivity of heterochromatin and heteropycnosis in *Microtus agrestis*. *Chromosoma*, **29**, 349–64.

Simon, Z. (1965). Lethal mutation hypothesis for the mechanism of action of cytostatic alkylating agents. *J. Theoret. Biol.*, **8**, 193–7.

Simson, J. V. (1969). Discharge and restitution of secretory material in the rat parotid gland in response to isoproterenol. *Z. Zellforsch. mikr. Anat.*, **101**, 175–91.

Sin, W. & Pasternak, J. (1970). Number and DNA content of nuclei in the free-living nematode *Panagrellus silusiae* at each stage during postembryonic development. *Chromosoma*, **32**, 191–204.

Sinnott, E. W. (1960). *Plant Morphogenesis*. New York: McGraw-Hill.

Sklarew, R. J., Pachter, B., Hoffman, J. & Post, J. (1971). Formation of splenic megakaryocytes in young rat. *Exp. Cell Res.*, **64**, 195–203.

Smith, B. J. & Wigglesworth, N. M. (1972). A cell line which is temperature-sensitive for cytokinesis. *J. Cell Physiol.*, **80**, 253–8.

Snow, M. H. L. & Ansell, J. D. (1974). The chromosomes of the giant cells of mouse trophoblast. *Proc. R. Soc. London B*, **187**, 93–8.

Sparrow, A. H. & Naumann, A. F. (1974). Evolutionary changes in genome and chromosome sizes and in DNA content in the grasses. In *Basic Mechanisms in Plant Morphogenesis. Brookhaven Symposia in Biology*, **25**, 367–89.

Sparrow, A. H. & Naumann, A. F. (1976). Evolution of genome size by DNA doublings. *Science*, **192**, 524–9.

Sparrow, A. H., Price, H. J. & Underbrink (1972). A survey of DNA content per cell and per chromosome of prokaryotic and eucaryotic organisms: some evolutionary considerations. *Brookhaven Symposia in Biology*, **23**, 451–93.

Speit, G., Mehnert, K. & Vogel, W. (1984). Induction of endoreduplication by hydrazine in Chinese hamster V79 cells and reduced incidence of sister chromatid exchanges in endoreduplicated mitoses. *Chromosoma*, **89**, 79–84.

Spierer, A. & Spierer, P. (1984). Similar level of polyteny in bands and interbands of *Drosophila* giant chromosomes. *Nature*, **307**, 176–8.

Spooner, M. E. & Cooper, E. H. (1972). Chromosome constitution of transitional cell carcinoma of the urinary bladder. *Cancer*, **29**, 1401–12.

Spradling, A. C. & Mahowald, A. P. (1980). Amplification of genes for chorion proteins during oogenesis in *Drosophila melanogaster*. *Proc. Natl. Acad. Sci. USA*, **77**, 1096–100.

Stadler, L. J. (1929). Chromosome number and the mutation rate in *Avena* and *Triticum*. *Proc. Natl. Acad. Sci. USA*, **15**, 876–81.

Steele, P. R. M., Yim, P. C., Herbertson, B. M. & Watson, J. (1981). Some flow cytofluorometric studies of the nuclear ploidy of mouse hepatocytes. II. Early changes in nuclear ploidy of mouse hepatocytes following carbon tetrachloride administration: evidence for polyploid nuclei arrested in telophase. *Brit. J. Exp. Pathol.*, **62**, 474–9.

Steinberg, M. L. & Defendi, V. (1982). Fusion-induced differentiation of SV40-infected human keratinocytes. *Exp. Cell Res.*, **139**, 369–75.

Steinbrück, G., Haas, I., Hellmer, K. H. & Ammermann, D. (1981). Characterization of macronuclear DNA in five species of ciliates. *Chromosoma*, **83**, 199–208.

Steinemann, M. (1978). Co-replication of satellite DNA of *Chironomus melanotus* with mainband DNA during polytenization. *Chromosoma*, **66**, 127–39.

Steinemann, M. (1981). Chromosomal replication in *Drosophila virilis*. III Organization of active origins in the highly polytene salivary gland cells. *Chromosoma*, **82**, 289–307.

Steinert, W., Pfitzer, P., Friedrich, G. & Stoepel, K. (1974). DNS-Synthese im Herzen von Ratten mit renalen Hochdruck bei Langzeitinfusion von ³H-Thymidin. *Beitr. Path.*, **153**, 165–77.

Stephen, J. (1973). Occurrence of polyteny, endopolyploidy and numerical variation of nucleoli in maize endosperm. *Science and Culture*, **39**, 323–4.

Stephen, J. (1974). Cytological investigation on the endosperm of *Borassus flabellifer*. *Cytologia*, **39**, 195–207.

Stewart, F. A., Denekamp, J. & Hirst, D. G. (1980). Proliferation kinetics of the mouse bladder after irradiation. *Cell Tiss. Kinet.*, **13**, 75–89.

Stich, H. F. (1960). The DNA content of tumor cells. II. Alterations during the formation of hepatomas in rats. *J. Natl. Cancer Inst.*, **24**, 1283–97.

Stich, H. F. & Acton, A. B. (1979). Can mutation theories of carcinogenesis set priorities for carcinogen testing programs? *Can. J. Genet. Cytol.*, **21**, 155–77.

Stockdale, F. & Holtzer, H. (1961). DNA synthesis and myogenesis. *Exp. Cell Res.*, **24**, 508–20.

Stöcker, E. (1966). Der Proliferationmodus in Niere und Leber. *Verhandl. Deutsch. Ges. Pathol.*, **50**, 53–74.

Stöcker, E., Schultze, B., Heine, W.-D. & Liebscher, H. (1972). Wachstum und Regeneration in parenchymatösen Organen der Ratte. *Z. Zellforsch. mikr. Anat.*, **125**, 306–31.

Stroeva, O. G. & Brodsky, V. Ya. (1968). Nuclear ploidy during growth and differentiation of the retinal pigment epithelium in newts and rats. *Zhurnal obshchei Biologii*, **29**, 177–85 (in Russian).

Stroeva, O. G. & Mitashov, V. I. (1983). Retinal pigment epithelium: proliferation and differentiation during development and regeneration. *Internat. Rev. Cytol.*, **83**, 221–93.

Stroeva, O. G. & Panova, I. G. (1976). Growth of the eye and the retinal pigment epithelium in the postnatal development of rats. *Ontogenez*, **7**, 170–7 (in Russian, English translation in *Soviet J. Dev. Biol.*, **7**, 141–6).

Stroeva, O. G. & Panova, I. G. (1983). Retinal pigment epithelium: pattern of proliferative activity and its regulation by intraocular pressure in postnatal rats. *J. Embryol. Exp. Morphol.*, **75**, 271–91.

Strom, C. M. & Dorfman, A. (1976). Amplification of moderately repetitive DNA sequences during chick cartilage differentiation. *Proc. Natl. Acad. Sci. USA*, **73**, 3528–32.

Strunnikov, V. A., Uryvaeva, I. V. & Brodsky, V. Ya. (1982). Two-mutational hypothesis of carcinogenesis and the protective role of somatic cell polyploidy. *Doklady Akad. Nauk SSSR*, **264**, 1246–50 (in Russian, English translation in *Doklady Biol. Sci.*, **264**, 285–8).

Sugar, J., Szentirmay, Z. & Decker, A. (1982). The significance of cytophotometry in the characterization of malignant transformation. *Acta Histochem.*, **28** suppl., pp. 59–68.

Sulston, J. E. & Horvitz, H. R. (1981). Abnormal cell lineages in mutants of the nematode *Caenorhabditis elegans*. *Dev. Biol.*, **82**, 41–55.

Summer, A. T. & Buckland, R. A. (1976). Relative DNA content of somatic nuclei of ox, sheep and goat. *Chromosoma*, **57**, 171–6.

Sun, N. C. & Chu, E. H. J. (1971). An improved method for chromosome preparation from mouse liver. *Can. J. Genet. Cytol.*, **13**, 612–17.

Sussman, M. (1973). *Developmental Biology: Its Cellular and Molecular Foundations*. New Jersey: Prentice-Hall.

Sutou, Sh. & Arai, Y. (1975). Possible mechanisms of endoreduplication induction: membrane fixation and/or disruption of the cytoskeleton. *Exp. Cell Res.*, **92**, 15–22.

Suzuki, J., Gage, L. P. & Brown, D. D. (1972). The genes for silk fibroin in *Bombyx mori*. *J. Mol. Biol.*, **70**, 637–49.

Suzuki, N., Williams, M., Hunter, N. M. & Withers, H. R. (1980). Malignant properties and DNA content of daughter clones from a mouse fibrosarcoma: differentiation between malignant properties. *Br. J. Cancer*, **42**, 765–71.

Swartz, F. (1967). Polyploidization of liver after partial hepatectomy in the dwarf mouse and hypophysectomized rat: effect of extended regenerative periods. *Exp. Cell Res.*, **48**, 557–68.

Sweeney, G. D., Cole, F. M., Freeman, K. B. & Patel, H. V. (1979). Heterogeneity of rat liver parenchymal cells. *J. Lab. Clin. Med.*, **94**, 718–25.

Swift, H. (1962). Nucleic acids and cell morphology in Dipteran salivary glands. In *The Molecular Control of Cellular Activity*, ed. J. M. Allen, pp. 73–125. New York: McGraw-Hill.

Szarski, H. (1976). Cell size and nuclear DNA content in vertebrates. *Internat. Rev. Cytol.*, **44**, 93–111.

Takanari, H. & Izutsu, R. (1981). Studies on endoreduplication. II. Spontaneous occurrence and cellular kinetics of endoreduplication PHA-stimulated tonsillar lymphocytes. *Cytogenet. Cell Genet.*, **29**, 77–83.

Tandon, P., Vyas, G. S. & Arya, H. C. (1976). Mechanism of *in vitro* gall induction in *Zizyphus jujuba*. *Experientia*, **32**, 563–4.

Tavares, A. S., Costa, J., De Carvalho, A. & Reis, M. (1966). Tumour ploidy and prognosis in carcinomas of the bladder and prostate. *Br. J. Cancer*, **20**, 438–41.

Tazima, J. (1964). *The Genetics of the Silkworm*. London: Academic Press.

Telfer, W. H. (1975). Development and physiology of the oocyte–nurse cell syncytium. In *Advances in Insect Physiol.*, ed. J. E. Treherne, M. J. Berridge & V. B. Wigglesworth, pp. 223–319. New York & London: Academic Press.

Teplitz, R. L., Gustafson, P. F. & Pellett, O. L. (1968). Chromosomal distribution in interspecific *in vitro* hybrid cells. *Exp. Cell Res.*, **52**, 379–91.

Terzi, M. (1974). *Genetics and the Animal Cell*. London: Wiley & Sons.

Theile, M. (1982). Mutagenesis in mammalian cells induced by oncogenic DNA viruses. *Biol. Zbl.*, **101**, 321–48.

Therman, E., Sarto, G. E. & Stubblefield, P. A. (1983). Endomitosis: a reappraisal. *Human Genet.* **63**, 13–8.

Thomas, C. & Brown, D. D. (1976). Localization of the genes for silk fibroin in silk gland cells of *Bombyx mori*. *Dev. Biol.*, **49**, 89–100.

Thompson, K. V. A. & Holliday, R. (1975). Chromosome changes during the *in vitro* ageing of MRC-5 human fibroblasts. *Exp. Cell Res.*, **96**, 1–6.

Thompson, K. V. A. & Holliday, R. (1978). The longevity of diploid and polyploid human fibroblasts. Evidence against the somatic mutation theory of cellular ageing. *Exp. Cell Res.*, **112**, 281–7.

Thompson, L. H. & Lindl, P. A. (1976). A CHO-cell mutant with a defect in cytokinesis. *Somat. Cell Genet.*, **2**, 387–400.

Thomson, K. S., Gall, J. G. & Coggins, L. W. (1973). Nuclear DNA contents of *Coelacanth* erythrocytes. *Nature (London)*, **241**, 126–7.

Tobler, H. (1975). Occurrence and developmental significance of gene amplification. In *The Biochemistry of Animal Development*, vol. 3, ed. K. Weber, pp. 91–143. New York & London: Academic Press.

Tobler, H., Smith, K. D. & Ursprung, H. (1972). Molecular aspects of chromatin elimination in *Ascaris lumbricoides*. *Dev. Biol.*, **27**, 190–203.

Tobler, H., Zulanf, E. & Kuhn, O. (1974). Ribosomal RNA genes in germ line and somatic cells of *Ascaris lumbricoides*. *Dev. Biol.*, **41**, 218–23.

Traut, W. & Scholz, D. (1978). Structure, replication and transcriptional activity of the sex-specific heterochromatin in a moth. *Exp. Cell Res.*, **113**, 85–94.

Tremp, J. (1981). Chromosome aberrations and cell survival in irradiated mammalian cells. *Radiat. Res.*, **85**, 554–66.

Troughton, W. D. & Trier, J. C. (1969). Paneth and goblet cell renewal in mouse duodenal crypts. *J. Cell Biol.*, **41**, 251–68.

Troy, M. R. & Wimber, D. E. (1968). Evidence for a constancy of the DNA synthetic period between diploid–polyploid groups in plants. *Exp. Cell Res.*, **53**, 145–54.

Tsanev, R. (1975). Cell cycle and liver function. In *Results and Problems in Cell Differentiation*, ed. J. Reinert & H. Holtzer, vol. 7, pp. 197–248. Berlin & New York: Springer Verlag.

Tschermak-Woess, E. (1956). Karyologische Pflanzenanatomie. *Protoplasma*, **46**, 798–834.

Tschermak-Woess, E. (1963). Structurtypen der Ruhekerne von Pflanzen und Tieren. *Protoplasmatologia*, 6/1.

Tschermak-Woess, E. (1967). Die eigenartige Verlauf der I meiotischen Prophasen von *Rhinanthus*, die Riesenchromosomen und das besondere Verhalten der Kurzenchromosomen und hochendopolyploiden Kernen. *Caryologia*, **20**, 135–52.

Tschermak-Woess, E. (1971). Endomitose. In *Handbuch der Allgemeinen Pathologie*, vol. 2, 2/1, ed. H.-W. Altmann, pp. 569–625. Springer Berlin & Heidelberg: Springer Verlag.

Tschermak-Woess, E. (1973). Somatische Polyploidie bei Pflanzen. In *Grundlagen der Cytologie*, ed. H.-W. Altmann pp. 189–204. Jena: Fischer Verlag.

Tschermak-Woess, E. & Hasitschka, G. (1953). Veränderung der Kernstruktur während der Endomitose, rhythmisches Kernwachstum und verschiedenes Heterochromatin bei Angiospermen. *Chromosoma*, **5**, 574–614.

Ts'O, M. O. M. & Friedman, E. (1967). The retinal pigment epithelium. I. Comparative histology. *Arch. Ophthalmol.*, **78**, 641–9.

Tultseva, N. M. & Astaurov, B. L. (1958). Enhanced resistance of polyploid silkworms and the general theory of biological effects of ionizing radiations. *Biofisika*, **3**, 197–205 (in Russian, English translation in *Biophysics*, **3**, 183–9).

Ullerich, F. (1966). Karyotyp und DNS-Gehalt von *Bufo bufo*, *B. viridis* and *B. calamita* (Amphibia, Anura). *Chromosoma*, **18**, 316–42.

Undritz, E. & Nusselt-Bohaumilitzky, K. (1968). Über das Auftreten diploider Megakaryoblasten, Promegakaryocyten und Megakaryocyten bei schweren Hämopathien, ein Rückgriff in die Phylogenese. *Schweiz. Wochenschr.*, **98**, 1686–7.

Uryvaeva, I. V. (1979). Polyploidizing mitoses and biological significance of polyploidy in the liver cells. *Tsitologia*, **21**, 1427–37 (in Russian).

Uryvaeva, I. V. (1981). Biological significance of liver cell polyploidy: an hypothesis. *J. Theoret. Biol.*, **89**, 557–71.

Uryvaeva, I. V., Arefyeva, A. M. & Brodsky, V. Ya. (1980). Modes of polyploidization of the mouse cardiomyocytes. *Byulleten exper. Biologii Meditsini*, **89**, 219–22 (in Russian, English translation in *Bull. Exp. Biol. Med.*, **89**, 209–12).

Uryvaeva, I. V. & Brodsky, V. Ya. (1972). Peculiarities of hepatocyte reproduction during liver regeneration in mice. *Tsitologia*, **14**, 1219–27 (in Russian).

Uryvaeva, I. V. & Faktor, V. M. (1971*a*). Rate of DNA replication in diploid and tetraploid nuclei of regenerating mouse liver. *Doklady Akad. Nauk SSSR*, **200**, 218–21 (in Russian).

Uryvaeva, I. V. & Faktor, V. M. (1971*b*). A method of hepatocyte chromosomes preparation. *Tsitologia*, **13**, 530–2 (in Russian).

Uryvaeva, I. V. & Faktor, V. M. (1974). ^{3}H-thymidine incorporation into cells of mouse regenerating liver by pulse and late labelling. *Arkhiv Anatomii, Histologii, Embryologii*, **66**, 54–61 (in Russian).

Uryvaeva, I. V. & Faktor, V. M. (1975). Growth fraction of liver, its composition by cell ploidy and changes during aging. *Ontogenez*, **6**, 458–65 (in Russian, English translation in *Soviet J. Dev. Biol.*, **6**, 396–401).

Uryvaeva, I. V. & Faktor, V. M. (1976*a*). Relationship between cell function and division. Resistance of liver to toxic effect of CCl_4 after partial hepatectomy. *Tsitologia*, **18**, 1354–9 (in Russian).

Uryvaeva, I. V. & Faktor, V. M. (1976*b*). Resistance of regenerating liver to hepatotoxins. *Byulleten exper. Biologii Meditsini*, **81**, 283–5 (in Russian, English translation in *Bull. Exp. Biol. Med.*, **81**, 322–4).

Uryvaeva, I. V. & Faktor, V. M. (1982). Formation of aberrant polyploid hepatocytes induced by the combination of alkylating drug dipin and partial hepatectomy. *Tsitologia*, **24**, 911–17 (in Russian).

Uryvaeva, I. V. & Lange, M. A. (1971). Binucleate cells in early postnatal development of mouse liver. *Ontogenez*, **2**, 26–32 (in Russian, English translation in *Soviet J. Dev. Biol.*, **2**, 18–22).

Uryvaeva, I. V. & Marshak, T. L. (1969). Proliferation of diploid and polyploid cells in regenerating mouse liver. *Tsitologia*, **11**, 1252–8 (in Russian).

Vakhtin, J. B. (1980). *Genetic theory of cell populations*. Leningrad: Nauka (in Russian).

Valeeva, F. S. & Kiknadze, I. I. (1971). Change in the mass of DNA in the nuclei of cells of the salivary gland of Chironomids following polytenization. *Ontogenez*, **2**, 406–10 (in Russian, English Translation in *Soviet J. Dev. Biol.*, **2**, 327–31).

Van Cantfort, I. I. & Barbason, H. R. (1972). Relation between the circadian rhythms of mitotic rate and cholesterol-7α-hydroxylase activity in the regenerating liver. *Cell Tiss. Kinet.*, **5**, 325–30.

Van Noorden, C. J. F., Vogels, I. M. C., Houtkooper, J. M., Fronik, G., Tas, J. & James, J. (1984). Glucose-6-phosphate dehydrogenase activity in individual rat hepatocytes of different ploidy classes. I. Developments during postnatal growth. *Europ. J. Cell Biol.*, **33**, 157–62.

Van Oostveldt, P. & Van Parijs, R. (1976). Underreplication of repetitive DNA in polyploid cells of *Pisum sativum*. *Exp. Cell Res.*, **98**, 210–21.

Van Parijs, R. & Vandendriessche, L. (1966). Changes in the DNA content of nuclei during the process of cell elongation in plants. I. Formation of polytene chromosomes. *Arch. Internat. Physiol. Biochim.*, **74**, 579–86.

Van't Hof, J. (1966). Comparative cell population kinetics of tritiated thymidine labeled diploid and colchicine-induced tetraploid cells in the same tissue of *Pisum*. *Exp. Cell Res.*, **41**, 274–88.

Van't Hof, J. & Sparrow, A. H. (1963). A relationship between DNA content, nuclear volume and minimum mitotic cycle time. *Proc. Natl. Acad. Sci. USA*, **49**, 897–902.

Vendrely, R. & Vendrely, C. (1956). The results of cytophotometry in the study of the DNA content of the nucleus. *Internat. Rev. Cytol.*, **5**, 171–94.

Vinikov, I. A. (1938). Growth and cell transitions in tissue cultures of outer pigment layer of retina. *Doklady Akad. Nauk SSSR*, **20**, 211–14 (in Russian).

Vlasova, I. E. & Kiknadze, I. I. (1975). Effect of cycloheximide on ^{3}H-thymidine incorporation into the chromosomes of *Chironomus thummi* salivary glands at different stages of larval development. *Tsitologia*, **17**, 518–23 (in Russian).

Vlasova, I. E., Kiknadze, I. I. & Sherudilo, A. I. (1972). The course of DNA replication during polytenization of chromosomes in ontogenesis of *Chironomus thummi*. *Doklady Akad. Nauk SSSR*, **203**, 459–63 (in Russian, English translation in *Doklady Biol. Sci.*, **203**, 159–62).

Von Denffer, H. (1970). Autoradiographische und histochemische Untersuchungen über das Teilungsvermögen von B-Zellen der Langerhansschen Inseln im Pancreas fetaler und neugeborener Mäuse. *Histochemie*, **21**, 338–52.

Vogel, H., Niewisch, H. & Matioli, G. (1969). Stochastic development of stem cells. *J. Theoret. Biol.*, **22**, 249–70.

Vorobyev, V. A. (1977*a*). Seasonal changes in the stomach epithelium of the starfish *Asterias amurensis*. In *Systematics, Evolution, Biology and Occurrence of Echinoderms*, ed. A. V. Zshirmunsky, pp. 14–15. Leningrad: Nauka (in Russian).

Vorobyev, V. A. (1977*b*). Somatic polyploidy in the stomach epithelium of starfish.

In *The Second Soviet Symposium on Cell Polyploidy*, pp. 25–7. Yerevan: Armenian Academy of Science Publishing Office (in Russian).

Vorobyev, V. A. & Leibson, N. L. (1974). Cytophotometric study of the stomach epithelium in the starfish *Asterias amurensis*. *Tsitologia*, **16**, 1222–7 (in Russian).

Vorobyev, V. A. & Leibson, N. L. (1976). Cytophotometric estimation of ploidy gradient in stomach epithelium of *Asterias amurensis*. *Tsitologia*, **18**, 451–7 (in Russian).

Wagner, G. (1951). Das Wachstum der Epidermiskerne während der Larvenentwicklung von *Calliphora erythrocephala* Meigen. *Z. Naturforsch.*, **6b**, 86–90.

Wake, N., Isaacs, J. & Sandberg, A. A. (1982). Chromosomal changes associated with progression of the Dunning R-3327 rat prostatic adenocarcinoma system. *Cancer Res.*, **42**, 4131–42.

Walbot, V. & Dure, L. S. (1976). Developmental biochemistry of cotton seed embryogenesis and germination. VII. Characterization of the cotton genome. *J. Mol. Biol.*, **101**, 503–36.

Walker, B. (1958). Polyploidy and differentiation in the transitional epithelium of mouse urinary bladder. *Chromosoma*, **9**, 105–18.

Walker, L. (1973). Syntheseprozesse an den Riesenchromosomen von *Glyptotendipes*. *Chromosoma*, **41**, 327–60.

Wang, R. J. (1974). Temperature sensitive mammalian cell line blocked in mitotis. *Nature (London)*, **248**, 76–8.

Wang, R. J. (1976). A novel temperature-sensitive mammalian cell line exhibiting defective prophase progression. *Cell*, **8**, 257–61.

Wang, R. J., Wissinger, W., King, E. J. & Wang, G. (1983). Studies on cell division in mammalian cells. VII. A temperature-sensitive cell line abnormal in centriole separation and chromosome movement. *J. Cell Biol.*, **96**, 301–6.

Wangenheim, K.-H. von (1970). Different effect of polyploidy on radiation response and on aging of differentiated cells. *Radiat. Bot.*, **10**, 345–9.

Wasserman, A. O. (1970). Polyploidy in the common tree toad *Hyla versicolor* Le Conte. *Science*, **167**, 385–6.

Watanabe, M. (1970). DNA synthesis in polyploid and binucleate hepatic cells in the regenerating rat liver of different ages. *Nagoya J. Med. Sci.*, **33**, 1–11.

Watanabe, T. & Tanaka, G. (1982). Age-related alterations in the size of human hepatocytes. A study of mononuclear and binuclear cells. *Virchows Arch. B*, **39**, 9–20.

Weinberg, R. A. (1983). Oncogenes and molecular biology of cancer. *J. Cell Biol.*, **97**, 1661–2.

Weinstein, R. & Hay, E. (1970). DNA synthesis and mitosis in differentiated cardiac muscle cells of chick embryos. *J. Cell Biol.*, **47**, 310–16.

West, J. D. (1976). Patches in the liver of chimaeric mice. *J. Embryol. Exp. Morphol.*, **36**, 151–61.

Weste, S. M. & Penington, D. G. (1972). Fluorometric measurement of DNA in bone marrow cells: the measurement of megakaryocyte DNA. *J. Histochem. Cytochem.*, **20**, 627–33.

Wheatley, D. N. (1972). Binucleation in mammalian liver: studies on the control of cytokinesis *in vivo*. *Exp. Cell Res.*, **74**, 455–65.

Whitlock, J. P., Kaufman, R. & Baserga, R. (1968). Changes in thymidine kinase and α-amylase activity during isoproterenol stimulated DNA synthesis in mouse salivary gland. *Cancer Res.*, **11**, 2211–16.

Whittaker, J. R. (1968a). Translational competition as a possible basis of modulation in retinal pigment cell cultures. *J. Exp. Zool.*, **169**, 143–60.

Whittaker, J. R. (1968b). The nature and probable cause of modulations in pigment cell cultures. In *The Stability of the Differentiated State*, ed. H. Ursprung, pp. 25–36. Berlin & Heidelberg: Springer Verlag.

Whittaker, J. R. (1970). The melanotic expression of embryonic pigment cells: regulation *in vitro* and *in situ*. In *Control Mechanisms in the Expression of Cellular Phenotypes*, ed. H. A. Padykula, pp. 89–108. New York & London: Academic Press.

Wigglesworth, V. B. (1966). Polyploidy and nuclear fusion. *Nature (London)*, **212**, 581–2.

Williams, D. L., Wang, S.-Y. & Klett, H. (1978). Decrease in functional albumin mRNA during estrogen-induced vitellogenin biosynthesis in avian liver. *Proc. Natl. Acad. Sci. USA*, **75**, 5974–8.

Wilson, E. (1925). *The Cell in Development and Heredity*. New York: MacMillan.

Wilson, J. T. & Frohman, L. A. (1974). Concomitant association between high plasma levels of growth hormone and low hepatic mixed function oxidase activity in the young rat. *J. Pharmacol. Exp. Ther.*, **189**, 255–70.

Wilson, J. T. & Spelsberg, T. S. (1976). Growth hormone and drug metabolism: acute effects on microsomal mixed-function oxidase activities in rat liver. *Biochem. J.*, **154**, 433–8.

Wilson, J. W. & Leduc, E. H. (1948). The occurrence and formation of binucleate and multinucleate cells and polyploid nuclei in the mouse liver. *Amer. J. Anat.*, **82**, 353–85.

Winter, W. A. (1974). Induction of DNA synthesis by isoproterenol in the rat urinary bladder epithelium. *Histochem.*, **41**, 141–4.

Wissinger, W. & Wang, R. J. (1978). A temperature-sensitive cell line defective in post-metaphase chromosome movement. *Exp. Cell Res.*, **112**, 89–94.

Wolf, B. E. & Sokoloff, S. (1976). Changes in the form of the polytene X chromosome in *Phryne (Silvicola) cincta* – causes and functional significance. *Chromosomes Today*, **5**, ed. P. Pearson & K. R. Lewis, pp. 91–108. New York & Helsinki.

Won Chul-Choi & Nagl, W. (1977). Patterns of DNA and RNA synthesis during the development of ovarian nurse cells in *Gerris najas* (Heteroptera). *Dev. Biol.*, **61**, 262–72.

Woodland, H. R. (1974). Changes in the polysome content of developing *Xenopus laevis* embryos. *Dev. Biol.*, **40**, 90–101.

Yanishevsky, R., Mendelsohn, M. L., Mayall, B. H. & Cristofalo, V. J. (1974). Proliferative capacity and DNA content of aging human diploid cells in culture: a cytophotometric and autoradiographic analysis. *J. Cell Physiol.*, **84**, 165–70.

Yarygin, N. E. & Yarygin, V. N. (1973). *Pathological and Adaptive Changes of the Neuron*. Moscow: Meditsina (in Russian).

Yarygin, V. N. & Felichkina, N. M. (1977). Some cytological patterns of the polyploid nerve cells. In *The Second Soviet Symposium on Cell Polyploidy*, pp. 127–9. Yerevan: Armenian Academy of Science Publishing Office (in Russian).

Zaborskaya (Uryvaeva), I. V. (1965). DNA cytophotometry in multinucleate cells of mesothelium. *Tsitologia*, **7**, 621–7 (in Russian).

Zaichikova, Z. P. (1976). DNA synthesis in nurse cells of the butterfly *Chrysopa perla*. *Tsitologia*, **18**, 438–44 (in Russian).

Zak, R. (1974). Development and proliferative capacity of cardiac muscle cells. *Circulat. Res.*, **35** suppl. 2, 17–26.

Zeligs, J. D. & Wollman, S. H. (1979). Mitosis in rat thyroid epithelial cells *in vivo*. I. Ultrastructural changes in cytoplasmic organelles during the mitotic cycle. *J. Ultrastruct. Res.*, **66**, 53–77.

Zhimulev, I. F. & Lychov, V. A. (1972). Functioning of salivary gland chromosomes in *Harmandia laevi* larvae (Diptera, Cecidomyiidae). *Ontogenez*, **3**, 194–201 (in Russian).

Zhinkin, L. N., Brodsky, V. Ya. & Lebedeva, G. S. (1961). Mitosis, amitosis and endomitosis in cells of multilayer epithelium in rats. *Tsitologia*, **3**, 514–21 (in Russian).

Zimmer, W. E. & Schwartz, R. J. (1982). Amplification of chicken actin genes during morphogenesis. In *Gene Amplification*, ed. R. T. Schimke, pp. 137–48. Cold Spring Harbor, New York: Cold Spring Harbor Laboratory.

Zimmermann, J. (1975). The initiation of melanogenesis in the chick retinal pigment epithelium. *Dev. Biol.*, **44**, 102–18.

Zutschi, U. & Kaul, B. L. (1975). Polyploidy and sensitivity to alkylating mutagens. *Radiat. Bot.*, **15**, 59–68.

Zybina, E. V. (1961). Endomitosis and polyteny in the giant cells of trophoblast. *Doklady Akad. Nauk SSSR*, **140**, 1177–80 (in Russian, English translation in *Doklady Biol. Sci.*, **140**, 797–800).

Zybina, E. V. (1963). DNA content in the nuclei of giant trophoblast cells. *Doklady Akad. Nauk SSSR*, **153**, 1428–31 (in Russian, English translation in *Doklady Biol. Sci.*, **153**, 1512–15).

Zybina, E. V. (1970). Peculiarities of polyploidization in trophoblast cells. *Tsitologia*, **12**, 1081–94 (in Russian).

Zybina, E. V. (1977). Structure of polytene chromosomes in mammalian trophoblast. *Tsitologia*, **19**, 327–37 (in Russian).

Zybina, E. V. (1980). Electron microscopic investigation of endopolyploid nuclei in the giant cells of the rat trophoblast. *Tsitologia*, **22**, 1284–90 (in Russian).

Zybina, E. V. (1983). Different ways of cell reproduction during mammal placenta differentiation. *Tsitologia*, **25**, 1103–19 (in Russian).

Zybina, E. V. & Grishchenko, T. A. (1970). Polyploid trophoblast cells in different parts of the rat placenta. *Tsitologia*, **12**, 585–95 (in Russian).

Zybina, E. V. & Grishchenko, T. A. (1972). Cytophotometrical estimation of polyploidy extent in decidual cells of rat endometrium. *Tsitologia*, **14**, 284–90 (in Russian).

Zybina, E. V., Kudryavtseva, M. V. & Kudryavtsev, B. N. (1973). Morphological and cytophotometrical study of the giant cells in rabbit trophoblast. *Tsitologia*, **15**, 833–40 (in Russian).

Zybina, E. V., Kudryavtseva, M. V. & Kudryavtsev, B. N. (1975). Polyploidization and endomitosis in giant cells of rabbit trophoblast. *Cell Tiss. Res.*, **160**, 525–37.

Zybina, E. V., Kudryavtseva, M. V. & Kudryavtsev, B. N. (1979). Distribution of chromosome material during fragmentation of giant nuclei in the rodent trophoblast. *Tsitologia*, **21**, 12–20 (in Russian).

Zybina, E. V. & Mos'yan, I. A. (1967). Sex chromatin bodies by endomitotic polyploidization of the trophoblast cells. *Tsitologia*, **9**, 265–272 (in Russian).

Zybina, E. V. & Rumyantsev, P. P. (1980). Formation of a complex plasmatic membrane and microfilament bundles during the completion of nuclear fragmentation in the trophoblast giant polykaryocytes. *Tsitologia*, **22**, 890–7 (in Russian).